CONTENTS

ENDANGERED MAIZE

ENDANGERED MAIZE

INDUSTRIAL AGRICULTURE AND THE CRISIS OF EXTINCTION

HELEN ANNE CURRY

UNIVERSITY OF CALIFORNIA PRESS

University of California Press
Oakland, California

© 2022 by Helen Anne Curry

Library of Congress Cataloging-in-Publication Data

Names: Curry, Helen Anne, author.
Title: Endangered maize : industrial agriculture and the crisis of
    extinction / Helen Anne Curry.
Description: Oakland, California : University of California Press, [2022] |
    Includes bibliographical references and index.
Identifiers: LCCN 2021016955 (print) | LCCN 2021016956 (ebook) | ISBN
    9780520307681 (hardback) | ISBN 9780520307698 (paperback) | ISBN
    9780520973794 (ebook)
Subjects: LCSH: Corn—North America—History. | Agrobiodiversity
    conservation—North America.
Classification: LCC SB191.M2 C867 2022 (print) | LCC SB191.M2 (ebook) |
    DDC 633.1/5—dc23
LC record available at https://lccn.loc.gov/2021016955
LC ebook record available at https://lccn.loc.gov/2021016956

Manufactured in the United States of America

30 29 28 27 26 25 24 23 22
10 9 8 7 6 5 4 3 2 1

For Andrew

The publisher and the University of California Press Foundation gratefully acknowledge the generous support of the Ralph and Shirley Shapiro Endowment Fund in Environmental Studies

FIGURES

ACRONYMS

CEC	Commission for Environmental Cooperation (Canada, Mexico, and United States)
CGIAR	Consultative Group on International Agricultural Research
CIMMYT	Centro Internacional de Mejoramiento de Maíz y Trigo/International Center for the Improvement of Maize and Wheat (Mexico)
FAO	Food and Agriculture Organization (United Nations)
GEM	Germplasm Enhancement of Maize (United States)
GM	genetically modified
IBP	International Biological Programme
INIA	Instituto Nacional de Investigaciones Agrícolas/National Agricultural Research Institute (Mexico)
IRRI	International Rice Research Institute (Philippines)
NAFTA	North American Free Trade Agreement
NGO	nongovernmental organization
RAFI	Rural Advancement Fund International
SAM	Sistema Alimentario Mexicano/Mexican Food System

| UPOV | Union Internationale pour la Protection des Obtentions Végétales/International Union for the Protection of New Varieties of Plants |
| USDA | United States Department of Agriculture |

INTRODUCTION

EACH YEAR THE WORLD'S FARMERS produce more than one billion tons of maize, or corn. Their annual collective harvest provides an estimated 20 percent of the world's caloric intake.[1] Maize seeds are sown nearly everywhere they can be coaxed to grow, from the rolling hills of Malawi to the river-basin plains of northeast China, from the temperate prairies of the United States Corn Belt to the tropical savannahs of central Brazil. They sprout in flat irrigated expanses, rocky hillside plots, and dry desert gardens. The end of each season sees engine-driven combines extract grain from some fields with relentless mechanical efficiency, most of it destined for animal feed and fuel. Elsewhere, flesh-and-blood harvesters do this work, ferrying ripe ears to homes and hearths and, ultimately, the table. In the twenty-first century, maize dominates farms and diets, and it does so across remarkably diverse cultural and ecological conditions.

It is also, by many accounts, in danger. A plant scientist managing an international gene bank in central Mexico curates thousands of samples of corn from around the world. "Maize's genetic diversity is unique," she emphasizes, "and must be protected in order to ensure the survival of the species and allow for breeding better varieties."[2] She and her colleagues recognize that the plant breeders developing tomorrow's corn varieties benefit from having access to diverse inputs, including the types in their collection. Since these distinctive local lines become rarer in farmers' fields as more uniform industrial lines proliferate, the gene bank's job is to perpetuate examples in cold storage. The Mexican government, motivated by similar concerns, runs its own maize gene bank. In recent years it has also sponsored community seed banks and designated local guardians of native, or criollo, maize, measures it hopes will keep disappearing maize types in cultivation. Scientists insist this support is essential. Subsistence cultivators are the only farmers who still grow many of Mexico's historical maize varieties, and their efforts ensure the crop's continued adaptation to changing needs and environments. Economic policies have made it difficult for them to carry on, however, and experts worry that farmers are "sowing and maintaining this diversity at an ever-increasing cost to themselves and their families."[3] North of the border, in the US southwest, a nonprofit safeguards maize varieties collected across Native American homelands, driven by a concern that every season fewer people grow the hardy desert varieties that long sustained agriculture in the region. Its collection saves seeds "for the proverbial 'rainy day,'" in this case, "when a crop can no longer be found growing in a farmer's field." Meanwhile, it strives to prevent the arrival of that rainy day by encouraging as many growers as possible to take up its treasured lines.[4]

The people behind these projects, and other conservation efforts like them, are divided on many issues. They nonetheless converge on a single, simple story. Maize diversity is disappearing, and action is essential.

Stories of endangerment are powerful. That's why we tell them. They are meant to provoke response and often to leverage change. "The very act of defining an entity as endangered entails the duty to find instruments and techniques to protect it," write the anthropologists Fernando Vidal and Nélia Dias.[5] With respect to wildlife and wild places, the threat of loss delivers nature reserves, inspires restoration programs, and prevents trade in endangered species.[6] Although less often in the public eye, dwindling domesticates like heirloom apples and heritage pigs are also targeted for conservation. Accounts of diverse agricultural plant varieties and animal breeds disappearing, possibly to extinction, have paved the way for seed and gene banks, rare-breed farms, local seed exchanges, and catalogs of culinary rarities. Because stories of plants, animals, peoples, and places under threat are influential—deployed as catalysts of change and also thrown up as obstacles to it—there's good reason to study them closely. That's the only way to judge whether they offer a useful understanding of what's happening in the world, to appreciate the kinds of responses they provoke, and to decide whether, in fact, we need different accounts.[7]

Consider the case of crop diversity, the central subject of this book. Since the late nineteenth century, an ever-expanding consensus has emerged—among scientists, policy makers, farmers, and eaters—about the diminishing biological diversity of the plants we grow for food. This idea has spawned eclectic and sometimes contradictory efforts to protect crop varieties understood as endangered, distinct undertakings that are nonetheless bound together by an insistence on imminent loss. Yet, as recent studies have shown, generalized reports of the disappearance of diverse crop varieties in the face of globalization and industrialization, itself an undisputed trend, have not always reflected local realities. The loss of crop diversity, the political ecologist Maywa Montenegro de Wit reminds us, "is more easily invoked than measured, more easily wielded than understood."[8] Blanket accounts of decline have tended to overlook the sites where diversity survives and thrives. These include the farms and gardens of small-scale and subsistence growers, people who have little, if anything, to gain from wholesale adoption of seeds designed for industrial production.[9] Tallies of diversity also tend to fixate on the loss of older types, overlooking the appearance of new ones. Farmers do not just adopt seeds, they adapt them, with the result that local lines and introduced varieties can mix to produce novel types. Professional breeders, too, create new combinations. As a result, the aftermath of displacing of local lines might see a plateauing of diversity rather than its relentless diminution.[10]

To identify patterns like these is not to deny the reality of loss or the importance of conservation. On the contrary, we know there has been an overall decline in the diversity of crops we cultivate for food, and we have many sound reasons to resist and reverse this trend.[11] Critical appraisals instead insist that existing accounts of crop diversity, tied to a powerful and generalized account of loss, don't always help the cause of conservation. Failure to see crop diversity where it survives or flourishes leads to missed opportunities for preserving it. It creates urgency for emergency off-site salvage to the detriment of longer-term investments that would keep crops, and cultivators, in place.

The recognition that a one-size-fits-all account does not, in fact, fit all circumstances points to the need to tell more tailored stories. Loss is historically specific. It proceeds at different rates and unfolds for different reasons, depending on the time, place, and crop in question. Accounts of loss therefore ought to attend to the local, embrace complexity, and allow for uncertainty. They should acknowledge that diversity can be, and is, made as well as destroyed. Experts working to conserve crop diversity often insist that making better ground-level observations, whether on farms, in breeding programs, at the market, or in gene banks, will lead to more appropriate solutions.[12] We can go further still. Tackling the challenges of conserving well demands an investigation of the origins of the master narrative and not just its fit. If stand-ard accounts of the loss of crop diversity aren't always a good reflection of what's happening in the world, then where do they come from? What has been, and is, really at stake when people talk about this disappearance?

In this book I draw on the history of maize to trace the origins of and motivations behind accounts of diversity's loss and to show how these shaped the methods and tools of conservation adopted by scientists and states. This research reveals interests and concerns that are often obscured, or deliberately masked, by simple declensionist tales. Like others who have studied the his-tory of conservation science and practice, I highlight the preoccupations that different people brought, and still bring, to the protection of threatened and endangered entities.[13] My research reveals how conservationists forged their methods for preserving crop plants—their modes of collection, classification systems, storage technologies, and negotiation tactics—around expectations of social, political, and economic transformations that would eliminate diverse communities and cultures. Until the 1970s few attempted to resist these trends. The introduction of technical concepts like "genetic erosion," the construction of cold-storage facilities, and the discussion of abstract imperatives such as enhancing global food security initially provided ways of obscuring lived human experiences and justifying urgent interventions over fair or reasoned ones. Wherever conservationists rely on similar ideas, tools, and strategies without questioning them, they risk perpetuating outdated narratives and, worse, the politics embedded within them.

In revising the history of today's crop conservation toolkit, I contribute to new understandings of endangerment and alternative strategies to protect and preserve. I join other scholars to ask what it might mean to abandon timeworn stories of imminent destruction and inevitable loss to focus on sustaining adaptation, embracing continuity with change, or surviving amid precarity.[14] We are so immersed in threat and endangerment with respect to the future of crop diversity that, as the historian Courtney Fullilove observes, "it's hard to conceive of a style of preservation that eludes the[se] specters."[15] But we can and should try. Undoing the limited endangerment narratives that pervade our conservation efforts is one place to start.

Present-day efforts to conserve crop diversity are often traced back to the late nineteenth century, when breeders in Europe helped to create a new view of plants and animals. Thanks in part to the emerging science of genetics, many came to think of individual organisms as bundles of heritable traits that could be disaggregated and reconfigured by enterprising breeders to create new and better forms. This meant that distinctive plants, whether barley grown in Germany, wheat from Turkey, or a Mexican farmer's maize, could be valuable as the source of new characteristics, even if they otherwise performed poorly. Marianna Fenzi and Christophe Bonneuil characterize this way of thinking as a "resourcist" vision of crop diversity, in which agronomists and breeders began to consider farmers' varieties and the wild relatives of crops as carriers of something a later generation would label "genetic resources."[16]

Achieving and maintaining control over these resources—typically by collecting them from farmers' fields and holding them at some research institution—became a major preoccupation of national and imperial governments needing to sustain agricultural productivity and project political power in the early twentieth century.[17] German, Soviet, and US institutions developed especially ambitious programs of collecting and disseminating crop diversity. Their plant explorers traversed the globe, often focusing on known diversity hotspots in East Asia, South Asia, the Middle East, North Africa, Mesoamerica, Andean South America, and the Mediterranean, establishing patterns that other agriculturally ambitious countries would retrace.[18] When scientists associated with these collecting enterprises started thinking of valuable farmers' varieties as likely to disappear, the resourcist worldview suggested seed bank storage as the ideal method of conservation. Seeds kept safe in storage would be accessible to the breeders who could make best use of them. Meanwhile breeders' products would continue their steady domination of farmers' fields.[19]

The resourcist vision and seed bank conservation, historically associated with US and European imperialism and by the 1970s aligned with increasingly powerful agribusinesses as well, eventually drew criticism from diverse quarters. These included plant scientists who prioritized continued evolution over static security in cold storage and representatives of national governments, especially in Africa, Latin America, and South Asia, which felt that they were being stripped of valuable products by more powerful countries and transnational corporations. They also included grassroots campaigners who saw crop diversity as a way to resist industrial agriculture or free-market capitalism. These critiques fueled different accounts of crop diversity. This was no longer simply a resource to be mined for power and profit but instead a contribution to the ecological and social well-being of societies. Farmers' varieties were now valued as biodiversity, ecosystem services, history, heritage, and culture. Novel strategies for their conservation accompanied these new perspectives, ranging from subsidies to farmers cultivating local varieties to community-organized seed libraries and exchanges.[20]

In their efforts to understand these transitions, historians have prioritized the circulation of ideas about crop plants and the ways in which they were and are valued. They have nonetheless also recognized that prevailing assumptions about farmers' knowledge and behavior similarly determined the trajectories of conservation. Christophe Bonneuil points out that state- and scientist-led conservation efforts in Europe and the United States, which emerged from the resourcist view of crop diversity, typically assumed that "local varieties would inexorably be swept away by breeders' cultivars." Researchers tended overwhelmingly to dismiss the potential role of farmers in making and maintaining varieties, especially when those farmers were peasant and Indigenous cultivators overseas.[21] Other historians note that the shift toward farmer involvement in conservation since the 1990s has been premised on a different narrative, one that sees farmers as knowledgeable, rational, and, crucially for the project of conservation, often interested in keeping diverse local varieties in cultivation.[22]

Here I place scientists' and policy makers' stories about farmers—about what farmers are doing, what they should be doing, and what they will do— front and center.[23] I argue that these narratives about people are more important to understanding the shape that conservation efforts have taken in the past 130 years than are the stories told about plants. The idea that crop varieties might be endangered has always been linked to the idea that a particular community, or way of life, is about to disappear. This connection has remained stable through successive attempts to preserve crop diversity, despite the variety of political projects they represented. It is *this* imagined extinction and its implications, and not that of plant varieties themselves, which has produced the approaches to conserving crop diversity we rely on today. By foregrounding this aspect of science's past, an important task for its future becomes clearer. The collection of more and better data about potentially disappearing crops, a need already widely acknowledged, must be accompanied by closer scrutiny of the narratives of progress and survival that surround their cultivators.

Although I tell a history of ideas about people and the consequences of those ideas for the seeds we save, the crops we grow, and ultimately the foods we eat, my way into this history is nonetheless a plant—ostensibly, one plant. I explore the conservation of crop diversity through the history of *Zea mays,* also known as corn or maize, especially as it's been bred, cultivated, researched, collected, stored, and consumed in Mexico and the United States since the turn of the twentieth century. A few crops would produce a story with similar contours.[24] None would so effectively reveal the events and ideas that provoked narratives about the loss of crop diversity and the responses those narratives encouraged. This is partly a result of the early and deep engagement of geneticists and evolutionary biologists in the intricacies of maize's history and life cycle. It is more obviously a consequence of the singular history of hybrid corn among twentieth-century agricultural crops and the unquestioned centrality of maize in cultures and economies across the Americas.[25]

Hybrid corn is an archetypical industrial crop, arguably the epitome of agricultural modernity in the United States. Its cultivation is characterized by uninterrupted fields, tended more by machines than people, and identical plants, tailored to chemical formulas and patented by corporations. It is, above all, big business. US farmers produce about one-third of the world's supply of corn, at an annual value of more than $50 billion.[26] The productivity of hybrid corn and the suite of technologies that surround it contributed to profound social, economic, and political changes in the twentieth-century United States. Between 1900 and 2000 the number of farms declined by 60 percent while the average farm size tripled, a trend especially pronounced across the swathe of midwestern agricultural heartland known as the Corn Belt.[27] Hybrid corn set a course for the seed industry, which underwent a similarly dramatic consolidation from myriad independent operations to a handful of transnational conglomerates. Hybrid corn influenced US foreign policy, from Cold War "food power" to the North American Free Trade Agreement, as the government looked to unload mountains of "cheap" grain sustained by farm subsidies. It changed diets, and not for the better. Inexpensive corn-fed beef and high-fructose corn syrup are often implicated in the steady rise of diet-related health problems like obesity, diabetes, and heart disease.[28]

The introduction of hybrid corn also entailed an unprecedented shift, from a diverse assortment of local varieties to a subset of breeders' lines based on a more limited genetic foundation, seen in a single crop, across a significant geographic expanse, in less than a generation.[29] It's this changeover that makes corn more than a case study for ideas about the loss of crop diversity. The sweep of hybrid corn across the midwestern United States in the 1940s, encouraged by science, industry, and government policy, provided the template for anyone imagining how later introductions of industrial crops in other places would unfold. Thanks to the interest of corn breeders and geneticists in having access to genetically diverse varieties, the rapid adoption of hybrid lines also inspired the first-ever large-scale crop conservation enterprise: a coordinated attempt to salvage all the corn varieties of the Western Hemisphere and keep these in perpetuity. Until the end of the 1960s, maize was the model for both the loss of diversity and what to do about this loss.

In the 1970s corn was shunted from this position as a result of the Green Revolution. This label—the Green Revolution—is often used to refer to an uptick in agricultural production in the late 1960s in parts of Asia, Latin America, and the Middle East. It's frequently attributed to the introduction of new varieties of a limited number of grain crops, namely, wheat and rice, which was achieved through investments in agricultural research and, crucially, driven by Cold War geopolitics. This specific Green Revolution profoundly reshaped many people's ideas about the direction and speed of global agricultural change. The phrase Green Revolution quickly came to have a general meaning too, referring to the "modernization" of agriculture through the adoption of new crop varieties, farm technologies, and cultivation practices, especially in countries where subsistence farming predominated.[30] Both Green Revolutions, specific and general, affected experts' ideas about crop diversity. The spread of "miracle" wheat and rice varieties became the new model for the rapid pace at which farmers' varieties could disappear. Meanwhile, the projection of a more general, global transformation in agriculture intensified anxieties about the loss of crop diversity, generating lasting international infrastructure for its conservation.

Although no longer the defining example of the phenomenon, hybrid corn remained at the center of discussions about dwindling crop diversity after the Green Revolution. Following a 1970 epidemic of corn leaf blight in the United States, it became the canary in the coal mine for genetic vulnerability in ever-more uniform industrial monocrops. When breeders called for the expansion and international coordination of intellectual property rights in plants, hybrid corn exemplified how the seed industry would wrest control over seeds, extracting profits from crop diversity at farmers' expense. When the introduction of new biotechnologies shifted the focus of international attention from the Green Revolution to the Gene Revolution, genetically modified (GM) corn featured centrally in debates about the safety and desirability of GM crops. In short, industrial corn occupied a central place in debates about crop diversity, even as the focus of these debates shifted from one decade to the next.

As pervasive as hybrid corn has become, it represents an almost insignificantly short span of maize's history.[31] Recent science places the origins of corn in south-central Mexico. There, some six thousand to ten thousand years ago, early agriculturists began the transformation of the wild grass teosinte into domesticated corn, *Zea mays*.[32] The spread of maize across the Americas over subsequent millennia was accompanied by many further transformations. Plants were shaped for and by different environments, from the chilly highlands of the Andes to the hot dusty deserts of Central and North America. Farmers adapted these for uses ranging from bread making to beer fermentation to religious rites to animal feed. When Christopher Columbus learned about corn, *Zea mays* had already diversified into more than two hundred distinct subpopulations.[33]

Corn is botanically plastic, proliferating shapes, sizes, colors, and textures and ranging widely over climates and geographies. It is also hugely productive, yielding more grain with less land and labor than other crops. These qualities have enabled it to sustain diverse peoples. In the ancient Olmec civilization of Mesoamerica, maize suffused daily life, not just as food but as god.[34] In some cultures humans shared the identity of maize as well. The *Popol Vuh,* which recounts the origin of the K'iche' people of today's Guatemala, describes the gods' creation of humans from white and yellow corn.[35]

Maize is still central to cultures of the region. In Mexico corn-based foods like tortillas and tamales dominate local cuisines, even as diets change. Maize is farmed on about one-third of the country's cropland and accounts for almost 60 percent in places without irrigation.[36] In the countryside a family might consume up to twenty-four kilos (fifty-three pounds) of maize per week; estimates for its overall contribution to rural diets run to 65 percent of calories or more. Although urban Mexicans have access to a more diverse array of foods, tortillas remain their staple too.[37] And what's on the dinner table is only the beginning. As the Mexican activist Gustavo Esteva describes, "Corn is present, even in the most unexpected forms, in most displays of Mexico's cultures today"—not just in cuisine but in art, language, dress, and even "everyday ways of thinking and behaving."[38] This is confirmed by Evaristo Polo, a farmer in Tlatlahuquitepec, Puebla. Polo is Nahua, and he voices a perspective shared widely among members of this Indigenous group, the largest in Mexico, when he says that maize "gives us life, the totality of everything, joy. We as Nahuas have said that maize is the one who shouts, who laughs, who talks, and who dialogues. . . . In any social or cultural event, the maize unites us."[39]

The history of hybrid corn unfolded against this backdrop, in which *Zea mays* was not one thing (a homogenized, industrial commodity) or even two (hybrid and nonhybrid) but many, profoundly diverse in form and meanings and deeply embedded in lives and cultures. As a result, the stories of loss and endangerment in which industrial corn was, and is, enmeshed have proved especially powerful.

Throughout this book I refer to certain kinds of corn as farmers' varieties, by which I mean distinct types developed and maintained by farmers through the seasonal acts of selecting and storing seeds to sow in the following season. These often have a long history in a particular place and, when they do, are also called landraces, another term I use here.[40]

Very few of the people who appear in the pages that follow would have used "farmers' varieties" to pick out certain kinds of *Zea mays*. Many, especially agronomists, breeders, and other agricultural scientists, identified farmers' varieties as primitive maize, indigenous strains, unimproved varieties, and, eventually, folk types. These identifiers served, then as now, to distinguish farmers' lines from breeders' varieties, which have often been called improved or modern varieties or, later still, high-yielding or elite. As the cultural hierarchies enacted in these labels became more apparent, those who celebrated or defended farmers' varieties adopted new descriptions: *nativo,* heritage, Mexican, Mayan, Tohono O'odham, and more.[41]

What corn is called is linked to the story that's told about it. When it comes to accounts of disappearing or endangered maize, no label has been more influential than that of "indigenous." The category of "indigenous corn" came into existence thanks to taxonomic, genetic, agronomic, and anthropological investigations. It was informed by researchers' ideas about race and progress, indigeneity and modernity.[42] When professional breeders and agronomists labeled some farmers' varieties as indigenous—or as "native" or "primitive" or any one of a number of similar terms—and their own varieties as modern or improved, they did so with reference to an imagined trajectory of agricultural progress. Through their efforts the improved would displace the unimproved, for the benefit of all. This logic shared in, and recapitulated, an existing colonial narrative that cast Indigenous peoples and cultures as, in the words of the historian Sadiah Qureshi, "the necessary victims of human racial competition." Corn varieties shared the fate of those who cultivated them, in that their "expected demise was both mourned and celebrated, and sometimes actively pursued."[43]

Although they did not invent the concept of indigenous maize and its analogues, Indigenous farmers, or farmers whose ancestry included Indigenous Americans, typically did create kinds of corn marked as indigenous. What's more, scientists' imposition of this category did not preclude its being subsequently reclaimed and remade by Indigenous and Native American farmers identifying certain kinds of corn as Indigenous and Native.[44] We have rich accounts of Indigenous and Native American views on maize, which continue to be deepened and extended.[45] My research complements these accounts, focusing on the ideas about indigenous maize espoused by professional scientists (a category that sometimes includes individuals who are Indigenous or Native) and the institutions that employed them. I tell the stories of how breeders, botanists, geneticists, and anthropologists encountered and engaged with Indigenous and Native peoples, knowledges, and maize varieties. I show how these experiences informed researchers' categorizations of diverse kinds of corn, including as "indigenous" or "native," and how they shaped efforts to preserve maize varieties considered in danger of extinction.

Understanding this history is essential, not because genetics, agronomy, and other scientific disciplines provided better or more important ways of knowing maize but because these were often aligned with powerful forces. States and industries relied on the information about maize generated through scientific research to create agricultural, economic, and political systems that systematically disadvantaged and exploited maize's most experienced cultivators. In sharing this history I hope to be an ally to the project of decoloniality, emphasizing the continuity of colonial, capitalist, and neoliberal systems of knowledge and governance and contributing to projects of resistance, repatriation, and reparation. In examining one totalizing account emanating from Western science—the inexorable loss of crop diversity—I document many displacements and effacements of Indigenous knowledges and cultures enacted by scientists in their pursuit of what most considered universal knowledge and agricultural modernity. This analysis aligns with the more direct defense of Indigenous food sovereignty, which, as the philosopher Kyle Powys Whyte has made clear, may rightly focus on the conservation of particular plant and animal species and varieties as critical to self-determination.[46]

In each of the chapters that follows, I highlight the history of a task that professional researchers considered essential to the conservation of crop diversity: collection, classification, cold storage, safety duplication, treaty negotiation, data generation, and, finally, cultivation. I connect these to the ideas about human cultures and agricultural change that inspired and shaped them at different moments and to the scientific, political, and economic currents of which they were a part. Although the story unfolds roughly chronologically, the tasks that I highlight did not emerge in sequence, each solving problems left open by an earlier approach. Each instead has its own history, some deeper and broader than others. Readers who follow the notes will know where to turn to learn more.

These tasks remain the basic elements of crop conservation today. As I researched this book, I spoke with many individuals who have dedicated their careers, and in some cases their lives, to the complex and important issue of saving crop diversity. I have deep respect for them. My effort to understand the origins of their and especially their predecessors' ideas and tools and to bring into view the political projects with which scientific programs historically have been aligned is not an argument against the conservation of crop diversity. I aim instead to deepen appreciation of the challenges they, and all of us, face in conserving diversity well. Emphasizing the urgency of loss may be important to gathering resources. It also leads to an overemphasis on salvage. As this book reveals, salvage has repeatedly failed as a conservation measure, not least because it ignores the sociality of crop species—that is, the interactions among plants and people that not only sustain diversity but generate it. In foregrounding this and related concerns, I hope to help those immersed in the study and conservation of plant genetic resources to assess anew the complexity of the work ahead.

In fact, I hope this book helps many people to appreciate that complexity. Stories of endangerment are meant to move us and often do. Repeated at a fever pitch, with prophecies of catastrophic consequences, they become declarations of crisis. Although often the motivation for change, crisis can also be a powerful tool for constraining and limiting it. The anthropologist Joseph Masco insists that in the contemporary world, "crisis blocks thought." Acceptance of crisis as the default condition, as has arguably happened with climate change, for example, can forestall deep engagement with the past, close off hard questions about the present, and diminish the possibilities for radical change in the future.[47] That's one way to understand what's happened in crop conservation, especially since the 1960s. Efforts by scientists and states to conserve diversity in crop plants historically have paired two dramatic crises: resource depletion and human hunger. They have linked the irreversible destruction of biological diversity with the relentless growth of populations. Insistence on these crises has enabled hasty response to prevail over reasoned redress. It's forced critics to respond in the same frenzied voice. Once crisis becomes the agreed mode, it's hard to opt out. The history of crop conservation is therefore not just about narratives of loss. It's also about acquiescence to crisis.

In writing the history of this acquiescence, I provide a counterpoint to it. I engage with the histories of crop diversity and with accounts of its endangerment and salvation as a way of enabling difficult questions about the present. Do we—as nations, communities, individuals—have the resources we need to survive? What do we want that survival to look like? How do we create systems that foster the futures we want? Important as they are, these questions typically get lost in the urgency that pervades discussions of saving crop diversity. To get past that urgency, we need to understand first where it comes from, to appreciate the origins and politics of endangerment narratives, and to understand how these shaped the task of conservation. With this past in hand, perhaps we can, collectively, tackle the important questions and in the process forge new stories—as well as new solutions.

[1] In its 1911 catalog Oscar H. Will and Company Pioneer Seed House and Nursery of the Northwest offered farmers in the northwestern United States the "descendants" of corn varieties cultivated by local Mandan, Hidatsa, and Arikara farmers and illustrated the bountiful harvest that would result from this "gift." State Historical Society of North Dakota 10190–14–1911.

ONE

COLLECT

IN JULY 1916 H. HOWARD BIGGAR, an employee of the Office of Corn Investigations of the US Department of Agriculture, set out on a two-month research trip. His journey eventually took him to fifteen Indian reservations across the mid- and northwestern United States. At each stop he asked residents about corn. What varieties did they know? How were they cultivated? Where were they stored? What were they used for? Biggar's most prized interviewees were elderly men and women, such as the men he knew as Seeking Land of the Crow Creek Reservation, "about eighty years old" and a "splendid informant," and Little Bald Eagle of the Rosebud Reservation, now seventy-three and able to "remember back about 65 years." Although he asked many questions, Biggar especially wanted to know what kinds of corn his interviewees could recall. Seeking Land remembered eight varieties, naming them by color and texture: white soft, white hard, yellow flint, small yellow flint, blue soft, red soft, spotted, and pink soft. Little Bald Eagle's list was shorter: white, spotted, red streaked, blue, and black. Wherever possible Biggar also gathered examples of these varieties. At a reservation in Minnesota, he encountered Clem Beaulieu, a trader of French and Anishinaabe descent. Beaulieu reported that for forty years he had grown a "pink hard corn" and "had never changed his seed." In other words, he had set aside ears from each harvest to use as seeds in the next season. Biggar enthusiastically purchased some corn from Beaulieu for the Office of Corn Investigations.[1]

Picking up samples of Beaulieu's pink hard corn and similar varieties was central to Biggar's mission. He and his colleagues from the US Department of Agriculture (USDA) were convinced that corn varieties traditionally grown by Native American farmers were nearing extinction. In the 1890s a handful of German breeders had been among the first to air concerns about farmers adopting "improved" varieties created through pure-line breeding and other methods. If farmers abandoned local varieties, as they were encouraged to do, it would spell the end of these lines. Some breeders thought this might be a problem, since farmers' varieties, even if not as productive, might nonetheless have useful qualities. They could, plausibly, be used by breeders to craft better varieties in the future. So they implored fellow scientists and governments to preserve farmers' varieties.[2]

In the United States some of the first farmers' varieties seen to be in danger of disappearing, with potential losses to breeders and growers and to the larger agricultural economy, were those cultivated by Native American farmers. But this was not because these farmers were understood to be transitioning to new varieties, even if some were doing exactly that. Rather, many people believed that Native American communities and cultures were vanishing. This explains Biggar's mission in 1916. He was to salvage Native American corn varieties before Native American farmers disappeared.

The idea that crop diversity might be endangered and that steps ought to be taken to save it depended initially on a set of closely linked assumptions, what Christophe Bonneuil describes as a "script," which mobilized subsequent action. Some elements of this script related to farmers' behavior, for example, the idea that they would always choose to adopt the "superior" products of professional breeders, with the result that "local varieties would inexorably be swept away." Another assumption was that farmers, seen to lack training and knowledge, had little or no role to play in either creating crop diversity or conserving it. These tasks could be undertaken only by breeders and other crop experts.[3] The early history of corn conservation in the United States reveals that the script of endangered crop diversity was often tied to yet another assumption, this one not about farmers' decisions but their very existence.

As was the case in settler colonies elsewhere, national expansion in the United States was achieved through the displacement and dispossession of Indigenous peoples, who were in turn cast as inherently fragile and destined to succumb.[4] In this context farmers' varieties that were also Native varieties were considered doubly doomed. Wherever it was an article of faith that Indigenous peoples were dying out, folding in confrontation with a supposedly more vigorous and virtuous culture, it was not just Indigenous crops that were thought inferior and therefore destined for extinction—but Indigenous farmers too.[5] This expectation shaped early conservation strategies and their outcomes. Although the rescue of assumed-to-be endangered Native American corn by researchers like Howard Biggar could have been a means of securing the future of Native economies and communities, it was not envisioned in these terms. "Indian corn" was a resource that would contribute to the expansion of settler cultivation. As a result, its successful conservation in collections only intensified its projected disappearance from fields.

THE OFFICE OF CORN INVESTIGATIONS

"The Indians are not growing corn to such an extent as in former times," Biggar noted in the log of his 1916 journey. In focusing on decline he both confirmed expectations of the endangerment of Native corn and legitimated his mission. Yet he arrived back at the home office with at least twenty-four varieties. When these were brought together with earlier collections and several samples sent by mail from sites he'd not visited, Biggar confidently declared that "all corn described by old Indians" was now safe in the hands of his employer, the USDA Office of Corn Investigations.[6]

He was probably wrong. By 1916 corn cultivation had a nearly four-thousand-year history in the lands that had become the United States. At the time of European contact, its cultivation had stretched from the arid Southwest, up through the Great Plains, east to the Atlantic, and north through today's New England. Differences in cultivation practices, soils, climate, and cultures inevitably meant diversification. Although early observations by Europeans focused on variations in the color of corn, this superficial differentiation had long since given way to settlers' appreciation of distinct uses and subtler qualities.[7] Even as Biggar wrote, contemporary ethnographers were charting the varieties grown among different Native American peoples, including their culinary and cultural significance.[8] Such diversity was unlikely to have been fully captured in the USDA's collections.

One explanation for Biggar's misapprehension lay in the changing place of corn in US culture. In the 1910s corn was already in the end stages of a process that Kelly Sisson Lessens characterizes as "de-Indianization." From the mid-nineteenth century, with corn established as a bedrock of the agricultural economy and key enabler of westward expansion, commentators "increasingly claimed corn more as a product of the U.S. nation and of white male farmers' ingenuity" than of the "physical and ritual labors" of Indigenous Americans, especially Native American women.[9]

This de-Indianization had gone hand in hand with expanding cultivation. In 1866, the year that the USDA began publishing statistics on agricultural production, US farmers harvested corn from some thirty million acres of farmland. Fifty years later this extent had more than tripled, to one hundred million acres.[10] With settlers now growing corn clear across the country, the number of recognized local varieties—farmers' varieties, or what today would likely be designated as landraces—soared. In this context corn's diversity seemed to be more a product of entrepreneurial settler farming than pre-Columbian cultivation. Still, this ostensible diversification had limits. Where Native American farmers used corn for food, animal fodder, and fiber, as well as in ceremonies, and created distinct types to suit these purposes, by the end of the nineteenth century many White farmers prized corn as feed alone. Qualities such as flavor and texture lost ground to the single metric that would come to define industrial corn: yield.[11]

Established in 1907 as part of the USDA's Bureau of Plant Industry, the Office of Corn Investigations both reflected and cemented this view of what corn should be. As one summary put it, the office sought "to find out how corn growers may produce larger yields per acre, of better quality and with less labor."[12] In its first decade staff undertook studies of corn breeding and seed storage, field tests of different varieties, and investigations of various methods of cultivation. If only US farmers were armed with better practices and better varieties, the expected yield of corn per acre would leap upward from the paltry national average of twenty-six bushels per acre—or so USDA staff routinely claimed.[13] The projection of agricultural progress through increased production was central to the still-young US agricultural research infrastructure at the turn of the century. State agricultural scientists possessed what one historian has described as an "unquestioned faith in the transcendent virtue of productivity," and corn researchers were no exception.[14]

The studies conducted by the Office of Corn Investigations, of necessity, generated a stock of corn varieties. Most would not have been the corn varieties of Native American communities but instead settler farmers' varieties, which often bore distinct names. Many of these had resulted from the efforts of ordinary growers who, by selecting within lines and occasionally crossing them and selecting again over many generations, had produced varieties they took to be novel.

In most cases these farmer-breeders worked with well-established types, modifying ever so slightly the dent, flint, flour, sweet, or pop corns that were familiar to Indigenous peoples of the Americas.[15] For example, an 1866 catalog of US corn varieties listed "New England Eight-rowed," a type common among Native American farmers of the Northeast, as the source of the copper-red flint variety King Philip. In comparison to its similar-looking antecedents, King Philip—named for Metacom, or King Philip, the Wampanoag sachem during a period of violent conflict between settlers and Indigenous peoples in the 1670s—was considered "one of the best field sorts in use." It was also distinguished from the variety Browne, "an eight-rowed sub-variety, improved from the King Philip by Mr. John Browne." By 1899 at least seven varieties in circulation answered to the same general description of a reddish-colored eight-rowed flint. The line between appropriation and invention was faint, if it existed at all.[16]

Not everything was old wine in new bottles, however. Through the lucky meeting of hard or flint corns from the northeastern United States and the softer dent corns established in the South, US farmers had unknowingly established a wholly new kind of corn, one not cultivated by Indigenous North Americans. This would eventually be called Corn Belt Dent, and it would come to dominate US corn production.[17]

Enterprising farmers subjected the emerging Corn Belt Dent varieties to selection and crossing in an effort to adapt them to local contexts and needs. A classic example, which illustrates the assortment of methods and ideas deployed by one of these self-taught breeders, is that of the variety Lancaster Surecropper. Around 1860 the Hershey family of Lancaster, Pennsylvania, received a new corn variety from a US Patent Office distribution, a small flinty corn that they soon preferred to more local options. The Hersheys, especially their son Isaac, took to mixing seeds of the flint corn with other varieties at regular intervals, combining as many as six different kinds over several decades. In 1910, when other farmers started taking an interest in his corn, Isaac stopped mixing and starting selecting for certain traits in each generation. He focused on ensuring that the variety would consistently ripen early, providing a "sure crop" even if summer weather were cut short. The eventual result was Lancaster Surecropper.[18] This trajectory was typical. Another popular variety with a similar origin story was Reid's Yellow Dent, reportedly the product of an Illinois farmer's unintentional cross in 1847 of a reddish flinty corn from Ohio with a small yellow corn found locally. This mixture was then "kept pure" by family members for decades, ultimately resulting in one of the country's most prized varieties.[19]

Samples of varieties like King Philip and Lancaster Surecropper likely formed the bulk of the collections held at the Office of Corn Investigations.

In the early 1910s the office had at least one sample of Reid's Yellow Dent on hand, along with other popular varieties like Boone County White, Hickory King, and Iowa Silver Mine.[20] Varieties created by the country's settler farmers were not the only types that interested USDA staff, however. They also investigated lines developed at state agricultural research stations, where an expanding pool of professional breeders worked to consolidate their expertise and deliver outcomes promised to taxpayers by producing better agricultural plants and animals.[21] Minnesota 13 was one exemplary product of such efforts. Breeders at the University of Minnesota's Agricultural Experiment Station had begun developing this variety in the 1890s, selecting for better yield and earlier ripening in the "common corn" of the area. By 1899 demand for the seeds from these selections outstripped the station's supply. Within a couple of decades, Minnesota 13 was recommended across fifteen states, reportedly the most popular variety in the northern reaches of the US Corn Belt.[22]

The Office of Corn Investigations also took keen interest in materials from other countries. The introduction of plants from abroad into the United States had been a concern of government officials since the country's founding. They began formalizing this work in the early nineteenth century. Introduction was initially delegated to the navy, later to the US Patent Office, and in 1898 to a new Office of Foreign Seed and Plant Introduction within the USDA. Over these decades exploration missions and volunteered donations had brought in corn from various corners of the world—Spain, Argentina, Peru, China, and elsewhere.[23] "The office is always on the lookout for foreign varieties of corn which seem to have unusual and valuable characteristics that might be of value in this country," the USDA reported. The "most useful" of these were said to be investigated closely, to determine whether and how they might be improved.[24] This was certainly the case with corn seeds sent to the Office of Foreign Seed and Plant Introduction from China in 1908. They ended up in the hands of Guy Collins, a botanist at the USDA Bureau of Plant Industry. Plants grown from the seeds were unlike anything Collins had seen or read about before. He logged its unique characteristics, pleased to announce that the "new type" of corn, thanks to its unusual qualities, "enhance[d] the possibilities of breeding." New traits meant new possibilities for expanding corn production and profit.[25]

In the late nineteenth and early twentieth centuries, the number of corn varieties known in the United States, whether well-established old types, recently developed varieties, or imported curiosities, leaped ever higher. An 1866 treatise, which claimed that "to repeat here the almost endless catalogue of existing varieties would be scarcely possible," nonetheless named thirty-four of the most widely used.[26] An 1899 survey identified some five hundred different varieties, and within two decades the number had risen to more than one thousand.[27] This proliferation reflected the biological plasticity of *Zea mays* as well as the ever-increasing scale of corn production in the United States and its central place in the national economy. In 1916, the year that Howard Biggar left on his cross-country tour of reservations, the USDA *Yearbook of Agriculture* estimated the annual average of corn production in the United States at 2.7 billion bushels—with the next largest record-keeping national producers, Argentina and Austria-Hungary, at just 250 and 210 million, respectively.[28] This figure, which, according to another estimate had "doubled in 40 years and quadrupled in 60 years," represented about one-fifth of the wealth produced on US farms, more than any other two crops combined.[29]

Proliferating varieties produced local headaches alongside national wealth. How was a farmer to know which of many varieties would be best suited to the soil and weather of his particular corner of the country? In the early twentieth century, the typical advice was that a farmer who needed seeds should buy or barter them in his own neighborhood. This way he could take advantage of the selective processes that had already shaped a variety to a particular place. One advice-giver in 1907 warned Kansas growers about the varieties brought in by seed merchants from eastern states. He declared that farmers would do better with the common "badly mixed in type and not pure bred" corn already found in Kansas, since it was "better adapted to our soil and climate than even the best and purest imported varieties."[30]

Even if one looked only locally for seeds, there was still the issue of varietal choice. Would a local strain of Reid's Yellow Dent grow well? Would one of Lancaster Surecropper be better? The comparative merits of different types could be hard to discern. For this reason the USDA and many state experiment stations took on the comparison of different varieties of corn as a basic service. These evaluations generated recommended lists for particular states or regions. The same Kansas growers encouraged to source seeds from close to home were also told to focus on 19 varieties "which have proven superior in productiveness" of 112 tested by staff of the Kansas State Agricultural College Experiment Station. The station staff did not stop at these recommendations but also worked to produce higher-yielding versions of the same varieties, hoping to improve Kansas corn production through the distribution of local "well-bred seed."[31]

The USDA Office of Corn Investigations meanwhile concerned itself with the productivity of the whole country. In 1908 its staff reported on "several hundred variety tests" conducted in cooperation with farmers across the United States. Through these tests, which typically consisted in planting out several varieties in similar conditions and comparing the eventual yields, they aimed to locate promising breeding materials and the best-adapted varieties for different locations.[32] Staff of the Office of Corn Investigations were also engaged in creating new varieties, chiefly through selection and hybridization. Recognizing the impossibility of USDA personnel meeting the needs of every imaginable corn niche, those in charge of the office imagined instead a process in which they produced reasonably good strains in each region of the country, which would then "be taken up by private corn breeders for more local adaptation."[33] Corn breeding was not just a national project: it was a nationwide one.

INDIAN CORN

Howard Biggar's exploration of Native American varieties in 1916 represented a modest contribution to these state-led efforts to increase corn productivity, most of which focused on creating locally adapted "pure bred" lines. Many contemporaries would have questioned the value of assessing corn from Indian reservations as part of the work of corn "improvement," given prevailing negative opinions of Native American farming among settlers and scientists. Biggar himself typified these opinions. He referred to interviewees who cultivated popular breeders' varieties such as Minnesota 13 as "very progressive," "enterprising," or "the newer types of Indians," labels that also indicated his opinion of farmers who continued to plant older varieties.[34] Not everyone was dismissive of Native agriculture, however. In addition to the many breeders who mixed and selected settler farmers' varieties, there were also some who saw value in those grown in Native American communities and who sought these as the foundation of future lines and future profits.

Guy Collins, the USDA researcher who had been so keen on the corn brought in from China, was one such breeder. His reasons for taking interest in the corn varieties grown by Native American farmers were similar to those that led him to rhapsodize about the unusual Chinese variety: he thought they might have characteristics useful to other US farmers.[35] Consider his response to a variety that was cultivated by Navajo, Hopi, and Zuni farmers of the arid Southwest, obtained by the USDA in 1912. Upon growing it out at the government research station, Collins discovered that the variety had an unusual root structure, with one elongated central root instead of a shorter, branching system. Traveling to Arizona and New Mexico to observe it in cultivation, Collins learned the value of this atypical form. There sandy soil predominated in the cornfields and spring rains came rarely, if at all. For the corn seeds to begin growing, they needed to reach the moist soil far below the sandy surface. Farmers ensured that they did, by planting their corn uncommonly deep.

As Collins observed, "When planted by the Indian methods, the Hopi and Navajo varieties of maize have been found superior to the more improved eastern varieties for these very dry regions." This finding might have surprised settler farmers. Many dismissed Native American corn varieties as inferior, unproductive, and undifferentiated "squaw corns," judging them to be of little more value than the Indian women, denigrated as "squaws," who typically cultivated them. For Collins, however, the observation of this perfectly attuned crop and cultivation system prompted a declaration that "a canvass of the varieties of maize grown by the Indians and a careful study of the agricultural practices of the different tribes"—that is, exactly the kind of work Biggar undertook three years later—"will disclose much of interest and value to American agriculture."[36]

Biggar encountered another noted enthusiast for Native American varieties on his travels in 1916. The seed merchant Oscar H. Will of Bismarck, North Dakota, ran what Biggar judged to be "practically the only seedhouse in the United States who [has] made a study of the squaw corns of the Indians."[37] The first example of this study had been Will's receipt of a "small quantity of squaw corn" that a friend had obtained from nearby Arikara farmers. "The ears were only about the size of a good sized finger and composed of many colors," Will later recollected, indicating that they were a far cry from the large and uniform types prized on the commercial market. He planted these seeds and selected only the white kernels from the resulting crop for planting the next season, a process that he reportedly repeated for ten years until he "got the color perfect." The end product was a variety he offered in his 1888 catalog as the white flint corn Pride of Dakota, "the earliest in cultivation."[38] As the earliest, it promised to be fully mature and ready to harvest before any other variety and especially before the cold, killing winter set in.

Pride of Dakota was better known in subsequent years by the synonym Dakota White Flint and famed for enabling, as a result of its quick

[2] Oscar Will poses in a field of Dakota White Flint corn, clutching a few prized ears, in 1911. The short-statured plants are hardly recognizable as corn to anyone accustomed to the much taller varieties that dominate in commercial production today. USDA Division of Cereal Crops and Diseases, Photograph Collection.

maturation, the cultivation of corn in challenging environmental conditions in the Northwest.[39] It was also, ostensibly, little different from its predecessor. As an agronomist from the Montana Agricultural College Experiment Station explained in 1915, the local Native American farmers "had a variety which they termed the Hard White." In the midst of the recurrent violence, dislocation, and disease outbreaks of preceding decades, farmers of the northwestern tribes had not always been able to maintain their varieties in the ways their ancestors had. By the time Will encountered seeds of the Arikara white flint corn, it had mixed with other corns grown in the region and, as a result, didn't produce a crop with uniform characteristics. The agronomist concluded that Will's efforts did little more than restore the originating variety.[40]

Will collected and experimented with many corn, squash, bean, and other vegetable varieties obtained from farmers at Fort Berthold, an Indian reservation near Bismarck that was home to the Mandan, Hidatsa, and Arikara peoples. He did the same with varieties he obtained from White farmers in North Dakota and from across the country. When he found something he thought might sell, he listed it. This didn't always require active searching or even adding much value through selection. Will reportedly received what would become one of his most famous catalog offerings, the Great White Northern Bean, from a Hidatsa man named Son of a Star, who had simply shown up at Will's business sometime in the 1880s with this gift.[41]

George F. Will soon joined his father, Oscar, in collecting and breeding, eventually surpassing him in both. George had left Bismarck in 1902 to study botany at Harvard College, but his interests led him to archaeology and anthropology and brought him back to North Dakota. While still an undergraduate, he participated in excavating the remains of a Mandan earth-lodge village near Bismarck. Soon he was studying Mandan culture and language as well.[42] George later returned to Bismarck to join the family business, taking over after his father's death in 1917 while continuing his studies of Native American culture. The two ventures were entangled. The family business led George Will to be particularly curious about corn, and in the 1910s this became a significant focus of his anthropological investigations. This research reached an apex in the years leading up to his 1917 book, *Corn among the Indians of the Upper Missouri,* which he coauthored with George E. Hyde, a fellow enthusiast of Native histories and cultures.

George Will dedicated *Corn among the Indians* to his father, "who in 1881 first perceived the value of the native varieties of corn from the Upper Missouri Valley, and who began at that time the work of selecting and breeding from them, to the lasting benefit of farmers of the Northwest." According to Will and Hyde's account, settlers had neglected "the native corns of the Indians" as they entered the region, preferring instead to cultivate dent corns brought from the East and leaving the local types "to degenerate and often to disappear." Will saw his efforts to promote corns of the region, like those of his father, as undoing the baleful consequences of this neglect. "The work of collecting seed of the old Indian varieties of corn has been very successful," he and Hyde reported. "Nearly all of the sorts formerly grown by the tribes along the Missouri, from the Platte northward, have been recovered. . . . The work of breeding and crossing these native corns will now be taken up again; and it is to be hoped that hardier and heavier yielding varieties for the Northwest may be ultimately produced in abundance."[43]

Will and Hyde's account shamed settlers of the Northwest for their ignorance and lauded efforts that, like their own, would offer these farmers a chance to recover from the error of their ways. In this sense it critiqued some of the devastation that accompanied White settlement. Yet its corrective fell

39th ANNUAL CATALOG

WILL'S

PIONEER SEED HOUSE
NURSERY & GREENHOUSES
OF THE NORTHWEST

GEHU
CORN
Earliest of All

OSCAR H. WILL

DAKOTA'S
FIRST SEED
SPECIALIST

WILL'S PIONEER BRAND

SEED SPECIALISTS FOR THE NORTHWEST

*Producers of the most hardy, drouth resistant
and earliest strains of Seed Corn, Field Seed, Gard-
en Seed and Nursery Stock for Northwestern Conditions*

OSCAR H. WILL & CO. Bismarck, N.D.

[3] The 1922 catalog of Will's Pioneer Seed House honored the late Oscar Will, founder of the company, and Scattered Corn Woman, "Dakota's First Seed Specialist," who provided Will with varieties of squash, corn, and beans. The 1914 catalog described Scattered Corn as "one of the very few surviving Mandan Indians." Image from the Biodiversity Heritage Library. Contributed by US National Agricultural Library. www .biodiversitylibrary.org

far short. Although the study valued the material culture of Native Americans in the form of corn varieties, it depicted them as passive possessors of this diversity and submissive or powerless in light of agricultural, economic, and social change. According to Will and Hyde, their corn had "degenerated" as a result of being ignored by settlers. Now it would be restored—not by Native American breeders but by entrepreneurial new arrivals like Oscar Will. They sketched a path to progress via the continued expansion of settler culture.

MANDAN MAIZE

George Will's key contacts for securing corn and learning about Mandan culture were residents of the nearby Fort Berthold Reservation, especially James Holding Eagle and his mother, Scattered Corn Woman.[44] The knowledge of Scattered Corn Woman, described by Will as "an elderly Mandan matron, whose father was the last Mandan corn priest," was particularly valuable for sorting out the history and names of Mandan corn varieties.[45] Another valuable source of both seeds and information was Maxi'diwiac, also known as Waheenee or Buffalo Bird Woman, a Hidatsa woman and the wife of Son of a Star. Both women reportedly provided Will with varieties that later featured in his seed catalog.[46]

Maxi'diwiac's knowledge was the target of a series of interviews by the Presbyterian minister and ethnographer Gilbert L. Wilson in the 1910s. Wilson, who, like Will, possessed a keen interest in Native American culture, began anthropological work around Fort Berthold in 1905 while working as a pastor in the town of Mandan, North Dakota. Maxi'diwiac's recollections of the agricultural practices of her community, translated for Wilson by her son Tsaka'kasakic (Edward Goodbird), featured as the central subject of Wilson's doctoral thesis at the University of Minnesota. Over four summers Wilson spent long days at Fort Berthold recording Maxi'diwiac's memories from the time of her childhood in a village named Like-a-Fishhook, which he then compiled and edited.[47]

Maxi'diwiac's stories captured the day-to-day life of people surviving amid violence and devastation. Early twentieth-century visitors to the Fort Berthold Reservation, like Will and Wilson—and Howard Biggar of the USDA too, who arrived on 18 August 1916 and encountered Wilson in the midst of continued studies—found the Mandan, Hidatsa, and Arikara tribes living together. This arrangement had been forged of recent disaster. All

[4] By the time of this 1916 photograph, Scattered Corn Woman of Elbowoods, North Dakota, identified in the original caption as "keeper of the Mandan corn," was acknowledged by White visitors as an expert on local crops. Howard Biggar of the USDA Office of Corn Investigations described her in 1916 as the "greatest corn authority in the Mandan tribe" ("The Corn Work of the Indians of the Middle West," ARS Records, 42). USDA Division of Cereal Crops and Diseases, Photograph Collection.

three groups resided along the upper Missouri River by the late 1700s and were well known beyond the Northwest through accounts of the explorers Meriwether Lewis and William Clark in 1804, the artist and author George Catlin in 1832, and others. Visitors and traders brought their diseases with them, and in 1837 a smallpox epidemic struck. Though not the tribes' first experience of the disease, it proved a particularly deadly outbreak. According to Wilson's 1917 account, the epidemic left fewer than 150 Mandan and about 500 Hidatsa. It is today thought that only 1 in 10 Mandan survived, 5 in 10 Hidatsa, and 6 in 10 Arikara.[48] Amid chaos and despair Mandan and Hidatsa survivors formed a new village, Like-a-Fishhook, at a site near to the Fort Berthold trading post. They were joined there by many Arikara in 1862.[49]

Maxi'diwiac had been born shortly after the smallpox epidemic, around 1839, and spent the next forty-odd years living in Like-a-Fishhook. As she recalled to Wilson, soon after arriving there, "the families of my tribe began to clear fields, for gardens." Clearing was the first step in recreating the agricultural and therefore the community life that the epidemic had disrupted. Maxi'diwiac's memory of clearing fields also marked the earliest knowledge of Native American agriculture in the region available to researchers like Gilbert Wilson. This "old woman expert agriculturist in one of the oldest agricultural tribes" of the Northwest was touted by Wilson's thesis adviser as an unparalleled resource on what he considered "primitive economic activity."[50]

Shared faith among anthropologists, ethnographers, and historians in this claim and its extension to a number of older tribe members made the reservation a popular site for researchers. "The Fort Berthold Reservation has become a sort of Mecca for those interested in Indian lore and customs," Biggar observed in 1916.[51] Among anthropologists and ethnographers like Gilbert Wilson, this value was heightened by the assumed fragility and fracturing of Native American life. As the historian Brian Dippie describes, anthropology in the United States was forged at the turn of the twentieth century as a "profession with a mission"—namely, to salvage these "vanishing" cultures. Ethnography "drew upon the belief in the Vanishing American and substantially reinforced it," he writes. Practitioners sought to preserve records and artifacts of cultures whose demise was insisted on as given. Within this anthropological worldview, Maxi'diwiac and her peers were increasingly rare relics of a "primitive" past.[52]

Wilson's ethnographic enterprise centered on Maxi'diwiac's knowledge of food and agriculture, and Maxi'diwiac's knowledge of these in turn centered

[5] In his efforts to document Native American agriculture in North Dakota, Gilbert Wilson photographed a woman demonstrating the use of a bone hoe in Maxi'diwiac's garden in 1912. Wilson's caption calls attention to the height of the fully mature corn plants. Squash grows to the right. Wilson Papers. Courtesy of the Division of Anthropology, American Museum of Natural History; and the Minnesota Historical Society.

on corn. Her account credited the Mandan, who had a longer history as settled agriculturists, with first introducing the Hidatsa to corn, after which it became as central to their survival as it was to their neighbors. Among the Hidatsa, tending to gardens, including the cornfields, was mostly women's work. Maxi'diwiac recounted how her grandmothers broke the soil with bone hoes and digging sticks prior to the arrival of iron hoes and axes; how she and her mothers prepared and planted the fields with sunflowers, beans, squash, and corn; and how they harvested these crops and prepared them to eat and store. When it came to corn, she was particularly thorough, identifying myriad customs associated with planting, such as watching to ensure that birds and boys didn't come after ripening ears. She described harvesting, threshing, storing, and especially preparing ears of different types and varying stages of ripeness for a wide range of corn-based meals.[53]

Although Wilson's doctoral adviser was happy for his student simply to have added to the "scanty knowledge" on Native American agriculture, he declared himself particularly proud that Wilson's study had "unexpectedly revealed certain varieties of maize of apparently great value to agriculture in the semi-arid areas west of Minnesota." Which of the corn varieties were perceived to be valuable remains unclear, as Maxi'diwiac had identified many. There had been nine "well marked varieties of corn" in her village, with names transcribed by Wilson as: *atą'ki tso'ki* (hard white), *atą'ki* (soft white), *tsï'di tso'ki* (hard yellow), *tsï'di tapa'* (soft yellow), *ma'ïkadicakĕ* (gummy), *do'ohi* (blue), *hi'ci cĕ'pi* (dark red), *hi'tsiica* (light red), and *atą'ki aku'hi'tsiica* (pink top). The varieties were not equally common. The gummy corn, which varied in color and was used solely for making a dish of boiled dumplings, was the least frequently cultivated. The soft white, which Maxi'diwiac declared "could be made into almost every kind of corn food," was the most popular. Another common variety was the hard white. Explaining that "perhaps at first, there was but one variety of corn, atą'ki tso'ki, or hard white; and that all other varieties have sprung from it," she declared that it "was very generally raised, nearly every family in the tribe having a field of it." Her own family grew the soft white and a hard yellow.[54]

In her interviews with Wilson, Maxi'diwiac elaborated the qualities and uses of the nine varieties of corn, emphasizing that they were understood, consumed, and managed as distinct types. Her settler neighbors, cultivating corn primarily as animal fodder, might only scarcely have grasped her appreciation of these. "If at night I were given to eat of hard white corn, or hard yellow or soft yellow, I could at once tell each from any of the others," she described. "If I were given mush at night made from these three varieties, each by itself, I could distinguish each variety, not by its smell, but in my mouth by taste."[55]

The village life in Like-a-Fishhook that Maxi'diwiac recounted for Wilson had come to an end in the mid-1880s. The Fort Berthold Reservation had been created in 1870, following from a series of US government decisions that relentlessly reduced the once-sizable joint Mandan, Hidatsa, and Arikara claim defined in the Fort Laramie Treaty of 1851. As one historian has summarized, "In a period of less than twenty years a tribal claim of more than twelve million acres had been whittled down to about a tenth of its original size, mostly by administrative fiat."[56] Then, beginning in 1884 and at the behest of Indian Bureau agents, families began to move from Like-a-Fishhook to separate homesteads or "allotments." The forced disintegration of village life and its communal underpinning was formalized with the Dawes Act of 1887, which authorized the survey and division of tribal lands into allotments. This was carried out at Fort Berthold and in many Native homelands across the United States.[57]

These transitions entailed profound changes for Native American agriculture. Government policies tried to replace community-oriented subsistence farming of traditional varieties with market-oriented production of commodity crops on individual farms, a transition that settlers and employees of various government agencies saw as "improving" or, later, "modernizing" Native American agriculture. As Maxi'diwiac explained, the US government did not just move people from the village to allotments in their bid to encourage different living and farming habits. Through the work of Indian Bureau agents, they also plowed up new fields. They issued seeds of wheat, oats, and unfamiliar vegetables and demanded the cultivation of potatoes. As a result, she summarized, "our old agriculture has in a measure fallen into disuse."[58] An agricultural tradition on the upper Missouri River that stretched back to the 1400s had been upended.

If interest in potentially useful corn varieties was what brought both Howard Biggar and George Will to Fort Berthold, this mission was given urgency by awareness of rapid change. They shared ethnographers' conviction that Native knowledge was disappearing and thought that the opportunity to make use of such knowledge in agricultural production would disappear along with it. Will and Hyde captured this urgency in introducing *Corn among the Indians:* "This work should have been undertaken fifty years ago when a great deal of material, now lost, was still available." They assumed that their efforts had missed extant varieties, types now too infrequently cultivated to be easily found, which they anticipated would "be lost with the death of the older Indians, who alone know of these things."[59] For Will this trajectory was no doubt confirmed by the surprise expressed by Scattered Corn Woman on seeing some of the varieties that he had collected. "Say Mr Will my mother was very glad to see the ear of corn you send. . . . She never thought you would have this Ear of Corn," her son reported in response to Will's delivery of some ears in 1917.[60] Biggar had a similar experience on his travels. When he told the Fort Berthold resident Red Bear that he had located a yellow-and-pink-striped corn on a few other reservations, Red Bear responded with interest. He thought this corn had disappeared completely from cultivation in Fort Berthold, a view that likely confirmed for Biggar the need to move quickly to successfully gather all the known varieties of Native American corn.[61]

On the reservation, meanwhile, many resisted the agricultural changes demanded by US government representatives. Maxi'diwiac reported that potato cultivation had become common only once the local agent realized that he could not expect the farmers of Fort Berthold to eat these and decided to purchase them for sale elsewhere. Other vegetables were similarly rejected. According to Maxi'diwiac, "We tried to eat some of these vegetables and did not like them and so let them stand in the field." The result, observed by Wilson, was that although there were some of the "white man's vegetables" grown on the reservation, "still there are many gardens in which are cultivated only the four vegetables [e.g., corn, squash, beans, and sunflower] originally used in this tribe."[62] Even as they experienced hardship and loss, residents of Fort Berthold did not simply acquiesce to outsiders' impositions. Inevitable vanishing was not their narrative.

It was instead the narrative of salvagers, whether ethnographers or agriculturists. It served to advance the economic and political interests of settlers and the state, and those of scientists too. Visions of looming and inevitable extinction—or, if not outright extinction, then inexorable assimilation—lay behind many attitudes toward and impositions on Native American people from the eighteenth century well into the twentieth and in some cases continuing to the present: forced removals, educational measures, social science investigations, and biomedical research.[63] The same before-it's-too-late mentality was common to many anthropological and ethnographic endeavors. The researchers that converged at sites like Fort Berthold Reservation in the 1910s to extract knowledge from residents—amateur ethnographers, professional anthropologists, local historians, linguists, and more—often viewed their enterprise as time-sensitive.[64]

For agriculturists like Biggar and Will, plants were as much in need of salvage as knowledge about them. It was this aspect that set their collecting apart from comparable efforts to assemble crop varieties undertaken in the United States at the time. Consider the activities of the Office of Corn Investigations, which included gathering varieties from state experiment stations, requesting corn from foreign consulates, and evaluating samples donated by missionaries abroad. These were not done with the thought that varieties would go extinct or that some unusual Chinese adaptations or impressive Mexican lines would be lost if not stockpiled immediately. Native American corn varieties were set apart: in the 1910s, it was collect now, or never.

HYBRID CORN

A later generation would wish that Biggar, Will, and others like them *had* been as concerned about the apparently robust and proliferating local varieties maintained by farmers across the United States as they were about "vanishing" Native American varieties. A sea change in corn breeding was underway in the United States, spurred on by increasingly sophisticated knowledge of plant heredity and the promise of greater profits to seed companies. The science and industry of hybrid corn lay ahead.

It's hard to understand the magnitude of this change without first appreciating the change in breeding practices that underlie the production of hybrid corn. In the late nineteenth and early twentieth centuries, most varieties were what we call today open-pollinated: populations maintained without exercising direct control over cross-fertilization. Mass selection was the chief method used to make these naturally variable populations more uniform. This "improvement" usually meant selecting seeds from individual plants with desirable traits, like a consistent color or extra rows of kernels, season after season, a breeding method compatible with open pollination. Sometimes breeders also controlled pollination to create inbred lines. By collecting and deliberately redistributing pollen from the corn's tassels to the silks of developing ears, farmers, experiment station workers, and other breeders ensured that "superior" plants fertilized themselves or siblings. This in turn made their desirable traits ever more prevalent in the subsequent population. Although it resulted in more uniform ears, this strategy also decreased plants' fitness over time, a phenomenon known as inbreeding depression, and eventually diminished yields. A pretty ear was not necessarily a profitable ear, which made inbred lines and corn shows that encouraged them controversial.[65]

Growing disdain for inbreeding among professional corn breeders took a sharp U-turn with the invention of techniques that took advantage of its capacity for generating uniformity while avoiding inbreeding depression. The secret lay in the phenomenon of heterosis, commonly known as hybrid vigor. Botanists and breeders had long been aware that, in many species, crossing two genetically distinct parents resulted in more vigorous offspring. More recently they had observed that crossing two inbred lines of corn typically produced an offspring more vigorous than either of its parents. Yet the plants grown from the seeds of this attractive first generation were nowhere near as productive. The hybrid vigor characterized only the first generation, a biological pattern that made the use of hybrid vigor in corn breeding tantalizing but infeasible. Farmers would not be able to grow the same seeds from one season to the next. Even if they could be convinced to buy a fresh supply of seeds each year, the cost of producing hybrid seeds was bound to be too high since the effects of inbreeding depression on each parent line led to low yields of the seeds necessary to make the hybrid generation.[66]

In the late 1910s a young breeder at the Connecticut Agricultural Experiment Station, Donald Jones, proposed a technical fix to this dilemma: Start with four distinct inbred lines (A, B, C, and D) instead of two. Cross A with B and C with D to create two different first-generation lines, AB and CD, each of which displays hybrid vigor. Because these first-generation lines do not suffer from inbreeding depression to the same extent as the parent lines, they can produce a reasonable amount of seeds from which to create a second hybrid generation, ABCD.[67] These seeds could then be offered for sale. The promise of this method was measured in yield per acre and especially in profits for seed companies that envisioned guaranteed return customers each year among farmers who transitioned to hybrids. In a comparatively short time, it led to a flurry in production of inbred lines. Starting with a sample of some open-pollinated varieties like Lancaster Surecropper or Reid's Yellow Dent, breeders self-pollinated (or sibling-pollinated with plants that shared parents) these over six or more generations. The resulting inbred lines then became the basis of the "double-cross hybrids" proposed by Jones.

As attention to double-cross hybrids grew in the United States, earlier methods of breeding corn were gradually neglected. An established practice of encouraging farmers to act as maintainers and improvers of their own corn varieties was all but abandoned. The production of hybrid lines would be the business of experiment stations and, in time, commercial seed companies.[68] The first company founded explicitly around the production and sale of hybrid varieties was the Hi-Bred Corn Company, launched in 1926 by the agricultural writer and publisher Henry A. Wallace and rebranded as the Pioneer Hi-Bred Corn Company a decade later.[69]

Through the promotional efforts of Wallace and others, aimed both at breeders and farmers, hybrid production and sales took off in the 1930s. In 1924, even before founding his company, Wallace had marketed a variety called Copper Cross, one of the first hybrids offered commercially. Copper Cross's competition in cornfields was the celebrated farmers' varieties of preceding decades: Reid's Yellow Dent, Lancaster Surecropper, Minnesota 13, and others. Twenty years later the cornfields had changed. "In the summer of 1944 one could drive across the state without seeing a single field of open-pollinated corn," reported the botanist Edgar Anderson of a trip through Iowa, a state famous for its corn harvests.[70] More recent historical work suggests Anderson's stark accounting was not far off. Hybrids represented less than 1 percent of acres planted to corn in 1933. By 1945 they accounted for 90 percent.[71]

Anderson had a privileged vantage point from which to observe changes in corn cultivation. In the late 1930s he was invited by Henry Wallace first to lecture at the Iowa headquarters of Pioneer Hi-Bred and then "to come back and study their inbreds and hybrids in any way [he] pleased." The result was a "gentleman's agreement" between the head of Pioneer's corn-breeding operations and Anderson, that each would take a continued interest in the other's work. This translated into several decades of no-strings-attached financial and technical support for Anderson and his students in their studies of corn.[72]

Although Wallace, the businessman, surely celebrated his friend Anderson's observation of hybrids' dominance across the Corn Belt, the expert in plant genetics saw cause for concern. It suggested to Anderson that "the whole genetic pattern of *Zea Mays* in the United States has been catastrophically overhauled."[73] He knew that the hybrids he saw on his drives across the Midwest had been created from various inbred lines of earlier open-pollinated varieties—but which varieties? Had they come from many different original types or just a few?

In an attempt to find out, Anderson picked six of the most popular double-cross hybrids on the market, which he thought accounted for about one-quarter of Iowa's annual corn production. Reconstructing the lineage of these varieties, he discovered that the parent inbred lines (of which there were twenty-four, four for each double-cross variety) derived from just three open-pollinated varieties: Reid's Yellow Dent, Lancaster Surecropper, and a variety called Krug. It was well known that Krug was closely related to Reid's Yellow Dent, created by a farmer who had combined that variety with another called Iowa Gold Mine.[74] As Anderson concluded, "We are faced with the remarkable fact therefore that much of the corn now being grown traces back mainly to two open-pollinated varieties." As if that were not striking enough, some widely circulating inbred lines, used in the development of different hybrids, appeared to be traced to the same original plant—even to a single ear of corn![75]

For Anderson this raised a host of alarming prospects. Geneticists and corn breeders did not know much about the origins of different Corn Belt

Dent varieties, nor what made one a better source of inbred lines than another. As a farmers' variety, Lancaster Surecropper had not been as successful as some others. Now as a source of inbred lines, it was highly prized. Researchers were also ignorant about what combinations of genetic material were most effective. These concerns led Anderson to wonder whether there was value in further study of the open-pollinated varieties that had once appeared on the market and in the field. It was this last point—the need to study earlier open-pollinated corn varieties to ensure better industrial corn in the future—that led Anderson to an unusual proposal. "It may be necessary for breeders of hybrid corn to subsidize highly skilled farmers as raisers of open-pollinated corn," he urged. This needn't be a permanent arrangement but one carried out only until "we have at least an estimate of when, where, and how the majority of the useful gene combinations in hybrid corn were accomplished."[76]

Although he thought it important to "move swiftly," given the rapid decline in open-pollinated fields, Anderson's proposal was not quite the salvage mission that Howard Biggar's or George Will's pursuit of Native American corn varieties had been. For the moment Anderson was less concerned about the corn varieties themselves than about the genetic knowledge that needed to be extracted from them. Yet there was an important similarity between his proposal and the earlier salvage efforts. He also imagined an inevitable, irreversible transition. Convinced that farmers would never value the old varieties of corn again, Anderson decided that they would need to be paid to grow it, as a service to breeders. George Will had also advocated preserving varieties through continued cultivation and similarly felt that growers would need to be convinced that this was worth their time and labor. Rather than sponsor farmers to cultivate Native American varieties, however, Will transformed these farmers' lines into breeders' lines, hoping they would find a secure place in the market. In both cases any future for these varieties would little resemble their past.

Later accounts painted Oscar and George Will as men ahead of their time, breeders who took care to preserve varieties not valued in the agricultural mainstream of the early twentieth century. "A focus on bioregionally-adapted heirlooms as a basis for place-specific agriculture is not at all new," observed an ethnobotanist in 1986. "In fact, it was in many ways pioneered over a century ago (1883) by horticulturist Oscar H. Will, whose son and grandsons continued to strengthen this tradition until 1959."[77] But focusing on the Wills' efforts to conserve and use crop diversity, even with their evident admiration of Native American knowledge and culture, obscures the conviction that motivated their collecting: the unlikeliness, or even impossibility, of Native American communities surviving. Collecting crops was essential only because guaranteeing a future for those who farmed them was inconceivable.[78]

Assessing the narrative of Indigenous extinction reveals the vicious cycle it set in motion for would-be corn conservators. Oscar Will's Pride of Dakota exemplifies how western expansion fueled interest in Native American varieties. Adopted by settler farmers, these varieties became a resource for further expansion, with all its ill effects for Native American communities: disease, dislocation, dispossession, and more. The Oscar H. Will seed company generated profits by fostering settler agriculture but in the process undermined the very agricultural systems—Mandan, Hidatsa, and Arikara—that the Wills and others saw as a key resource for agricultural development. Making "Indian corn" suitable for settler cultivation, even if it resulted in the continued farming of Mandan corn or other Native American varieties, would never address the root causes of these varieties' perceived endangerment.

By the 1940s the imagined fate of Native American varieties appeared to be the actual fate of other open-pollinated varieties, whether settler farmers' varieties or breeders' lines, as hybrid corn rose rapidly to dominance. At least one observer, Edgar Anderson, thought that attention should be given to keeping these in circulation, as Will had done for Native American varieties. Anderson, again like Will, did not upset the prevailing order by suggesting the rush to hybrids be resisted. He instead recommended the creation of a new farming professional—paid conservator—who would be compensated for the costs of staying out of the hybrid revolution. Anderson, who often swam against the tide in his scientific research, in this case moved with the prevailing currents of agricultural industrialization. He embraced a view common in crop conservation, assuming that farmers who did not adopt the latest breeders' varieties were bound to suffer economically.

Until the late 1940s the dwindling number of varieties in large-scale cultivation was worrying to only a handful of people invested in corn breeding. The scale of threat to corn diversity imagined by US breeders was soon to change, however, as they imagined a worldwide rollout of hybrid corn. Mexico, the site of corn's domestication and a hotbed of corn diversity, grabbed attention first, thanks to state projects that aimed to transform

Indigenous farmers and campesinos, or peasants, into "modern" agriculturists. Some experts worried that for peoples whose cultures were organized around subsistence production, and specifically around *maíz,* these state programs portended the end of distinct ways of life—and with it the end of distinct kinds of corn.

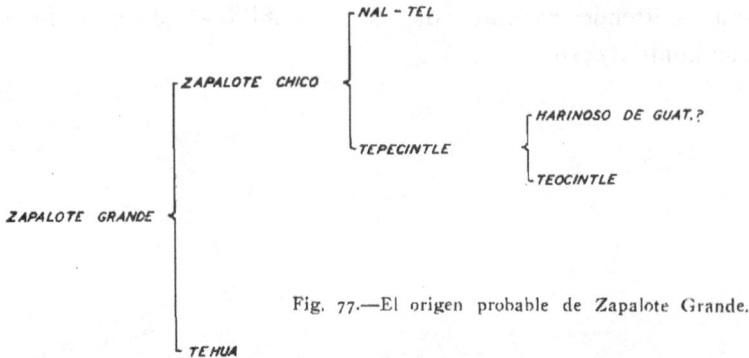

ZAPALOTE GRANDE ⟨

⟨ ZAPALOTE CHICO ⟨

NAL - TEL

⟨ TEPECINTLE ⟨

HARINOSO DE GUAT.?

TEOCINTLE

TEHUA

Fig. 77.—El origen probable de Zapalote Grande.

[6] In 1951 scientists working in Mexico published an account of the country's "races of maize," which classified common types of corn and established evolutionary relationships among them. They considered the race Zapalote Grande *(center)*, one of the "prehistoric mestizo races," to have originated in crosses between the race Zapalote Chico *(left)* and a type with thick ears, yellow kernels, and many rows, such as the race Tehua. From Wellhausen, Roberts, and Hernández Xolocotzi, *Razas de maíz*, 141. Reproduced with permission of Secretaría de Agricultura y Desarrollo Rural, Mexico City.

TWO

CLASSIFY

"THE SOUTH AND SOUTHWEST OF MEXICO constitute one of the great culture hearths of the world," the influential geographer Carl Sauer declared in 1941. He had been studying the history, culture, and landscape of Mexico for fifteen years, research that led him to conclude that the historical peoples of the region had possessed "an economic complex that is one of the great achievements of mankind." Evidence of this civilizational attainment lay in the agriculture of those who still lived in the region. "Of maizes there are in the west not only a great many kinds of the dent variety but many flour, sugar, pop, and flint corns, which have never been collected or classified," Sauer reported. "The economic botanist does not yet know the wealth of maize, beans, chili, squash, upland cotton, amaranth, and tomatoes that marks the hill lands behind the west coast."[1]

Sauer's praise for the Indigenous cultures of Mexico aligned with a view adopted among Mexican intellectuals and pursued in state policies after the Mexican Revolution of the 1910s. In forging a new nationalist narrative, influential Mexican thinkers rejected an earlier elite perspective that saw little room for Indigenous peoples—a large, visible, and diverse part of the national population—in a modernizing, Europeanizing Mexico. In the 1920s many instead espoused *indigenismo*. They celebrated Indigenous cultures, especially their pre-Colombian attainments, and articulated a vision in which contemporary Indigenous peoples would be integrated into a unified mestizo nation.[2]

Even if he shared elements of the *indigenista* valorization of Mexico's diverse peoples and cultures, Sauer was pessimistic about some of the economic projects that aligned with an integrationist perspective. Agricultural programs, in particular, seemed to ignore the value of local expertise. Sauer thought the "scientific agriculture" of Latin American countries possessed a "dangerous imitativeness" of US agriculture, not least in its "contempt for native crops and methods."[3] By 1940 he was convinced that state agricultural programs would eliminate diverse crop varieties grown by Mexico's peasant and Indigenous farmers, just as hybrid corn had displaced farmers' varieties in the United States.[4] His concern was soon shared among a number of US scientists interested in the genetics, evolutionary history, and improvement of corn. By collecting and classifying samples of Mexican maize, they sought to understand and preserve this resource in advance of what they considered its inevitable decimation.

For much of the twentieth century, breeders, agronomists, and plant explorers in the United States and Europe were quick to extrapolate from their countries' agricultural experiences in projecting what would happen in other contexts. This was especially true when it came to crop diversity. "The perceived genetic erosion of food production in industrialized countries became testament to the future of agricultural production elsewhere," write Susannah Chapman and Paul Heald.[5] This assessment captures perfectly the assumptions of Sauer and his contemporaries about the effects of implementing corn-breeding programs in Mexico. Their evidence for what would happen came more from their observations of agriculture in the United States than of transitions in Mexico. It also explains their response: they tried to mobilize collection and conservation of Mexican farmers' varieties, assuming their extinction was already in progress.

But this remains only one aspect of the story. The prevailing narrative of Mexico's evolution as a nation, and especially the trajectory of Indigenous peoples within it, also shaped early efforts to understand and conserve Mexican corn diversity. In the 1920s and later, successful integration of rural Indigenous peoples into the Mexican nation—that is, their melding into a society whose ethnic identity would be, simply, mestizo—seemed to depend in part on standardizing and industrializing corn. Mexico's pursuit of modernity meant the "modernization" of Indigenous peoples, a process that demanded traditional agricultural methods and crop varieties, corn chief among them, be abandoned for "improved" ones. In other words, the vision for Mexico's economic, political, and racial evolution was tied to that imagined for corn. This lent unique urgency to efforts to conserve Mexican corn diversity and sharply constrained the forms it was imagined to take.

Maíz occupies a central place in Mexican history and culture. For centuries peoples of Mexico have identified themselves as being, literally, of corn, and it continues today to be synonymous with Mexican identity and nationhood. As the anthropologist Guillermo Bonfil Batalla once described, in Mexico "maize, society, culture and history are inseparable. Our past and our present have their basis in maize. Our life is based in maize. We are people of maize."[6] Over millennia farmers cultivating corn across Mexico's diverse physical and cultural landscapes produced tremendous diversity in the crop: tall in stature and squat, with broad floury kernels and tiny flinty ones, fast maturing and slow, drought tolerant and irrigation dependent.[7] This vast diversity was perhaps the unifying characteristic of Mexican maize—and, in the early twentieth century, scientists in the United States had yet to recognize it. Once they did, they almost immediately began to understand it as a valuable resource, not only for improving agricultural production but also for understanding its history.

Mexican researchers started classifying the nation's maize varieties in the nineteenth century. A survey published in 1846 began by determining which of the types cataloged by European cultivators were also grown in Mexico.[8] A more throughgoing account appeared in 1913, based on reports and samples of corn submitted to the government's Central Agricultural Station (Estación Agrícola Central) by dozens of farmers across the country. The results of this survey enabled the agronomist Eduardo Chávez to describe and photograph fifty-six varieties grown in Mexico. These included dark-purple *maíz morado* from Yucatán, wide-kerneled *maíz ancho* from Guerrero, rain-fed white *maíz pepitilla* from San Luis Potosí, and many others.[9]

In the 1920s domestic efforts to chart the diversity of corn across Mexico were overtaken by those of Soviet researchers from the All-Union Institute of Applied Botany and New Crops in Leningrad. The institute's director, the geneticist and plant breeder Nikolai Vavilov, had developed a theory about the "centers of origin" of crop diversity that linked the prehistoric site of a particular crop's domestication with the region of its greatest contemporary genetic diversity.[10] As part of his ambition to investigate the centers of origin of all the world's crops, Vavilov orchestrated an agricultural survey of the Americas and Caribbean in 1925 and 1926. One team traveled through Mexico, Guatemala, and Colombia, with a quick pass through Panama,

Venezuela, Cuba, Curaçao, Trinidad, and Barbados, while a second journeyed to Peru, Bolivia, and Chile. The two teams gathered about 2,800 samples of corn on this expedition, which they used to identify concentrations of diversity in Mexico and Peru. They ultimately settled on Mexico as corn's center of origin—that is, the place where domesticated *Zea mays* had first appeared. Their 800 samples of Mexican maize confirmed, above all, the "extreme diversity" of forms found in that country. Among dent corns, for example, the Soviet scientists declared "a variety of forms and types not met with anywhere else in the globe."[11]

The Soviet agricultural tours across the Americas were not only, or even chiefly, an academic enterprise. Soviet agriculture depended in part on crops originating in the Americas, like corn and potatoes. Plant exploration sustained the state-led agricultural modernization efforts of the All-Union Institute of Applied Botany and New Crops, providing Soviet breeders with the resources they needed to create and disseminate new varieties.[12] According to one of the Soviet explorers, history had shown that many introductions of American crops to Europe had come from a very few specimens—in some cases, as few as one. By comparison, he explained, a crop species in its place of origin "exhibits infinite diversity," which "would open great possibilities to plant breeding." The thousands of samples that he and his fellow explorers obtained were prized resources for the agricultural future as much as keys to an evolutionary past.[13]

Soviet studies of the origin and diversification of crops are said to have inspired Carl Sauer's research on crop domestication and evolution.[14] Trained at the University of Chicago, where he studied the physical geography and patterns of settlement and land use in the midwestern United States, Sauer had moved to California in 1923 to take a position at Berkeley. Soon he began traveling to Mexico, studying a landscape he considered uncharted territory among geographers. Equal parts historian, anthropologist, and geographer, Sauer's effort to acquaint himself with the lands and peoples he encountered in Mexico immersed him in colonial and pre-Hispanic history and culture. His studies—perhaps in concert with his reading of Vavilov—also led him to consider the origins and history of Mexico's staple crops. He thought these might shed light on Indigenous cultures and land use, past and present. Sauer started collecting samples of corn, beans, and squash on his travels, often sharing them with colleagues who could conduct further botanical analyses.[15]

Sauer prized these still-cultivated grains and vegetables as clues to life in pre-Columbian America. In 1941 he wrote to Edgar Anderson at the

Missouri Botanical Garden, offering to send the geneticist samples of Mexican corn that an archaeologist colleague had just collected. The corn had come from remote communities, high in the mountains and far from markets, some of which Sauer claimed had "never been visited by an American or European before." He hoped that Anderson, who was also studying the history of corn, would share any insights useful to Sauer's studies of "cultural diffusion." Sauer's description of these samples, which had needed to be "packed on muleback considerable distance to bring them out" was as much an account of his methods and assumptions as it was of the corn itself. He considered remote places like these mountain communities ideal sites to recover pre-Columbian cultures, believing them to be disconnected from wider networks of economic exchange and isolated from outside influences. For Sauer, visiting these communities was akin to stepping into the archive or visiting an archaeological site, enabling the recovery of "archaic forms" of corn and other crops from living farmers.[16] These would help him tell the history of agriculture.

By the 1930s Sauer was convinced that the window of opportunity for this kind of collecting, and the research dependent on it, was closing. He was particularly gloomy about the effects of new state-sponsored agricultural programs and the breeder-improved varieties they promised.[17] The presidential administration of Lázaro Cárdenas del Río (1934–40) had set some of these programs in motion, as part of reforms that aimed to improve conditions for rural Mexicans. With grinding poverty and continued oppression by wealthy landlords generating instability in the Mexican countryside, Cárdenas and other reformers understood that the future of the Mexican state depended on profound transformation.

The most important of the Cárdenas-era interventions was land reform. By the early twentieth century, ownership of land in Mexico had become concentrated among a rich elite, often in the form of vast, private haciendas that comprised lands earlier taken from campesinos, or peasants, many of whom were also Indigenous. Landless campesinos eked out their existence by contributing labor, forced by circumstance, to the haciendas. Righting this inequality through the redistribution of land was a central demand of the Mexican Revolution and had been promised in the constitution of 1917. Several postrevolution presidential administrations made piecemeal efforts at redistribution, but land reform ground to a halt in the early 1930s. Cárdenas reinstated it, eventually going far beyond his predecessors in converting tens of millions of acres of expansive, and often foreign-owned, haciendas into ejidos, lands owned by the state and collectively farmed and managed by peasants.[18]

In his bid to transform the countryside, Cárdenas also invigorated agricultural science. He directed funds to long-neglected agricultural experiment stations, expanded extension activities, and founded a new biological research center, the Biotechnical Institute (Instituto Biotécnico). These ventures were led by a growing number of professional agronomists, or *agrónomos,* including graduates of the decades-old National School of Agriculture (Escuela Nacional de Agricultura). Many agronomists, and some Mexican leaders as well, thought that domestic agricultural woes could be solved by emulating foreign research programs. They blamed low productivity—which reduced campesinos to poverty, created unrest in the countryside, and threatened the country's ability to feed its urban laborers—on farmers who grew crops using traditional techniques. In a bid to transform these circumstances, they pursued intensification and mechanization on the model seen in the United States and targeted campesinos through various interventions.[19]

Among the many perceived agricultural failings of peasants, poor seed choice featured prominently. The historian Joseph Cotter suggests that campesinos were "chastised" for relying on local varieties or selecting seeds "improperly" and urged to buy certified seeds of imported varieties. For some agronomists good seeds were fundamental to agrarian reform, essential for the government to achieve its cultural and economic goals. That was certainly the case for the agronomist Antonio Rivas Tagle, who equated good seeds with better diets and therefore healthier bodies and better intellectual development. "It is evident that the selection of the richest seed will result in the improvement of our population," wrote Rivas.[20]

Carl Sauer, meanwhile, thought the clamor for scientific agriculture in Mexico, including the pursuit of "good" seeds, would soon undermine the country's rich crop diversity. In 1935 he began looking for funding to support "the collection, preservation, propagation and genetic investigation of native American cultivated plants," namely, corn, beans, squash, and cotton.[21] He also invited collaboration from colleagues who shared his interests. One of his correspondents, the botanist and breeder Paul Mangelsdorf, was so alarmed after hearing from Sauer that he immediately brought Sauer's concerns about "disappearing" varieties to the attention of the US Department of Agriculture. "[Sauer] states that as a result of activities upon the part of the Mexican Department of Agriculture, these old local strains are being replaced by so-called improved varieties and that they will soon disappear,"

Mangelsdorf reported. He urged the head of corn breeding at the USDA to take some action on the matter.[22]

Sauer needed a collaborator who actually cultivated corn to preserve the Mexican varieties he saw as endangered, and Mangelsdorf, then leading a corn-breeding program at a Texas agricultural experiment station, was in a position to help. Sauer offered to start collecting maize varieties on his next journey to Mexico if Mangelsdorf would maintain them. But Mangelsdorf thought that both collecting and maintaining corn varieties would be better managed by his USDA colleagues. As he pointed out, between its programs of plant exploration and plant breeding, the federal government could undertake these tasks and also pass samples on to researchers for study, breeders for evaluation, and even to the Smithsonian Institution in Washington, DC, for display. Although pressed by Mangelsdorf to consider this possibility, the USDA demurred.[23]

PRIMITIVE CORN

Sauer's proposed preservation project was never realized. Paul Mangelsdorf, however, went on to establish a collection of corn varieties from not just Mexico but across the Americas. When Sauer reached out to him in 1940, Mangelsdorf was already saving samples of maize from far-flung places. He was less interested in big questions about human history and culture than in the history of the plant itself. What were the ancestors of corn, and how had these combined or transformed to create the modern crop plant? Though his objectives differed, his research materials were very much the same as Sauer's. He, too, prized samples of farmers' varieties thought to have generations-long histories in a particular site and hunted down types he considered "native" or "primitive" among remote places and peoples.

Old debates about the origin of corn gathered new energy in the early twentieth century, as botanists, breeders, and geneticists became more knowledgeable about the genetics of corn, the diversity of its forms, and the nature of its closest relatives. Vavilov's theories added fuel to the fire. Early efforts to resolve the where and how of corn's evolution had emphasized comparative morphology and sometimes relied on archaeological finds or textual evidence in natural histories and explorers' accounts. By the mid-1920s more experimentally minded researchers focused on sorting out the genealogy of corn through hybridizations and cytological analyses.[24] Many agreed that corn had its closest relatives among two genera: the wild grasses of the genus *Tripsacum* and another weedy grass from Mexico and Guatemala called teosinte, then classified in the genus *Euchlaena*.[25]

Since no corn had ever been discovered growing wild, the secret of its origin was thought to lie among these relatives, especially teosinte. Corn and teosinte had the same number of chromosomes, and the two readily hybridized to produce fertile offspring. Some botanists thought they ought to be placed in the same genus—as they are today—and further proposed teosinte as the wild ancestor of corn. The main problem with this hypothesis was that the female flowers of teosinte and the seed spike that results from them look nothing like those of corn. How could a tiny spike of six to twelve seeds, which readily shatters to disperse them when ripe, have transformed into the immense ear of hundreds of seeds that cling patiently to the cob awaiting a farmer?[26] A wild corn intermediate between teosinte and *Zea mays* might have lent greater credence to this idea, but no such thing was known to exist.

This ambiguity left the door open to further hypothesizing, and Mangelsdorf was among those who strode confidently through. He made his first contribution to the debates while working as a breeder in Texas. He had recently completed his graduate studies with two leading maize geneticists, Edward East at Harvard and Donald Jones (of double-cross hybrid fame) at the Connecticut Agricultural Experiment Station. Although his main responsibilities in Texas lay in crop improvement, Mangelsdorf was eager to establish a genetics research program. Searching for a topic in Texas cornfields, his eye landed on *Tripsacum*. He had seen species of *Tripsacum* years before and been struck by some of their similarities to corn. He'd thought then that perhaps the two plants had been placed too far apart on botanists' evolutionary tree. Now he saw stands of wild *Tripsacum* growing adjacent to cornfields, and this got him wondering again. He invited a cytologist, Robert Reeves, to join him in a cytogenetic study of corn and its relatives, one that combined visual analysis of their chromosomes with experimental hybridizations meant to illuminate patterns of inheritance.[27]

Mangelsdorf and Reeves soon arrived at their own hypothesis about the evolution of corn. They imagined that *Tripsacum* and an ancestral *Zea* shared a common remote ancestor but had diverged and evolved independently, *Zea* in South America and *Tripsacum* to the north. The first teosinte had emerged when these two branches of the family tree recombined at a later point, somewhere in Guatemala. Successive hybridizations between teosinte and *Zea* in Central and North America had produced further teosintes and *Zea* varia-

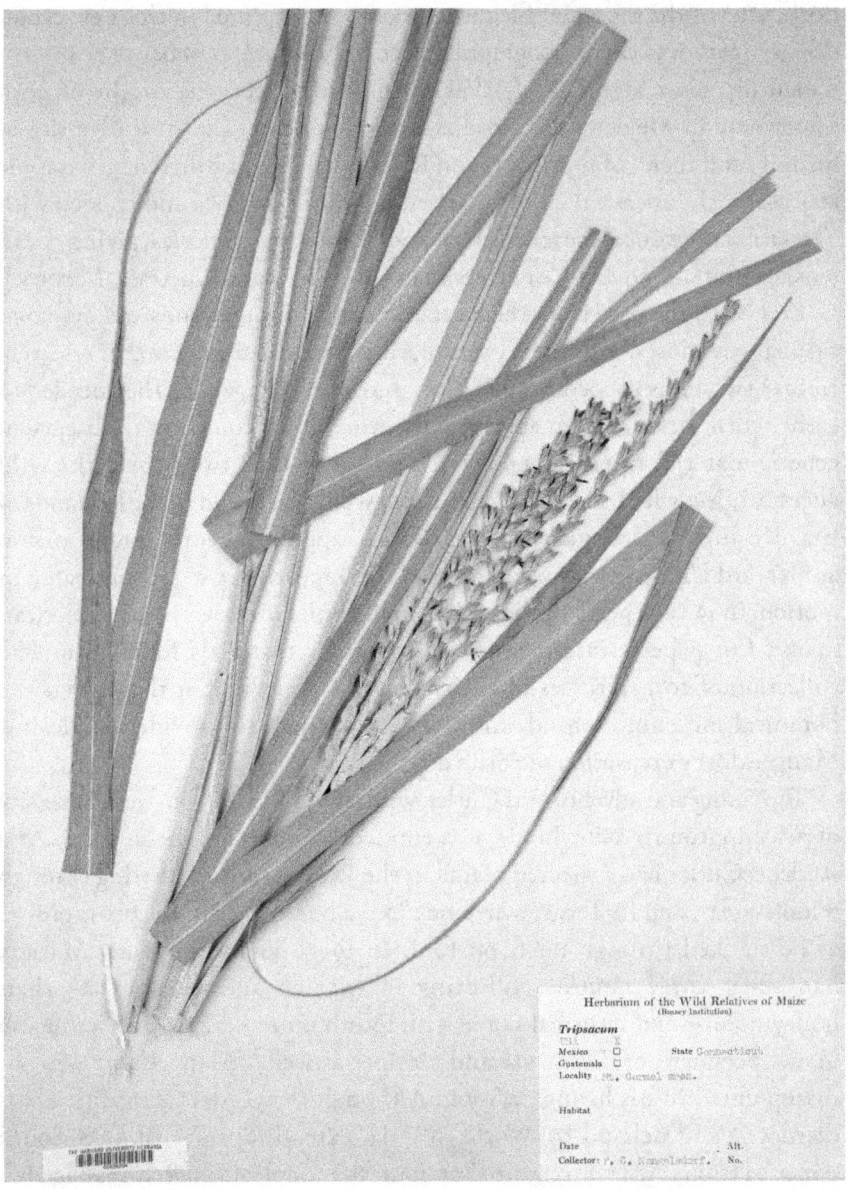

Herbarium of the Wild Relatives of Maize
(Bussey Institution)

Tripsacum

Mexico ☐
Guatemala ☐   State Connecticut
Locality Mt. Carmel area.

Habitat

Date   Alt.
Collector: P. C. Mangelsdorf.   No.

THE HARVARD UNIVERSITY HERBARIA

[7] Early in his career Paul Mangelsdorf preserved this specimen of *Tripsacum* from Connecticut in the Herbarium of the Wild Relatives of Maize at Harvard University's Bussey Institution. Courtesy of the New England Botanical Club.

tions. One of the most significant revisions encompassed in this new evolutionary story was corn's geographic place of origin. A number of studies—including, most significantly, Vavilov's—had placed the origin of corn somewhere in Mexico or Central America, based on the great diversity of forms found there. Mangelsdorf and Reeves imagined a different geographic trajectory: the ancestral *Zea* had arisen in South America and crossed with *Tripsacum* only once it reached cultivation in Central America, giving rise to teosinte and subsequent *Zea* diversity in Guatemala and especially Mexico.[28]

Even with the data from their experimental hybridizations and cytological analyses, most of this theory was speculative, demanding further research. In 1938 two tasks topped Mangelsdorf and Reeves's agenda. They needed to gather corn varieties from across South America to confirm the absence of genetic material from *Tripsacum*. They also wanted to discover the wild ancestral *Zea,* which the theory predicted would be found in the lowlands of Brazil, Bolivia, or Paraguay.[29] Mangelsdorf's appointment to a professorship at Harvard University in 1940 provided the opportunity to set this search in motion. In 1941 he hired the botanist Hugh Cutler to survey South American maize. On paper Cutler's task was to gather materials for a "complete collection of corn varieties of Latin America" to exhibit at the university's botanical museum. Behind this was the mandate to provide samples for Mangelsdorf's excavation of corn's deep past.[30]

The young and adventurous Cutler, with a recent PhD from the University of Washington in Saint Louis, was eminently suited for the mission. As a student, Cutler had collected plants in the US Southwest, boating through remote areas, and in Texas, where he hired a burro to haul his plant presses as he trekked through fields on foot. In 1940 he and his wife, Marian, had spent three months collecting in Mexico and Guatemala—their honeymoon—and returned home with four hundred ears of local corn and many specimens of *Tripsacum* and teosinte as well.[31] Mangelsdorf was not disappointed in his hiring decision. Although the Cutlers' expedition was disrupted and delayed by World War II, extending their time in South America from one year to more than four, the couple ultimately managed to ship hundreds of samples of maize back to an expectant Mangelsdorf.

The Cutlers' letters and diaries show just how hard it was to gather samples of these farmers' varieties, which Hugh Cutler and Mangelsdorf usually referred to as "primitive types" or "native corn." All three of the countries slated for the Cutlers' trip had experienced significant political and social upheaval in the 1930s. Paraguay and Bolivia had been at war from 1932 to

[8] Hugh and Marian Cutler, around 1941, pack ears of corn to ship from Cochabamba, Bolivia, to Paul Mangelsdorf at the Harvard University Botanical Museum. The stack of paper in the right foreground is probably a collection of pressed plants destined for herbaria, while a few of the Cutlers' experimental maize plants lean in from the left. Bill and Elisabeth Cutler, private collection.

[9] Hugh and Marian Cutler bought this canoe with the intention of floating a few hundred kilometers down the Paraguay River, stopping to collect plants along the banks. "We had not counted on the abundance of crocodiles," Hugh told Mangelsdorf in a letter dated 9 December 1941. Here a man (presumably Hugh) is managing a stack of pressed specimens that also traveled in the canoe. Bill and Elisabeth Cutler, private collection.

1935, over disputed and reportedly oil-rich territory in the semiarid lowlands of the Gran Chaco—a state conflict often described as the bloodiest in South American history. Transportation infrastructure remained thin across large swathes of rural territory, and government control was often similarly patchy. Hugh Cutler griped that towns marked on maps turned out, upon arrival, to be "a single Indian *tolda* [tent or shelter] or shack or a pile of mud that remains from a Chaco War shack or a Jesuit Mission, or a cabin of some old colonist."[32] He planned for trips to take more or less time depending not on what he discovered in the field but whether transportation options panned out. "I will be gone about two months in all," he estimated of one journey, "54 days if I make all connections and can get the trucks and trains and make the return from Cochabamba to Corumba by plane, and 72 days if I make none of the connections and have to depend on horse, oxcart and foot, with a little canoeing thrown in now and then."[33]

These uncertainties weren't really the hard part, however. Finding the right kind of corn was. Cutler avoided collecting at markets, although it would have saved him time to bag many samples in one spot. Like Sauer and others, he thought that the most interesting varieties didn't circulate through markets. Better-off farmers often used commercial seeds imported from other regions or countries or mixtures of local varieties and imports. These would not reveal the ancient history of corn. Even smaller-scale farmers producing for central markets were seen as embedded in networks of exchange that had altered their agricultural practices, including their choice of seeds. Collectors like Cutler therefore targeted subsistence farmers, in particular Indigenous farmers, distanced from commercial exchange and believed to be cultivating the same varieties their ancestors had grown. Cutler envisioned a great diversity of unknown types scattered among remote farming communities and made it his goal to access these.

"I am trying to secure maize directly from the tribes, or from reliable sources," Cutler wrote from the field in 1942. This led him to ever-more inaccessible locations, which in turn meant leaving Marian behind to process their photographs, prepare specimens and slides, and tend plots of experimental corn. "The only opportunity to get into the Chaco was on a raiding party, but the 'sport' of shooting Indians doesn't appeal, even if there might have been maize in their camps," he reported to Mangelsdorf, explaining why there would be no samples from that region. In other cases known sites of interest required intermediaries. Cutler had heard that the Nambikwara people were thought "very primitive, and . . . still in a stone-age civilization" and yet also were settled agriculturists, which made their communities an inviting destination for collecting. He couldn't get there himself, however, and the missionaries he asked for samples failed him.[34]

Cutler's search for "primitive" corn, premised on the idea that it would resolve the question of corn's evolutionary past, looked a lot like anthropologists' earlier hunt for "primitive" peoples as clues to human evolution.[35] Cutler associated "primitive" corn, assumed to reveal the historical evolution of the species, with "primitive" peoples, thought to belong to an earlier stage in human evolutionary history. What Cutler encountered on his travels, however, was not a static—and, ultimately, mythic—corn of prehistory but the messy, dynamic corn of the present. Although he dismissed anything recently imported, Cutler enthusiastically followed the trail of almost any kind of corn he considered "native." By this he meant either local in origin or cultivated by Indigenous farmers or both. He took an approach typical of midcentury ethnobotany, quizzing farmers about local names for different varieties, cultivation strategies, or culinary uses and systematizing this information along with his own botanical evaluations.[36]

Cutler assiduously gathered information from Indigenous informants. Often he was interested in learning local classifications, in part because this would help him assess whether he had found all the extant varieties in a particular area. "According to an Indian near San Benito, the types of maize in that district are Morocho (Huillcaparo), Chuspilla (only yellow in this district altho others white are known), Arrocilla (Uchuchilla), Blanco, Culli," he jotted during his travels in Bolivia. Here he annotated the local names used by his informant with further classifications supplied by a Bolivian botanist, an act of calibration that made local categories legible within a larger botanical enterprise. Cutler supplemented this taxonomy when he learned that *lagua* was the name given to soup made with the variety Huillcaparo, while *api* referred to a sour porridge that tasted best when prepared with Culli. The notes grew so capacious that they could be used as recipes. "Huminta is mashed green [immature] corn, preferably white, with fat of pigs, salt or sugar, and cooked in [an] oven on plates or within husks of corn tied with husks and cooked in water like tamales of Mexico," rambled one observation, before concluding, "Cheese, pepper, chopped meat and seasoning are sometimes added . . . and these are the best kinds of humintas."[37] Everywhere he turned, Cutler encountered corn in motion. It not only moved across space—on mules, in markets, and as meals—but also transformed over time through cultivation, exchange, and breeding.

Hugh and Marian Cutler returned home in 1945 without wild corn, but the mission had hardly been fruitless. In fact, Mangelsdorf marveled at what they achieved. "Your collection is very impressive," he told Hugh, having finally had a chance to spread it all out at once on tables in the Harvard Botanical Museum. "It not only represents an enormous amount of work on your part and Marian's, but it is I suspect, the most comprehensive and valuable collection of South American maize which has ever been made."[38]

BREEDERS' RESOURCES

Although studies of human culture, domestication, and crop evolution generated collections of corn diversity, agricultural ambitions generated many more. By the 1920s genetic science and breeding practice had transformed the perceived value of genetically distinct crop varieties, making collections of these a critical component of successful breeding programs. Soviet scientists of the 1930s may have been the first to attempt comprehensive collections of corn across Latin America for this purpose, but they were not the last.

By the twentieth century the economic pursuit of the plants and knowledge of Indigenous peoples of the Americas already had a long history. Plant hunting and exchange, centered on the circulation and cultivation of potentially profitable species, had motivated global exploration and fueled the growth of European empires since the sixteenth century.[39] In the twentieth century botanical exploration increasingly adopted a new goal: the deliberate amassing of genetic diversity. Earlier generations of plant hunters had typically searched for wholly novel crops and flowers or superior individuals of known valuable species. These were cultivated at botanical gardens and experiment stations, sent abroad to launch colonial plantations, or sold as novelties in domestic markets. In the twentieth century plant hunters focused on crops increasingly sought to encompass a wide range of variation within a given species, sometimes amassing hundreds of examples of just one kind.[40] This shift corresponded, in part, to changing ideas about heredity. The refinement of breeding strategies went hand in hand with the development of genetic theories emphasizing the permanence of inherited traits, which were linked to a physical gene, as well as the possibility of predicting these over successive generations. Breeders now assessed individual plants or animals as assemblages of genes that could be shuffled by skilled breeders into desirable combinations. This understanding identified diverse varieties, regardless of their overall performance, as potential sources of new traits.[41]

In the first decades of the twentieth century, breeders and state agricultural institutions assembled increasingly large varietal collections, seeing these as storehouses of useful traits and therefore the wellsprings of future productivity and profit. By the 1930s interest in farmers' locally adapted varieties, or landraces, fueled a growing number of state-sponsored international collecting missions, especially in Germany, the Soviet Union, and the United States.[42] Collecting ventures were also profoundly shaped by Vavilov's ideas about the nature and location of crop genetic diversity. Elaborations on his theory of the centers of crop origins and diversity essentially generated maps to guide anyone hoping to discover wild relatives and ancestors or diverse varieties of particular species to mine for genetic traits.[43]

Collecting trips often served multiple purposes. Consider Vavilov's world circuits, which simultaneously fueled his evolutionary theorizing and provided materials for breeders at his institute; or those of Hugh Cutler, who was as attentive to the agronomic possibilities of South American corn varieties as to the evolutionary data they might provide. In Bolivia, where Cutler found "so many types of corn," he hoped to pick up some that could be used "in producing quick-growing, high-yield varieties for the states."[44] Mangelsdorf also managed to combine his interests in the origin of corn with more practical endeavors. In 1941, around the same time that he sent Cutler to South America, he was invited by the Rockefeller Foundation to join a survey of Mexican agriculture, touring farms and research institutions and meeting government ministers and scientists.[45] Although the purpose of the mission was to advise the foundation about potential philanthropic opportunities in Mexico, Mangelsdorf took the opportunity to study corn varieties in the various regions they visited and to make collections of *Tripsacum* and question local farmers about it. Recalling the trip in an interview in the 1960s, he went so far as to declare that the plant hunting had been his primary motivation for participating.[46]

The 1941 survey was the first of many Rockefeller-funded corn-collecting adventures for Mangelsdorf. The agricultural program that developed on the heels of the survey is often identified as a pivotal moment in the history of agriculture, the starting point of efforts that eventually brought about the so-called Green Revolution.[47] It was also a decisive moment in the collection of Mexican corn diversity. The program opened the door to collecting, comparing, and classifying varieties at scales that Mangelsdorf and others had yet only dreamed of, in the name of increasing Mexican agricultural production.

Historians dispute the precise events that led the Rockefeller Foundation to become involved in Mexican agriculture. Many accounts foreground the role of US vice president Henry Wallace—the same Henry Wallace who had founded the Pioneer Hi-Bred Corn Company and been a keen early promoter of hybrid corn. He was now a national political figure, having served first as secretary of agriculture under Franklin Delano Roosevelt and then, from 1941, as US vice president. In 1940, visiting Mexico for the inauguration of Manuel Ávila Camacho (1941–46), Wallace became convinced that the Mexican government would be open to philanthropic intervention and saw this as a way to improve the country's agricultural and economic outlook.[48]

Intervening in Mexico would also sustain US national security interests. The sweeping reforms of Ávila Camacho's predecessor, Cárdenas, had strained US–Mexico relations. Nationalization of the oil industry and land reform had involved expropriation of assets held in the United States. In keeping with its "Good Neighbor" policy toward Latin America, Roosevelt's presidential administration did not interfere, but it did keep a watchful eye on the nation's southern neighbor. US officials worried that continuing popular unrest might result in either a communist uprising or a fascist coup. Anxious to keep Mexico on a trajectory amenable to US interests, administrators like Wallace sought ways to stabilize or improve the Mexican economy as well as the circumstances of peasants and laborers.[49]

The Rockefeller Foundation was inclined to support this endeavor. In early 1941 foundation officers met to discuss an aid program for Mexico. One of the central sticking points was what kind of initiative would be most effective. Was it one that attempted to educate farmers and improve practices through extension, or one that conducted research and sought new technical solutions? The survey mission that Mangelsdorf joined attempted to resolve that question for the foundation. They reported that research should come first and urged the foundation to build scientific knowledge and capacity rather than focus on extension and farmer education. And because they judged Mexican science, like Mexico as a whole, to be "underdeveloped," they encouraged the foundation to send US scientists to Mexico rather than disperse grants to researchers at existing institutions.[50]

In February 1943 the Mexican government and the Rockefeller Foundation established a joint agricultural research program along these very lines. At a new Office of Special Studies (Oficina de Estudios Especiales), Rockefeller staff and Mexican researchers employed by the Ministry of Agriculture and Livestock (Secretaría de Agricultura y Ganadería) would work to raise the levels of production of Mexico's staple crops, especially corn, wheat, and beans. Those negotiating the parameters of this cooperative program agreed that better crop varieties were essential to enhanced agricultural productivity. It followed that establishing breeding programs topped the to-do list of the Office of Special Studies. This was the case especially for Mexico's most important crop, maize. Although central to life in Mexico, its production in the twentieth century had never been stable. Poor farmers working difficult terrain with no irrigation were particularly susceptible to bad seasons. In years of low production, shortages led to price spikes, leaving many Mexicans, rural and urban alike, hungry. If corn production could be increased, including through the development of more productive or resilient varieties, this would be a direct route to improving agriculture and social and economic conditions more generally.[51]

Not everyone agreed that a corn-breeding program run by scientists from the United States was desirable. Asked for his opinions about the possibilities of a Rockefeller Foundation agricultural assistance program in Mexico, Carl

Sauer was pessimistic. "The possibilities of disastrous destruction of local genes are great unless the right people take hold of such work," he maintained. "Mexican agriculture cannot be pointed toward standardization on a few commercial types without upsetting native economy and culture hopelessly. The example of Iowa is about the most dangerous of all for Mexico."[52] He thought the homogenized heart of the Corn Belt, by then replete with uniform and high-yielding double-cross hybrids, was the wrong model to emulate.[53] Edgar Anderson was similarly skeptical that a bunch of US scientists had much to offer. Imagining himself "a Mexican wealthy enough to endow a cultural mission to West Virginia," Anderson drafted facetious explanatory notes for a program of "Mexicanization of American hill-lands" that would assist US citizens in the cultivation of steep inclines. "Three Mexican missionaries will be selected annually one for West Virginia, one for Mississippi and one for Arkansas," he proposed.[54] Although critical of blithe assumptions about US superiority, Anderson's satire tellingly did not single out Iowa as in need of Mexican aid but instead Appalachia, the Deep South, and the Ozarks, regions of the United States stereotyped as backward, impoverished, and ignorant.

The prioritization of corn breeding also made some Mexican geneticists and plant breeders uneasy. A few already had their own programs underway but were now asked to collaborate with Rockefeller Foundation staff. These Mexican corn-breeding efforts had their roots in programs of the 1920s and 1930s. The agronomist Pandurang Khankhoje, a political refugee from India, began improving varieties of corn and other crops at the National School of Agriculture in the mid-1920s.[55] A few years later staff of the Biotechnical Institute started a corn-inbreeding program. This process would "purify" what they perceived to be the messy genetic heritage of the varieties cultivated by many subsistence farmers and create "standardized" varieties. Meanwhile, the agronomist Eduardo Limón, director of a school of agriculture in Michoacán, had by the 1930s started a corn-breeding program that focused on developing hybrid varieties.[56]

In 1940 Limón's program became the model for a new national corn-breeding initiative. Although the Biotechnical Institute closed that year as a result of administrative reorganization, a new Office of Experiment Stations (Oficina de Campos Experimentales) opened. Its first director, Edmundo Taboada, instructed several state experiment stations to launch programs like that of Limón's at Michoacán. Limón, meanwhile, was relocated to the station at León, Guanajuato, to lead a hybrid corn program there. These new corn-breeding efforts started with samples of typical landraces from the region covered by each of the stations. Collected landraces in turn became the source material for inbred lines, which would be used to create double-cross hybrids specific to local conditions. Within a few years Limón had thousands of potential inbred lines in development at León.[57]

At the joint Mexico–Rockefeller Foundation Office of Special Studies, the new corn-breeding program, directed by the Oklahoma-born and Iowa-trained plant breeder Edwin Wellhausen, also needed collections of local corn varieties and inbred lines to get going. Mexican scientists at the Office of Experiment Stations provided the program with an initial set of inbred lines. Meanwhile, its assortment of farmers' varieties originated in independent collecting missions. After Wellhausen's arrival in September 1943, he, Mangelsdorf, and the Office of Special Studies director, J. George Harrar, "took trips to Chapingo, Tlalnepantla, Tehuacán, Toluca, Ozumba, Puebla, Querétaro, León, Morelia, etc., in order to collect corn varieties."[58] By the end of the year, the Office of Special Studies had "accumulated" what it described as the "very large number" of more than 200 corn varieties. Four years later that very large number had more than septupled, to 1,500.[59]

Bringing this collection together required the labor of Mexican agronomists and especially students of the National School of Agriculture in Chapingo. Early collections were made along road sides, as Wellhausen and his colleagues traversed the countryside, but after 1943 collecting was more systematized and its labor distributed. Wellhausen dispatched students to gather varieties from different regions, providing them with "bags and tags and money to travel." From there they relied on buses and sometimes burros to travel to what Wellhausen called the "hinterlands."[60] Collecting presented difficulties, especially when it involved visiting remote locations and negotiating with impoverished farmers for access to a precious resource. As described by the botanist Efraím Hernández Xolocotzi, a Mexican scientist on the Office of Special Studies staff, "to collect the genetic variation of maize in a given community one has to be persistent and to use a great deal of tact." If a corn were a ceremonial variety, for example, it sometimes could not be sold at any price, as he recalled had been the case for a kind grown by Huichol farmers in Nayarit. In another community, this one high in the mountains, growers refused to sell their corn because, if they ever ran short, they had to descend two thousand meters to buy it—and then make the arduous journey home via the steep mountain trails. "I practically stole two ears of each type, since I could not return empty-handed," Hernández remembered.[61]

The diverse Mexican corn varieties gathered by Hernández, Wellhausen, and others were an essential resource for the corn program at the Office of Special Studies. One strategy for getting more productive maize to Mexican farmers was simply to find the best existing farmers' varieties and circulate them more widely. The collected varieties also provided starting materials for the development of inbred lines and open-pollinated populations to use in making "synthetic hybrids," whose seeds could be replanted each year.[62] As Mangelsdorf explained to the head of the Mexican Ministry of Agriculture, amassing varieties from across the country had produced a new tool for research on breeding as well: the collection. With all the country's maize diversity in a single place, the staff could identify "the basic types of maize from which the modern varieties are derived" and chart their expected characteristics. This classification in turn provided knowledge that would allow them to better predict what varieties would perform best in a given region through visual examination alone, or so Mangelsdorf claimed. The research would inform breeders' selection of the initial lines for inbreeding and enable a "more intelligent combining of inbred strains" in making hybrids.[63] The collection, considered as a whole, was as important as the samples within it.

CLASSIFYING MEXICAN MAIZE

After a visit in February 1947, Mangelsdorf declared the collection at the Office of Special Studies as "probably the largest which has been made in a single country."[64] Initially gathered to provide materials for the corn program, the collection eventually became the basis for an ambitious evolutionary study. This project, a landmark still celebrated today, demonstrated the power of a new method of classifying corn and generated an influential evolutionary account of *maíz* in Mexico.

Edgar Anderson was among the first to insist that a detailed study be made of the collection of corn stockpiled at the Office of Special Studies. In June 1946 he suggested that "one of W[ellhausen]'s Mexican boys measure his collections and work up the data." A year and a half later, he was still hot on the issue, believing that Wellhausen had no further plans to study the varieties now occupying his storeroom.[65] Anderson's eagerness was linked to his interest in seeing the evolutionary history of corn definitively resolved. Inspired by Mangelsdorf's theory of the origin of corn and no doubt encouraged by his continued association with the Pioneer Hi-Bred Corn Company,

[10] Edgar Anderson measures the kernel width on a sample of maize, circa 1948. Further specimens of the same type sit on the table, suggesting that he is in the process of creating an average measurement for the lot. Courtesy of Missouri Botanical Garden, Saint Louis.

Anderson had sketched out a new classification system for *Zea mays*. This would be a "natural" one based on evolutionary relationships and would supplant the "artificial" one derived from kernel characteristics (i.e., flint, dent, sweet, pop, etc.) that was typically used.[66] Working with Hugh Cutler, he began by classifying corn of the US Southwest. He then moved on to corn of

Mexico and Guatemala, relying on materials collected by Cutler, Sauer, and others. The resulting publication, in which Anderson and Cutler delineated several "races" of corn and their relationships to one another, illustrated Anderson's approach to classification.[67]

The category of race is most familiar today in its applications to human groups, but it has a longer history of use in the classification of populations of domesticated species such as breeds of animals or crop varieties.[68] Anderson was versed in this notion of race and its related use in indicating a distinct subpopulation of a wild species through his work as a botanist, but he nonetheless drew inspiration for classifying races of corn from physical anthropologists' efforts to classify human populations. He and Cutler described the "problem of races and their recognition"—that is, the challenges of clearly defining what races consist in and determining the boundaries among them—as being essentially the same between corn and people, a correspondence that perhaps encouraged their adaptation of techniques from physical anthropologists.[69] When it came to defining races in maize, they settled on a "loose" definition, still cited by maize biologists today, of "a group of related individuals with enough characteristics in common to permit their recognition as a group."[70] Sorting out race in corn required data, however, and there still were not enough published to facilitate their work. Anderson felt this was especially true when it came to categorizing the corn of Mexico. "It is indeed as if one were called upon to discuss the physical anthropology of Europe before any of the technical papers upon that subject had been published," he reflected.[71]

As he expanded his study of Mexican corn varieties, Anderson generated a great deal of data on these: typical ear widths, kernel widths and thicknesses, degrees of denting in the kernel, row numbers, type of striation on the husk, and still others. His measurements and observations, repeated systematically across collections, formed the basis of the large data set whose depths he plumbed for race-revealing patterns. Yet by early 1948 he still felt that what he had was insufficient to do justice to the problem, a view that spurred his repeated inquiries about Wellhausen's collection at the Office of Special Studies.

Anderson didn't have long to wait. By August 1948, with the samples assembled in the Office of Special Studies collection numbering around two thousand, Wellhausen reported that he and his colleagues were "now working very hard bringing together the data collected . . . on our different types of corn."[72] One motivation for this flurry of activity might have been managerial. It was difficult to maintain so many distinct samples as living seeds that could be grown out when needed. Figuring out which samples were likely to share an evolutionary past—and therefore present-day genetic material—would allow Wellhausen to decide how many he could discard and still have a collection representative of the diversity of Mexican corn. Wellhausen also understood that the information gleaned in the process would be valuable to fellow geneticists and breeders. Together with several Office of Special Studies colleagues, he began planning a technical publication on the corn of Mexico and invited Mangelsdorf to join in its preparation.[73]

A six-week blitz in the winter of 1948–49 produced most of the text of *Razas de maíz en México,* a 240-page book that identified and described the key characteristics of twenty-five races *(razas)* of corn in the country. Although the listed authors were Wellhausen and his colleagues Lewis Roberts and Efraím Hernández Xolocotzi, in collaboration with Mangelsdorf, the study depended on a wider pool of labor. "Roberts had all these Mexicans tabulating the data and so forth," Mangelsdorf later remembered, just barely acknowledging the many agronomists who contributed physical and intellectual labor to the project, including the eighteen explicitly acknowledged for collecting and measuring varieties and working up the data. Given the scale of the effort—growing plants, taking measurements, making slides, preparing maps, and more—Wellhausen and the other authors surely needed the assistance.[74]

There had been earlier attempts to classify and characterize the types of corn found in Mexico, not just Anderson's recent start but also the work of Mexican agronomists and the Soviet surveyors. Most of the races identified in *Razas de maíz en México* corresponded to types already recognized among scientists—and among farmers too. But the authors attempted, in addition to classifying types, to discern their relationships to one another. The book therefore included an overarching evolutionary account in which distinct races were clustered together based on their assumed time, place, and manner of origin. According to this evolutionary history, "Ancient Indigenous" races of Mexican corn had arisen directly from a "primitive" pod corn in Mexico, while the similarly prehistoric "Pre-Columbian Exotic" races had been introduced from Central and South America. "Prehistoric Mestizos" were the product of hybridizations between the "Ancient Indigenous" and the "Pre-Columbian Exotics" or their recombination with teosinte. A fourth cluster of races, "Modern Incipient," were those that had "developed since the Conquest and which [had] not yet reached a state of racial stability." These were more genetically variable from one region to another. Instability was not necessarily a problem, however. The type *Celaya,* crucial to production in the central lowland region known as the Bajío, "Mexico's counterpart of the Corn Belt," numbered among the study's "incipient" types.[75]

Ironically, although the determination of races of maize borrowed some methods of physical anthropology, the emphasis on quantitative measures and statistical analyses helped the authors of *Razas* to write humans out of their story. This was not a cultural geography, uniting knowledge of land and peoples in the manner preferred by Carl Sauer, nor was it an example of ethnobotany, with its attention to the classification systems and naming conventions used by local peoples. In this new evolutionary account, races of maize moved and transformed through space and time as though autonomous entities.

Still, though not explicit, ideas about human movement and mixing remained powerfully inscribed in the overarching narrative of *Razas.* The authors presented an account of fixed indigenous maize populations transformed by racial mixing with newly arriving populations to produce valuable "incipient" but still unstable racial types. This story fit contemporary Mexican racial politics as well as the assembled data on corn. The Mexican government's nationalist rhetoric had long espoused the integration of isolated Indigenous peoples into the social and economic life of the country. But from the 1920s the *indigenismo* philosophy rejected straightforward assimilation, at least rhetorically, and instead emphasized the valuable qualities that Indigenous people would contribute to Mexican society. Effective integration would occur through racial mixing, *mestizaje,* a process that would ultimately produce a stronger, more unified nation.[76] Thanks to the authors of *Razas,* it was now possible to see evidence of the benefits of *mestizaje* in corn too. One need look no further than the success of the "modern incipient" race *Celaya.*

The goal of the classification effort that underpinned *Razas* had not been to produce a palatable political narrative about racial mixing, but it did aspire to assist in another, related, evolutionary vision. It was intended to generate knowledge that would facilitate the practical work of the Office of Special Studies, whether paring down the corn collection to a more manageable size or enabling better-informed breeding decisions. As such, it was a contribution to the development and distribution of "improved" corn varieties in Mexico.[77] It was thus inseparable from the Rockefeller Foundation and Mexican government's agreed project to remake the Mexican countryside, a project that aimed to transform peasant, often Indigenous, cultivators using "primitive" varieties or methods into "modern" farmers with breeders' varieties and better inputs. In planners' minds this project would succeed when these farmers and their local corn no longer dominated the landscape.

With this transformation written into their mandate, staff of the Office of Special Studies faced the prospect of being handmaidens to the extinction of maize varieties. This was disconcerting. Mangelsdorf, in a new foreword written for the 1952 English translation of *Razas de maíz en México,* placed a word of caution amid his otherwise celebratory account of the maize classification and its hoped-for uses. "Maize is the basic food plant in most of the Americas and its diversity . . . is one of the great natural resources of this hemisphere," he assessed. He reminded readers that loss of "any part" of corn's diversity would reduce the options of future breeders, especially in coping with emerging diseases and environmental change. He concluded with an exhortation for conservation: "The modern corn breeder . . . has a responsibility not only to improve the maize in the country in which he works but also to recognize, to describe and to preserve for future use, the varieties and races which his own improved productions tend to replace and in some cases to extinguish."[78]

Mangelsdorf's warning extrapolated from the agricultural experiences of some industrialized countries—notably, the United States—to the rest of the world. He had no evidence from Mexico, for example, to support his claims. This was a move typical of midcentury scientists concerned about the loss of crop diversity. But Mangelsdorf was also engaged in, even instrumental to, a project that had as an explicit goal the replacement of local varieties. He contributed to a vision of agricultural modernization in Mexico in which success meant turning local corn into modern breeds so that poor peasants could become modern farmers. He joined others at the Rockefeller Foundation and beyond in forging a powerful account of the contribution of agricultural science to the transformation of rural economies in general: a story about the evolution of the countryside and the people who lived there and the role of both in a modernizing nation. In other words, Mangelsdorf's projection of the loss of crop diversity also drew on powerful narratives about cultural transformation and the imperative for this at local, national, and international scales. This was not just about what would happen to crop plants. It was about what would, and must, happen to people. In Mexico, it was about what would happen to Indigenous people.

Although the Office of Special Studies boasted a large maize collection, it had been gathered in service of the breeding program, with little discussion of the ongoing care and continuing renewal of individual samples. It is almost certain that the staff initially counted on recollecting anything that got lost along the way rather than attempting a costly and time-consuming regeneration program. Collecting materials and preserving materials were two distinct enterprises.

Now preservation appeared to be an important consideration, as Mangelsdorf emphasized. But how could one go about preserving Mexican varieties? They could not be saved through continued cultivation, given the vision of modern agriculture that prevailed. Within the Mexican Ministry of Agriculture and its various research stations, including the Office of Special Studies, it was assumed that farmers would be better off—and that Mexico would be better off—if breeder-developed varieties made it into the hands of peasant and Indigenous farmers still dependent on older types. Otherwise Mexico's evolution as a nation could not be realized. It also seemed agriculture could not be modernized *without* these same varieties, however, now that they had been identified as essential inputs into agricultural research programs. The most obvious answer was therefore to make preservation a concern of professional breeders. This was, unsurprisingly, the group that Mangelsdorf addressed in his admonition. Perhaps breeders in the United States could be convinced that Mexican maize diversity was as essential to the future of the Corn Belt as to the farm economy of Mexico.

Mangelsdorf had a hard time selling the project. By 1950 he had made several efforts to launch a corn preservation effort. These had started with his informal proposal to the USDA in 1940, which had been inspired by Carl Sauer's reports, and continued with the "complete collection" of corn varieties he had attempted to create at the Harvard Botanical Museum with the help of Hugh Cutler and others. In less than ten years he had amassed more than one thousand samples, which he kept in cold storage in hopes of extending their life span. In 1948, still hoping to expand these efforts, he sent off "A Proposal for the Collection, Classification and Maintenance of Corn Varieties from Latin America" to be considered by the USDA Division of Plant Exploration and Introduction. Mangelsdorf pitched this project in terms of US agricultural needs. It was not merely academic, though of course classification would be an important component. It was about US corn production. "When selection in United States varieties reaches a point of diminishing returns it will become necessary to introduce small amounts of germ plasm from exotic varieties," Mangelsdorf predicted, with "exotic" here referring to apparently endangered varieties originating south of the US border.

In other words, the US hybrid corn industry would someday need his cobbled-together materials.[79] Although he emphasized breeding, Mangelsdorf surely had in mind his own passion for sorting out the history of corn. This reconstruction depended on his having access to varietal diversity, especially as found in Mexico and in Central and South America.

Regardless of Mangelsdorf's motivation, he captured a concern that was increasingly shared among corn breeders, geneticists, and others knowledgeable about the diversity of corn found in the Americas. In this early moment would-be salvagers like Mangelsdorf and Sauer saw US-based collections as the best way to secure the future of endangered Latin American varieties and make these available to botanists and breeders alike. Within a few years their approach had shifted. When a group of US researchers, including Mangelsdorf, finally made headway in launching a hemisphere-wide corn conservation effort, their vision for how best to save varieties and the arguments they presented for doing so reflected greater biological knowledge and increasing political savvy. It produced the first long-term preservation facilities for corn diversity—not in Washington, DC, as Mangelsdorf had imagined in 1940, but in Mexico, Colombia, and Brazil.

[11] A researcher examines a sample stored in the corn seed bank of the Office of Special Studies in Chapingo, Mexico, in 1954. Each jar ideally contained seeds from a few ears harvested at the same time from the same field. Around the time the photograph was taken, the collection was expanding rapidly, thanks to the Committee on Preservation of Indigenous Strains of Maize. Courtesy of Rockefeller Archive Center.

THREE

PRESERVE

ALTHOUGH PAUL MANGELSDORF'S 1948 PLEA for USDA sponsorship
of a Latin American corn collection fizzled, an opportunity soon arose to
participate in almost exactly the project he had imagined. In 1949 a fellow
botanist approached him with the idea of launching a large-scale corn-
collecting effort through the National Research Council of the US National
Academy of Sciences. The eventual Committee on Preservation of Indigenous
Strains of Maize, or Maize Committee, which convened from 1951 to 1960,
set out to gather and preserve in perpetuity all of the "indigenous" corn varieties of the Western Hemisphere. A dramatic account of agricultural transition by Friedrich Brieger, a German émigré maize geneticist in Brazil, had
generated the committee's initial momentum. Brieger reported that the
diverse farmers' varieties of Latin America were rapidly vanishing. Corn varieties "developed in various civilized areas" had been brought in, and, according to Brieger, "the natives are taking to these strains and abandoning the
ones they have grown for countless centuries." In ten years there would be
"practically nothing left."[1]

The committee that formed on the heels of Brieger's intervention boasted
an all-star lineup of US corn experts, including Paul Mangelsdorf, Edgar
Anderson, and Hugh Cutler. Several other maize geneticists and breeders,
two USDA representatives, and an archaeologist rounded out the committee.[2] Over the span of a decade, these men and many collaborators in North,
Central, and South America generated immense collections of corn, many
gathered directly from farmers, and developed national and regional racial
analyses modeled on *Razas de maíz en México*. Their ideas and assumptions
gave shape to a preservation enterprise that in turn patterned many later crop
conservation efforts.

The agricultural scientists who forged the idea of diverse crop varieties as endangered genetic resources in the early twentieth century wielded a compelling story about the trajectory of agricultural change. Their account of impending extinction assumed an inevitable industrialization of agriculture and emphasized the power of plant breeders and public programs to realize this transformation.[3] No place, and no people, would escape "modernization" through the adoption of new methods and tools and especially the expanded cultivation of breeders' varieties. US scientists who assessed the future of corn cultivation across the Americas in the 1950s didn't just extrapolate from their experiences of hybrid corn in Iowa and the rest of the midwestern United States. They embraced a far more encompassing narrative about a necessary and ultimately inexorable agricultural transformation across the hemisphere.

The Maize Committee found further motivation for its salvage mission in members' understandings of indigeneity. The historian Courtney Fullilove describes how Nikolai Vavilov's theory of centers of origin of diversity, which encompassed both crop and linguistic diversity, "set plant genetic resources conservation on a path of privileging indigeneity as the locus of biodiversity."[4] Plant collectors sought, and still seek, sites thought to be originary and undisturbed. These are qualities that, within a narrative of inexorable agricultural modernization, define these sites and the indigenous varieties within them as always under threat—indeed, as inherently vulnerable. Just as "fragility fundamentally defines the 'Indigenous'" within settler colonial worldviews (to quote Lorenzo Veracini), so too have indigenous crop varieties, from the moment of their invention as a category, been understood as fragile, vulnerable, and vanishing.[5] The "indigenous strains" prized by the Maize Committee were no exception. These were conceived as static relics that would be either wholly displaced in the relentless advance of breeders' varieties or diminished through genetic mixing with introduced varieties.

These understandings of agricultural modernization and indigeneity, and of the relationship between them, shaped the strategies pursued by the National Research Council's Maize Committee in its quest to preserve the "indigenous strains of maize" of the Americas. From their funding sources to their collaborators, their field sites to their conservation methods, the US corn experts who formed the committee adhered to a vision in which the farmers' varieties cultivated by—but, crucially, not improved or transformed by—Indigenous peoples were valuable resources whose extinction was a foregone conclusion. The continued availability of these varieties as a resource for breeders, and by extension industrial agriculture, therefore necessitated purpose-built facilities. These would be overseen by the only people considered capable of managing and manipulating crop diversity: professional plant breeders and geneticists, especially those from or trained in the United States.

The founding of the Committee on Preservation of Indigenous Strains of Maize in 1951 extended concerns about the loss of corn diversity to encompass all of the Americas. As the Cold War deepened and US leaders endeavored to maintain a global balance of power against the Soviet Union, the social and economic stability of Latin American countries took on ever-greater significance. Members of the Maize Committee successfully capitalized on these circumstances to secure funding for their project. But once the money was in hand, they faced a more perplexing problem. How, exactly, could endangered corn diversity be preserved in perpetuity?

"Throughout Latin America and in parts of the United States, there are countless strains of maize which are in danger of dying out because of the inroads of corn introduced from the outside," declared the committee's first detailed plan of action. The committee predicted that these types of corn would survive only another ten to thirty years, absent some intervention. Adhering to a view of relentless agricultural change, the proposal emphasized that farmers' varieties had to be collected now, or they could never be.[6]

As it sought resources to undertake this rescue mission, the Maize Committee transformed an established view of this loss as a threat to industrial corn productivity and thus the US economy into a vision in which dwindling corn diversity also threatened regional and even global security. Yes, access to new breeding materials was important in the United States, where corn breeders were said to be already "beginning to turn to Mexico and Central America" in search of genetic variation for the Corn Belt. But in fact all countries of the Americas stood to benefit from the preservation of corn diversity. Farmers' varieties were essential to future corn breeding in the countries where they originated, an input needed "to improve the lot of the peoples of Latin America by increasing their food supply." And adequate food was more than a basic need or even a human right, as the recently adopted Universal Declaration of Human Rights suggested it might be. It was a weapon that would let the United States advance on the Cold War battlefront. "We must look forward to the possibility of a long world struggle in which the Western Hemisphere must become more self-sufficient. We must also remember that Communism thrives on food shortages, on conditions of undernourishment," the Maize Committee emphasized. "Improving food sources" in Latin America, where "no commodity is half so important as maize," was therefore essential to securing democracy abroad.[7]

The Red Scare rhetoric would not have been lost on bureaucrats in the newly established Technical Cooperation Administration, which received the Maize Committee's proposal for a hemispheric corn collection program in 1951. The administration was responsible for delivering on US president Harry Truman's 1949 inaugural address, in which Truman called on the United States to make "the benefits of our scientific advances and industrial progress available for the improvement and growth of underdeveloped areas." This was the fourth element, "Point Four," in Truman's four-pronged strategy for promoting peace and democracy—and capitalism—around the world and for keeping communism firmly at bay.[8] Evidently persuaded by the pitch, the Technical Cooperation Administration awarded $85,000 to the Maize Committee to organize the collection and "permanent preservation" of corn varieties from across North, Central, and South America.[9]

This left only the problem of how, exactly, to pull it off. While geopolitical concerns shaped the Maize Committee's bid for funding, more immediate practical needs shaped its plans for securing maize diversity. Operating on the assumption that varieties were rapidly disappearing and that the arrival of modernity foreclosed any possibility of these surviving in cultivation, they considered assembling collections their first priority. This was comparatively straightforward, a task that plant hunters had been pursuing for centuries. Some of the Maize Committee members were themselves successful collectors. Arranging for permanent preservation was by comparison a puzzle. Even under good storage conditions, any seeds the committee collected would have to be regenerated from time to time, perhaps as often as every five to ten years. Because corn cross-pollinates freely, keeping seed samples "pure" would also demand tightly controlled plots in which plants from one sample in the collection were pollinated only by other plants grown from the same sample. This might not be too onerous if one could rely on the fact that each individual corn plant produces abundant seeds and therefore efficiently replenish a stock by growing just a couple plants. Unfortunately, this approach would lead to a sharp narrowing of genetic diversity. Instead of maintaining the full range of genetic variation found within the many plants originally gathered into a sample, it would be reduced to that found within one or two individuals. As a result, it might be necessary to grow twenty or a hundred or more plants, just to renew one sample.[10] Any effort at preservation would therefore require storage facilities *and* a plan for renewing collections.

Mangelsdorf, who had experience managing a collection of corn varieties from outside the United States, was quick to point out that the Maize Committee would have to establish storage facilities in or near the places where corn seeds were collected. Although he had once thought that seeds from elsewhere in the Americas could be brought to the United States, stored, and regenerated whenever the seed supplies dwindled or began to decay, by 1949 he knew otherwise. He'd failed at several attempts to grow out and harvest fresh seeds from many varieties in his collection, even when he tried to do so at a field site in Cuba. This led him to conclude that many of the samples collected by the committee "will have to be grown in the countries in which they originated."[11] Still involved with the Rockefeller Foundation, he suggested the philanthropy as an ideal collaborator for this aspect of the project, better positioned than the USDA to offer assistance and expertise in Latin America.

By the late 1940s the Office of Special Studies in Mexico was just one element in a growing portfolio of Rockefeller Foundation grants and fellowships supporting Latin American agriculture.[12] This work was, as many historians have pointed out, attuned to the aspirations of the US government. Intervening in agricultural production promised political stability by promoting the welfare of impoverished farmers and the expansion of capitalism in the form of commercial cultivation and agriculture-related industries.[13] From the perspective of the Maize Committee, the foundation was a useful ally because it had staff and grantees on the ground who could act as collectors and collaborators. With Mangelsdorf as an intermediary, the Maize Committee approached the Rockefeller Foundation, hoping to incorporate the foundation's operations in Mexico and Colombia into its developing plans.[14]

Whereas the Office of Special Studies in Mexico had been around for the better part of a decade by 1950, the Colombia site was comparatively new. Colombian scientists and politicians eager to spark agricultural development in the 1920s and 1930s had laid groundwork—in the form of agricultural research and extension infrastructure—that enabled the Rockefeller Foundation to see Colombia as a promising place to invest.[15] A series of foundation fellowships and grants in the mid-1940s paved the way for a full-blown cooperative agricultural program, activities that unfolded despite an inauspicious political climate. By the late 1940s the Colombian country-side was embroiled in a bloody conflict, known simply as La Violencia, that escalated in 1948 after a deadly riot in Bogotá. The tumult only galvanized the Rockefeller Foundation, which continued to see agricultural research as a way to resolve rural unrest.[16]

In 1950 the foundation provided funds to launch a new agricultural pro-gram in Colombia and appointed two scientists from the Mexican program to lead it. These men traveled south to open the new Office of Special Research (Oficina de Investigaciones Especiales) at Medellín in May.[17] Patterning its activities on those of its Mexican predecessor, the office imme-diately hired a collector, the botanist Víctor Patiño Rodríguez. Trained in agriculture and horticulture in Colombia and Brazil, Patiño had started his career as an educator in Bugalagrande. After traveling across South America in 1933 and 1934 to collect plants for the Colombian government, he had returned to Colombia, where he continued to work as an agricultural explorer and became founding director of an agroforestry research station. Political violence had forced him leave the station in 1950, just as the Office of Special Research was getting started. The office hired Patiño to help with, among other tasks, collecting local varieties for use in the breeding program.[18]

The existing emphasis on collections at the Colombian office and its Mexican counterpart facilitated the Maize Committee's entreaties for col-laboration. Foundation administrators agreed that establishing collections of diverse corn varieties was important. At a minimum it would be useful for the breeding programs in which they were already invested. They easily grasped how they could be useful to the Maize Committee in return. The director of the Office of Special Studies in Mexico thought that agreeing to help the Maize Committee would "immediately make available to them the most important maize material now in existence, at no cost."[19] The costs from there on out, however, remained to be agreed on. Foundation higher-ups were particularly keen to correct any assumption that the foundation would pick up the tab for what was sure to be the most expensive part of the endeavor, the ongoing—indeed, indefinite—maintenance of the maize collections as living seeds.[20]

To understand why this was such a concern, it's helpful to consider who else was attempting to save seeds and how. By the 1940s many agricultural scientists harbored concerns about the loss of crop diversity resulting from international trade in seeds and the spread of commercial varieties. As government-sponsored missions fanned out to gather seeds from centers of diversity in the 1930s, typically as resources for state-directed breeding, calls for dedicated conservation programs became more common. In Europe representatives to a 1931 international meeting of plant breeders highlighted the need for the "conservation of local varieties" and called on governments to "bring together and preserve" all those remaining. This group imagined that "enlightened farmers and institutes" would "cultivate old local varieties, in the regions of origin and according to old cultivation methods." Mass collecting missions of the 1930s and 1940s instead generated centralized stockpiles at state agricultural institutions but little in the way of explicit conservation measures.[21]

In the United States some administrators at the US Department of Agriculture shared the concerns of European breeders. Although plant exploration and introduction were long-standing activities of the USDA, it hadn't devoted resources to maintenance of collected materials. This task was left to breeders, who tended to keep just those seeds or plants they thought immediately useful, abandoning the rest. Although a USDA plant collector might bring home fifty different samples of wheat from a journey to Europe, if USDA wheat breeders judged only a handful to be useful, these alone would be kept—and even then, not necessarily for very long. By the 1940s this approach seemed risky. Established channels of plant exchange with colleagues now in the Soviet bloc were sometimes cut off amid growing Cold War tensions. This shift away from free exchange, combined with a growing consensus that it was possible, even likely, that local varieties would disappear when replaced by commercial lines, meant that plant collectors might not be able to recollect items once readily available. Breeders therefore needed to think about both present needs and all possible future needs, which for some scientists at the USDA meant they ought to be saving all collections. The problem was that keeping collections indefinitely demanded time, labor, and facilities, which in turn required funds simply not available in the USDA budget.[22]

In 1946, driven by fears of a postwar glut of surplus crops and ensuing depression, the US Congress approved legislation expanding agricultural research, including "the discovery, introduction, and breeding of new and useful agricultural crops, plants, and animals, both foreign and native."[23] This influx of money led to a cooperative program between the USDA and state agricultural experiment stations for the introduction, testing, and maintenance of "plant germplasm"—that is, gathering and studying the seeds, tubers, and other plant genetic materials that would be the basis of future research and especially breeding. Four regional plant introduction stations, tasked with introducing and developing new agricultural crops, supplemented an existing national plant introduction station at Glenn Dale, Maryland.[24] Providing for maintenance remained a problem. By 1950 the committee charged with organizing this cooperative system had endorsed the need for a national seed storage facility that would take on the problem of maintaining plant seeds and stocks for the entire network. Even with the additional support of industry, it took another six years to secure the necessary funding from Congress.[25] In short, the USDA was still sorting out the "permanent preservation" of its collections of diverse plant materials in the 1950s. If the National Research Council's Maize Committee wanted to ensure the long-term survival of its collections, it would have to be equally inventive, establishing "seed centers" where corn samples would be sent, sorted, stored, and regenerated on a regular basis.

A first question was where to establish such centers. As Mangelsdorf had observed, successful regeneration of seed required facilities in or near the places where samples had been collected. The Office of Special Studies facilities in Mexico and those of the equivalent office in Colombia were considered the best locations for storing, respectively, the maize of Mexico and Central America and of the northwestern and Andean regions of South America. With the Rockefeller Foundation signed on as a collaborator, the Maize Committee also had access to all the seed samples already collected in Mexico and Colombia, the help of the foundation staff in overseeing collection and regeneration, and the use of its land and laboratory facilities. These resources were not freely donated. In a reversal of the typical flow of funds, the Rockefeller Foundation accepted money from the US Technical Cooperation Administration to pay for equipment and supplies as well as the salaries and stipends of additional collectors.[26]

The coverage provided by the Rockefeller Foundation in Mexico and Colombia left only the lowlands of eastern South America—Argentina, Brazil, Paraguay, and Uruguay—in need of a facility. Here Friedrich Brieger, based in Piracicaba, Brazil, at an agricultural college of the University of São Paulo, seemed a natural collaborator. Brieger had emigrated to Brazil in 1936, one among many Jewish scientists who landed at the country's new universities after being expulsed from Nazi Germany.[27] Familiar with the problems of corn genetics and breeding from a stint under Edward East at Harvard—the same geneticist who trained Mangelsdorf and Anderson—Brieger had immediately started experimenting with Brazilian corn, especially kinds collected from nearby Indigenous communities.[28] His on-the-ground assessments had set the Maize Committee in motion, and his existing collection of Brazilian corn varieties could serve as the foundation of a larger one. The University of São Paulo was soon lined up to host the third seed center, with Brieger placed in charge of overseeing its operations. Here, too, Technical Cooperation Administration funds underwrote the costs of collecting and equipment for storage.[29]

The Maize Committee also made arrangements for seed storage closer to home. The remnant farmers' varieties that the Maize Committee hoped to locate in the United States and Canada would go to a regional experiment station in Ames, Iowa, that already had responsibility for maintaining US corn stocks. Meanwhile, a small duplicate sample of every single collection made would be sent to the Plant Introduction Station in Glenn Dale, Maryland, where the USDA had recently built a modest temperature- and humidity-controlled facility to preserve breeding materials.[30]

The Maize Committee referred to this duplicate collection at Glenn Dale as "standby storage." Today this would be called "safety duplication" or, more colloquially, a backup. Even though they knew that most of the seeds collected in Latin America would be difficult, if not impossible, to regenerate in the United States, committee members felt that it was important to have duplicate seeds. There were all sorts of contingencies that worried them, especially given their understanding of life in Brazil, Colombia, and Mexico: unstable governments, uncertain electrical connections, incompetent administrators, poorly trained scientists, and more. "It is easily possible for some accident to destroy one or more samples in one of the centers," explained the chair of the Maize Committee. This meant that "every strain should be preserved in two centers, and one of those places should be in the United States."[31] This imagined geography of security would have long-lasting influence.

COLLECTING INDIGENOUS MAIZE

While arrangements for improved storage, especially refrigeration equipment, were being sorted out for the seed centers, collecting got underway. The committee's eventual successes in gathering thousands of samples depended on their recruitment of knowledgeable collectors and networks of agronomic expertise already in place across Central and South America. With many participants engaged in the hunt for "indigenous maize," conflicts inevitably arose, including over the very fundamentals of the project—for example, what corns counted as "indigenous" and what it would mean to preserve these adequately.

Víctor Patiño, already employed at the Office of Special Research in Medellín to collect Colombian varieties, extended his activities across the region overseen by the Colombian seed center. Patiño was an experienced collector, adept at acquiring the information and assistance needed to carry out his assignments. Traveling fifteen thousand kilometers over 108 days in Ecuador and Peru between May and September 1952, he returned with 420 samples of corn and hundreds of beans and other plants besides. Along the way he had the help of numerous institutions and individuals. The Agricultural Directorate (Dirección Técnica de Agricultura) in Ecuador, for example, had lined up five agronomists, each helping in a different province of the country. Patiño located another two collectors on his own. His experiences in Peru, and in Venezuela a few months later, were similar.[32] Thanks especially to Patiño's efforts, by the end of 1952 the Colombian center held 1,189 samples from Colombia, 199 from Ecuador, 221 from Peru, and 250 from Venezuela. Trusted collaborators had agreed to continue collections in Peru and Venezuela, freeing Patiño for trips to Bolivia and Chile and follow-up work wherever needed.[33]

While Patiño built up the collections of the Colombian seed center, other collectors were doing the same for the center in Mexico. The agronomist Alfredo Carballo Quiróz of Costa Rica, an alumnus of the Office of Special Studies, organized collecting missions across Central America. Edwin Wellhausen arranged for letters introducing Carballo to be sent to key agricultural institutions across the region. He supplied Carballo with blank forms for recording information about samples, along with equipment such as collecting bags and labels.[34] Unlike Patiño, Carballo did not venture much into the field to fulfil the requests of the Maize Committee. He instead made arrangements for independent collectors in each country to ship seeds to the Office of Special Studies via an arrangement with Pan American airlines.[35] By the end of 1952, hundreds of corn samples from across Central America had landed at the office, and many more were expected. Wellhausen tallied about 3,400, including materials in the collection prior to the organization of the Maize Committee. The vast majority of these—more than 2,500— were of Mexican origin. Still, there were a few hundred samples from

[12] Researchers associated with the Rockefeller Foundation's Central American Corn Improvement Program examine races of maize in 1959, possibly in Guatemala. The exemplars on the floor, which the ruler suggests are soon to be measured, were drawn from a larger collection, some of which is visible on the shelves behind these men. Courtesy of Rockefeller Archive Center.

Guatemala, another site of significant corn diversity, about forty from Nicaragua, a couple dozen each for El Salvador and Panama, and eleven from Costa Rica.[36]

At both the Colombian and Mexican seed centers, the Maize Committee effort was facilitated by its access to the resources and networks of the Rockefeller Foundation and the existing infrastructure for managing collections. This was not the case at the Brazilian center. Bureaucratic wrangling had slowed the University of São Paulo's approval of the program, which in turn delayed the arrival of funds to support collecting. Brieger and an assistant spent the first part of the grant period reviewing existing materials and deciding on collecting priorities.[37] In 1952 Brieger used university funds to travel to Argentina and Uruguay for a cursory survey of corn diversity. He also met with potential collaborators, eliciting offers of assistance from some and provoking irritation in others. Mangelsdorf despaired that on this first mission, Brieger had "inevitably rubbed the people in these countries the wrong way" and urged the National Research Council office to send someone "to indoctrinate and orient Dr. Brieger."[38]

The frustration was mutual. Brieger griped that the US-based Maize Committee did not understand the realities he faced on the ground, in terms of state bureaucracy, funding shortfalls, and lack of facilities and equipment, while the Maize Committee complained that Brieger simply did not produce what he had promised.[39] Brieger did have a point about the Maize Committee's blindness to his circumstances. One of his first requests after being asked to lead a center was that the committee not provide refrigeration equipment. He already had a unit meant to provide cooling and humidity control, but it caused more problems than it solved. When the electricity went off or some other complication arose and the unit was opened to investigate the issue, warm air would rush in and the humidity would skyrocket. Humidity control was, in Brieger's assessment, more essential than temperature management, and he wanted only "sufficiently large air tight containers" and "efficient drying material" for the new center.[40] The Maize Committee, unmoved, ordered him a refrigerator.[41]

Brieger also provoked disagreement when he questioned the methods the Maize Committee intended to use when regenerating samples. Brieger was not just worried that genetic diversity be preserved to its fullest extent, a concern he shared with most of the geneticists involved. He also wondered whether and how the "technique of the Indians" in maintaining their own varieties ought to be perpetuated. He offered the example of Guaraní farmers, who relied on three plots, one to maintain their typical field corn, a second for a popcorn, and the third for their ceremonial corn. The last of these involved planting two distinct strains, to obtain the particular color pattern of the mixed offspring each year. "In such a case . . . it would be advisable, to maintain the two ceremonial races in the same way, and not in separate lots," Brieger ventured.[42] He understood that Indigenous farmers worked deliberately to generate and maintain diversity. Many of his US colleagues disagreed, however, as would become more apparent with time. And even if they did see a role for Indigenous peoples in maintaining varietal diversity, members of the Maize Committee did not acknowledge that the preservation of Indigenous methods was necessary for the preservation of Indigenous corn. In fact, it was essential to their salvage mission that it not be.

Brieger's early insistence on paying attention to the practices of Indigenous farmers may also have emerged from his understanding that collecting indigenous strains meant collecting corn from Indigenous peoples. He targeted

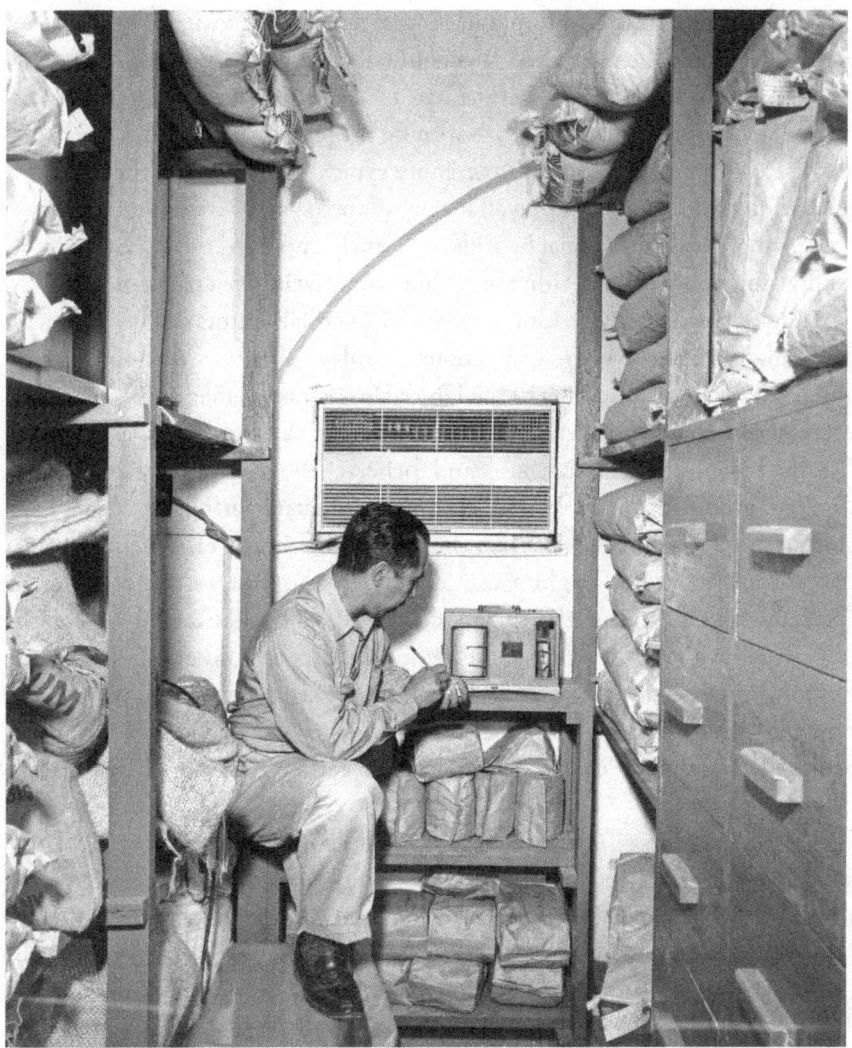

[13] A Rockefeller Foundation staff member checks an environmental monitor in the seed storage room, momentarily enjoying the breeze from the air-conditioning unit, in the 1960s. As the original caption suggested, researchers celebrated the power of "simple artificial air conditioning" to prolong the "vitality of seed" in the tropics. Courtesy of Rockefeller Archive Center.

Indigenous groups more than some of his counterparts at other sites, and he more consistently identified samples as coming from peoples as well as places. This was still apparent in his thinking in 1956, long after his "indoctrination" in the ideas of the Maize Committee, when he offered the following categories to encompass the material under examination at Piracicaba: races of the

Guaraní people, races of Humahuaca (a region in Argentina), races of the Kaingang people, races of the Calchaquí tribe (Diaguita people), and commercial corn of the Paraguay-Paraná Basin.[43] He explicitly associated most of his indigenous maize samples with specific peoples of the region. Other collectors, and the Maize Committee, more typically interpreted the remit of indigenous strains as including all locally adapted varieties grown in a country, leaving aside commercial hybrids or recently imported types. In this view "indigenous" meant something more like "originating in a particular area."

Despite this, the Maize Committee and its collaborators did prize collections made directly from Indigenous peoples. This was obvious in the approach taken to collecting in the United States and Canada. Hugh Cutler was dispatched to the southwest to collect among the Papago (today Tohono O'odham), Havasupai, Navajo, and others. The committee assembled further samples from extant collections originating in Native American communities. George Will, still at the helm of the Oscar H. Will seed company, donated twenty-eight varieties, which the committee considered "probably the best and the widest collection of Indian corns in North America." Another seventy came from the North Dakota Experiment Station, which included "original Indian corn of that State." To include the northernmost reaches of corn cultivation in the Americas, the committee acquired twenty-five "original Indian or the earlier grown open pollinated corns" of Canada.[44]

This association of endangered corn with peoples and cultures still considered endangered linked the Maize Committee to earlier salvage efforts—materially so in the case of recollecting George Will's collections and others like it. It also revealed slippage in their use of the term "indigenous" to describe the varieties of greatest value to their mission. This sometimes meant "local," but it could also mean "from Indigenous peoples." In either case it was clear what was *not* indigenous: breeders' products and commercial lines, all those "improved" and "modern" varieties. Indigenous was defined—invented even—through its juxtaposition with industrial corn.

PURITY AND "MONGRELIZATION"

While its collaborators in Brazil, Colombia, and Mexico organized collecting within their respective regions, the Maize Committee made arrangements to search for varieties cultivated by farmers in the Caribbean. The experiences of two collectors it engaged in this task, William Brown and William Hatheway, reveal the committee's guiding assumptions, especially about the presumed existential threats to indigenous varieties and who was best positioned to assess and salvage these. Above all, they highlight the ambiguous identity of indigenous strains of maize.

Many of the Indigenous peoples of the Caribbean islands had cultivated corn before 1492. In the colonial period the islands were transformed end to end. Early Spanish colonizers had forced Indigenous peoples like the Taíno of today's Puerto Rico and Cuba into labor in the encomienda system, eroding existing agricultural patterns. Later colonial authorities, whether Spanish, British, Dutch, or French, intensified the exploitation of land and people, not least through the development of sugarcane plantations. The labor of enslaved Africans first supplemented and then replaced that of Indigenous groups as their numbers diminished in light of imperial coercion, brutality, and disease. In short, pre-Columbian ways of life had been disrupted early and vigorously across the Caribbean.[45] For the Maize Committee, finding Indigenous farmers who had maintained varieties intact across centuries could not be the primary goal in this region, though it remained an aspiration. Instead the committee hoped to identify local varieties that had not mixed too much to prevent clear racial identification. These could still be considered "indigenous strains."

The Maize Committee's official emissary to the Caribbean was one of its members, William Brown, a geneticist on the staff of the Pioneer Hi-Bred Corn Company. Though he represented the oldest hybrid corn seed company in the world, Brown was no ordinary corn breeder. He had been a graduate student of Edgar Anderson's at the University of Washington in Saint Louis. Hired by Pioneer in 1945, its first employee with a doctorate, he was offered free reign to conduct "fundamental studies" on corn. Brown's earliest contributions as a Pioneer staff member were cytological investigations—and not improved varieties—of Corn Belt Dent.[46]

Brown's mission for the Maize Committee was to survey "the local varieties of maize currently in use" in the Caribbean, with "local" standing in here for "indigenous" and signifying locally adapted farmers' varieties rather than hybrids or imports. Brown was not to collect at each location on his itinerary but instead make arrangements for others to do so, wherever "local personnel had sufficient training and interest." As this suggests, he was tasked with surveying humans, though in this case not Indigenous peoples but instead local agronomists, breeders, and other agricultural workers. Wherever he did not find the right kind of expertise or training, he was to sample what he could on his own. Brown spent seven weeks in early 1952 traveling among seven Caribbean islands and Suriname (then also called Dutch Guiana), during which time he gathered seventy samples on his own while making arrangements for further collecting.[47] He dutifully surveyed both people and maize on his travels, noting where he thought experiment stations had "good men" or were doing "good work" alongside his observations of what corn was being grown where.[48]

Brown deemed the corn varieties cultivated by poor farmers with little access to external markets or agricultural advice to be of most interest to the committee. For example, he was impressed by several varieties maintained by Haitian farmers. He felt that the country was "primitive beyond belief" and the desk-bound experiment station workers knew "little of the corn or the country." Most farmers grew small crops of corn in plantings mixed with other grains, beans, and vegetables. "Although they sometimes exchange seed among themselves they seldom get anything from the markets," Brown reported. "I suggest this maize is as critical as any in the Caribbean."[49] His next stop, Puerto Rico, presented a different picture. As he noted, "Varieties in Puerto Rico mean very little—much [are] mixtures. Farmers buy seed from market which may include introduced varieties." As a result, he judged the maize "of no value for botanical and historical purposes."[50] For Brown the value of a given variety was dependent on the degree of "primitiveness" he attributed to communities in which it was grown, even if he did not identify these as Indigenous, and on their isolation from commercial markets.

Brown decided later that the mixing he observed among corn in Puerto Rico was characteristic of the whole Caribbean. This posed a challenged for identifying clear races in preparation for their preservation and future use. "The problem of classification of West Indian maize is a particularly difficult one, not because of the number of races found in the area but due to the absence of barriers to hybridization and the resulting mongrelization of races," Brown summarized. He was talking about races of corn, but the mixing of corn was linked to the mixing of peoples. Before 1492 this had arisen from the free travel of Indigenous Americans, together with their corn, from island to island. As Brown described, "Spaniards, English Colonials, Orientals, African negroes as well as various peoples from both North and South America" arrived later, contributing to the aggregation of peoples and plants. These arrivals had "some influence upon the agricultural history of the area"—an egregious understatement given the wholesale transformation of many Caribbean islands to sugar plantations—with the result that "the area is a melting pot not only of peoples but of their cultivated plants as well." Brown concluded that "if primitive races ever existed in the West Indies most have long since been drastically altered though hybridization."[51] In other words, true indigenous strains of maize could be considered extinct.

In her study of twenty-first-century genomics, science and Indigenous studies scholar Kim TallBear describes a process by which the drawing of genetic boundaries around Indigenous populations inevitably led, and leads, to the conclusion that these are in decline. As TallBear notes, first "scientists worry about indigenous peoples vanishing because they view them as storehouses of unique genetic diversity." That diversity is thought particularly useful for exploring deep human history. The genetic signatures of Indigenous individuals determined to be "admixed," in that their ancestry includes peoples not genetically understood as Indigenous, are considered less useful. They are therefore deemed "not sufficiently indigenous" and excluded from the population under scrutiny. As TallBear concludes, "If admixture is on the rise, the indigene is, by genetic definition, vanishing."[52] TallBear's observations about contemporary human genetics apply equally to the process of extinction Brown imagined in Caribbean maize. Where "primitive" races of corn were seen as unchanging populations with a particular genetic composition, any subsequent alteration of the genetic profile automatically constituted extinction-in-process.

There were further problems with Brown's assumptions. Characterizing local mixing of seeds and lines as "mongrelization," he presented this change as a failure of farmers he generally deemed too ignorant to maintain obvious types. But mixing was also a deliberate strategy. Varieties originating from the Caribbean were known to have spread widely, which suggested not just the transit of peoples but also the qualities of the corn. As Brown described, because of "its wide adaptability and exceptional vigor," corn of the Caribbean had spread across the American tropics and parts of Asia, Europe, and the United States. In other words, "mongrelization" was in many cases linked to improvement through recombination. Back at Pioneer Hi-Bred in Iowa, Brown experimented, for decades, with materials from his Caribbean collections. Corn of the Caribbean was, in his words, "probably one of the world's most important sources of maize breeding materials," and he aimed to make use of it—including through hybridizations he might have deemed "mongrelizations" had he seen them in process among Caribbean farmers.[53]

[14] Corn of the Caribbean awaits assessment by researchers in the early 1960s. William Brown of the Pioneer Hi-Bred Corn Company led the study of Caribbean maize for the Committee on Preservation of Indigenous Strains of Maize. His high esteem for some of these varieties no doubt informed the Rockefeller Foundation's assertion in the original caption that the "races of maize of the Caribbean" promised to contribute to the "further improvement of maize throughout the world." Courtesy of Rockefeller Archive Center.

In 1953, shortly after Brown's visit to the Caribbean, the Harvard University student William Hatheway arrived in Cuba, hoping to correct prior accounts of the history of corn on the island. He was not officially commissioned by the Maize Committee but contributed to its activities through his doctoral supervisor, Paul Mangelsdorf. Hatheway's hopes for finding "indigenous strains" in the hands of Indigenous peoples, or at the very least "pure" races of corn, reveal just how problematic the assumptions underpinning these concepts could be when used as a guide to collecting. Upon his arrival in 1953, Hatheway immediately encountered a host of problems, from bureaucratic wrangling to bad weather. Even more troubling, when he finally hit the road, he couldn't find any of the corn he was looking for. "Today I established what must be very nearly a world's record for futility," he wrote after an initial survey. He had traveled 230 kilometers, many on rough tracks; forded three rivers in the station jeep; and had only "2 handsful of grain, all of it mongrel" to show for his effort. His search "for relatively pure old races of Cuban corn" already seemed like a bad idea. "I don't believe that we will ever find towns or districts in which a pure race is grown to the exclusion of almost everything else," he wrote to Mangelsdorf. "No one here has heard of such a thing, nor in my experience have I seen anything approaching it."[54]

Hatheway's first strategy for locating "pure races" was to follow reports of Indigenous islanders. Disappointed by the results of his initial foray, he headed for the region west of Guantánamo, "where there are said to be Indians (!) and lots of corn." What he found was not what he anticipated. Although he encountered "people of undoubted Indian blood," he was disappointed to discover that "they were growing the same mongrelized corn as everyone else."[55] As a result, Hatheway started to think that the hunt for isolated human populations maintaining remnants of pure races of corn was wrongheaded. Everyone, it seemed, had access to the market, so perhaps it was more reasonable to look for people who intentionally preserved uncommon varieties for their unique qualities. This appeared to be the case for three varieties of popcorn he found in a single community, a discovery that renewed his confidence. When Hatheway made a preliminary evaluation of his collections a few weeks later, he assessed these popcorn varieties as "the only primitive corns in Cuba which are still kept relatively pure." Two further varieties, an orange flint corn called *Argentino* and a dent variety with very thin cobs, he considered "close to the primitive type."[56] Like Brown, Hatheway dismissed other farmers' varieties he'd seen as too new or too mixed to be of much interest.

As Hatheway's studies of Cuban corn progressed, this initial assessment completely unraveled. On a return trip to Cuba in 1954, he and his wife, Merilyn, conducted interviews to supplement his field collections. Keen to learn more about the primitive popcorn races and the "older" *Argentino,* he asked farmers about these specifically. The answers he received were certainly informative. An experienced farmer told him that the white popcorn had arrived in packages from the United States after World War I. A nearby storekeeper said the packages were from the Canary Islands but otherwise agreed they were recent arrivals. Hatheway also learned that the yellow popcorn, which he thought might be an ancestor of *Argentino,* had only been around five or ten years. Meanwhile, *Argentino* derived from a similar flint corn imported from Argentina between 1917 and 1928 and sold widely to local farmers.[57] Hatheway's disappearing primitive types were in fact recent commercial introductions. His interest in endangered indigeneity had led him to impose this narrative on corn that possessed a very different history.

These and other findings led Hatheway to a dim view of the usefulness of scientists naming corn races and to a far greater appreciation of cultivators as informants. The first shift did not stop him from describing races in his PhD dissertation, which was soon published as *Races of Maize in Cuba,* with funds from the Maize Committee.[58] The second shift caused a temporary hang-up in publication, however. William Brown, who peer reviewed the manuscript, objected to Hatheway's emphasis on farmers' knowledge. Brown was of the opinion that the vast majority of farmers "knew very little about their maize."[59] Hatheway in contrast thought they knew so much that their seed-saving practices and naming conventions should be used by scientists in delineating races.[60]

Aside from this dispute, Hatheway's and Brown's assessments of indigenous corn shared much in common. In his dissertation Hatheway dwelled at length on contemporary conditions in Cuba, the most dominant of which was the lack of "old" races of corn. He associated this pattern in corn populations with those of human populations, observing that, just as the people who "claimed Indian ancestry" had the same houses and clothing as their neighbors, "mostly of European or African descent," everyone shared the same corn, and it was usually a "mongrel" of recent origin. From his perspective it was equally apparent that the mixing of peoples, genetic or cultural, eliminated "true" Indigenous people and that the mixing of corn left little for the student of indigenous types to collect.[61]

When it came to understanding why "pure" types had given way recently to "mongrels"—the only evolutionary history to which his data gave access— Hatheway offered many explanations. He had observed farmers' poor knowledge of corn biology, the effects of cross-pollination in particular. He had heard from seed dealers that growers were "becoming more and more careless in [their] farming methods." Most important, however, he had driven on the "wide two-lane concrete highway" that cut across the island and enabled "modern tractor-trailers" to transport fruit and grain to the markets of all the big cities.[62] Though he may not have contributed much to the Maize Committee's ambition of rescuing varieties in danger of extinction, Hatheway's explorations in Cuba nonetheless provided the kind of narrative that drove them forward. In Cuba, he thought, the trappings of "modern" life had diminished the Indianness of the Indians and eliminated indigenous corn.

LIVING CULTURES AND COLD STORAGE

Brown and Hatheway's disagreement over the value of farmers' knowledge was one of many tensions to arise within the work of the National Research Council's Maize Committee. Disagreements proliferated as the project progressed. From the field Víctor Patiño pointed out that the boundaries established for the three seed centers, in most cases based on national borders, often made little biological sense. The Rockefeller Foundation complained when it got wind of Patiño being described as a foundation employee by a newspaper in Lima. The director of the agricultural college in La Molina, Peru, grumbled that he had not received sufficient information nor been invited to collaborate. Another Peruvian agronomist decided that turbulent Colombia was too politically unstable for his nation's seeds to be trusted to a facility in Medellín and began arranging his own long-term storage. And all of this was just in Peru![63]

A second phase of the Maize Committee project, dedicated to studying and classifying the corn collections, brought still other tensions to the fore—perhaps more, even, than collecting had. This phase of the project began in 1955 and consisted primarily in publishing a series of books on the races of maize of the Americas. It was funded by an additional three-year grant from the US foreign aid office.[64] Some of the tensions were, like the irritations arising during collecting in Peru, about institutional and professional hierarchies. Friedrich Brieger, chronically behind schedule, and not always because of bureaucratic snags, complained that the Mexican and Colombian centers had advantages that he did not. Unsurprisingly, he thought that more money should be directed to him in Piracicaba. He was particularly aggrieved that the second phase of the project did not involve funds for collecting, of which his center still had much to do.[65] Meanwhile, the staff of the Office of Special Research in Colombia rebuffed an attempt by the committee to send its experts Edgar Anderson and William Brown to South America to study the material from Venezuela, Ecuador, Bolivia, and Chile. The scientists in Colombia insisted that they, as collectors, had priority.[66]

Other tensions were more intellectual but nonetheless politically charged. The question about what a race of corn was and how to determine one intensified as authors with different opinions on the subject were recruited for the national and regional studies.[67] The role played by Indigenous Americans in creating varieties of corn was a further flashpoint. Some researchers dismissed the idea that Indigenous peoples had the knowledge and skills needed to intentionally create new strains. The authors of *Races of Maize in Central America,* Edwin Wellhausen, Alejandro Fuentes Orozco, and Antonio Hernández Corzo, noted the strong correlation of Guatemala's extraordinary diversity in corn with the presence of Indigenous groups. Clearly these groups were responsible for many diverse races. Yet the authors insisted that Indigenous farmers were not breeders. "That the Guatemalan Indian also played a conscious role in the creation of new races of maize is doubtful," they contended, allowing Indigenous cultivators only the successes of maintaining types and perhaps making them more uniform.[68]

Other accounts challenged this view, elaborating on the methods and aims of pre-Columbian and more recent communities of Indigenous Americans. Describing the Incan Empire at the height of its agricultural attainments, the authors of *Races of Maize in Peru* identified this as a period in which the farmer "exercised his best breeding ability" in developing "through selection" more productive maize.[69] Brieger and his colleagues in Brazil similarly referred to the "efficient breeders" and "breeding programs" found among Indigenous farmers past and present. They further observed that the task of "preserving and perpetuating strains," casually dismissed in *Races of Maize in Central America* as a routine task, in fact "requires a considerable skill," which "has been carried out and maintained successfully and constantly through many hundreds or even thousands of years."[70]

The debate over the agency and skills of Indigenous farmers had practical implications for the Maize Committee's preservation enterprise. Its US-based leaders saw themselves, by and large, as collecting biological materials. These could be extracted from the communities in which they were found, placed in cold storage, and conserved in perpetuity by agricultural technicians. But what if the corn varieties were not just biological materials but cultural products, as the historical narratives advanced by some of the *Races of Maize* contributors suggested? What if these varieties were not static or near-static relics shaped mainly by slowly changing environments but were instead dynamic, subject to human design and changing cultural needs? If corn diversity had not been passively stewarded but actively developed and maintained, then surely cold storage was only a partial conservation measure. Local cultures and especially agricultural traditions ought to be considered for protection as well, or at very least studied more closely. As historians have described, scientists built the edifice of *ex situ* conservation of plant genetic resources—that is, the practice of conserving seeds off-site, typically in cold storage—around the assumption that professional breeders and geneticists alone possessed the knowledge needed to efficiently and effectively manipulate genes. Christophe Bonneuil confirms scientists' systematic devaluation of the "creative and innovative agency" of farmers and their depiction of crop diversity "as mostly resulting from 'natural' mechanisms and 'unconscious' or 'unguided' practices."[71] The Maize Committee adhered to this perspective, even while some of its collaborators disagreed.

Despite these diverging understandings of Indigenous knowledge, all who surveyed and classified corn collections for the Maize Committee agreed on the endangered status of the diversity they sought. The corn varieties they observed and cataloged were considered in danger of disappearing in cultivation, if not gone already. "Some of the varieties reported on here and preserved in the 'germ plasm bank' at Medellín are no longer readily obtainable," came the bleak assessment from Bolivia.[72] In Venezuela, where reportedly "only small groups of Indians remain now, and even these are for the most part culturally and racially much modified," the research team was less sanguine about the success of the salvage mission. "It now seems only too obvious that the Venezuelan collections were made about 20 years too late," they wrote.[73] The experts offering these views did so from a shared vantage point. By and large they were agricultural scientists associated with state agricultural programs. Many were engaged in the task of creating and promoting breeders' varieties as their primary occupation. Agricultural change was not just something they observed. It was also something they enacted. It's therefore possible, even advisable, to read their accounts as both congratulatory updates and sobering reports of extinction in progress.

Aggregated in Maize Committee analyses, these accounts provided further cause for celebration, in that they underscored the importance of the committee's work. By 1960 it had established three seed centers in Brazil, Colombia, and Mexico and filled these facilities with more than twelve thousand samples from across the hemisphere. It had installed the equipment thought most essential in preserving the collections. It had recruited collaborators from many countries, which had not only facilitated the project but also spread word of its work to agronomists, geneticists, and breeders and inspired the creation of still other collections. In the final phase of work, it had arranged for its amassed corn to be studied and for this research to be published and distributed. Now any researcher wishing to work with the increasingly rare "indigenous strains" of corn that had been collected would have an entry point. The projected disappearance of Indigenous peoples and cultures would not disrupt the professional-led projects of evolutionary research, plant breeding, or agricultural development.[74]

One significant problem remained. The Maize Committee had never arranged for ongoing storage and renewal of the seeds they had collected. Regenerating seed was a time- and resource-consuming chore, and no one was lined up to do it. Nor was there any money available to persuade an institution to take it on. Although committee members had once envisioned raising a permanent endowment, possibly by soliciting donations from the hybrid corn seed industry, they had never pursued this with conviction. The Maize Committee had taken precautions against the loss of material, most notably with the duplicates shipped to the United States for standby storage. These materials had made their way in fits and starts to the USDA facility in Glenn Dale, in the form of small packets containing about two hundred seeds of each sample. The existence of these duplicates meant that, in theory, any collection lost from a seed center in Latin America could be restored.[75]

In 1951 an optimistic USDA official offered the Maize Committee the opportunity to cache additional seeds in a yet-to-be-built national seed storage facility in Fort Collins, Colorado.[76] Plans had taken shape for this in the late 1940s, as the US system for plant introduction expanded. The envisioned facility would accept deposits of potentially useful seeds from botanists and breeders around the country, store these under the best possible conditions for extending shelf life, and arrange for regeneration if and when viability began to drop off. This service was desperately needed, advocates argued, because breeders could not be expected, or trusted, to keep up all the potentially useful seed stocks gathered by plant collectors or developed in the course of breeding programs. Valuable resources, collected at taxpayers' expense, were being allowed to die, simply because busy breeders did not have the time or inclination to maintain them. In the early 1950s USDA scientists, supported by lobbyists representing agricultural industries, pressed the US Congress to make seed storage a national priority. The expenditure was

[15] A seed scientist demonstrates pullout shelving in the storage rooms of the newly built—and mostly empty—US National Seed Storage Laboratory in 1959. US National Laboratory for the Preservation of Genetic Resources, Fort Collins, CO.

finally approved in 1956, and within two years a new US National Seed Storage Laboratory was ready to accept deposits.[77]

From the Maize Committee's perspective, this was none too soon. In April 1957, well before the National Seed Storage Laboratory opened its doors, the committee had already approved a procedure whereby the seed centers in Brazil, Colombia, and Mexico would each prepare one-peck samples (about nine liters of seeds), representing the races identified in the ongoing studies. These would be created by mixing seeds from several samples of a given race to create a composite that represented that race. The centers would then ship these to the National Seed Storage Laboratory for long-term storage.[78] A solution to their preservation problem had arrived at last.

Or had it? The National Seed Storage Laboratory was not an ideal long-term solution for the Maize Committee. For starters the laboratory did not have the resources to regenerate samples. It was designed chiefly to safeguard collections on behalf of breeders and other research programs, which would retain responsibility for renewing collections. Moreover, irrespective of this arrangement, it would not be possible in many cases to regenerate the tropical landraces collected by the Maize Committee in the United States. Still, the seeds sent from the Latin American seed centers to Fort Collins would be kept under the best possible conditions. This process could significantly extend their shelf life. As a result, the Maize Committee members believed it would provide greater security for the collections held in Brazil, Colombia, and Mexico.

With the decision to move its endangered corn seeds to Fort Collins for long-term preservation, the Maize Committee expressed—and instantiated—its imagined hierarchies of knowledge (professional scientist versus peasant or Indigenous farmer), of professional expertise (US scientists versus Latin American scientists), and of security (US institutions versus Latin American institutions). Its embrace of long-term cold storage also exposed its underlying convictions about the nature of indigenous crops and their relation to Indigenous peoples. The committee considered indigenous maize varieties not only inherently vulnerable in light of agricultural modernization but also static. These were essentialized, much like the Indigenous peoples they were associated with and yet not attributed to. Indigenous varieties were bounded such that any change to them constituted loss.

The work of the Committee on Preservation of Indigenous Strains of Maize had profound long-term consequences. It extended an existing vision of the fragility of corn diversity in light of rapid agricultural change to encompass the whole Western Hemisphere. It persuaded many people, politicians and agriculturalists alike, that this fragility threatened economic security and social stability across that same region. Once equipped with the funds and personnel to start a salvage mission, it generated thousands of samples and dozens of studies of corn diversity. It made institutional arrangements to secure its collections as best it could, establishing regional seed centers where it hoped collections could be easily accessed by researchers and regenerated when necessary. It lined up the best available facilities for duplicate collections, lest any disasters befall the seed centers. This network of seed centers and standby storage, a novel institutional assemblage that the committee imagined would provide maximum security for collections, established a model for many seed and gene banks in the decades that followed.

These were major accomplishments, yet the Maize Committee concluded its work with uncertainties about its own long-term significance. Its samples, gathered with urgency and advertised as precious resources for the future of peoples and especially economies across the Americas, were left to the interests and whims of institutions with many commitments besides conservation. A decade earlier that might well have meant their abandonment or disintegration, not only for the lack of facilities capable of long-term seed storage but also as a result of the circulation of concerns about crop extinction only within limited professional domains. But new international environmental concerns lay just over the horizon. Dawning recognition of the "genetic erosion" of crop species and worries about a "population bomb" would soon generate international institutions dedicated solely to preventing the loss of diverse crop species.

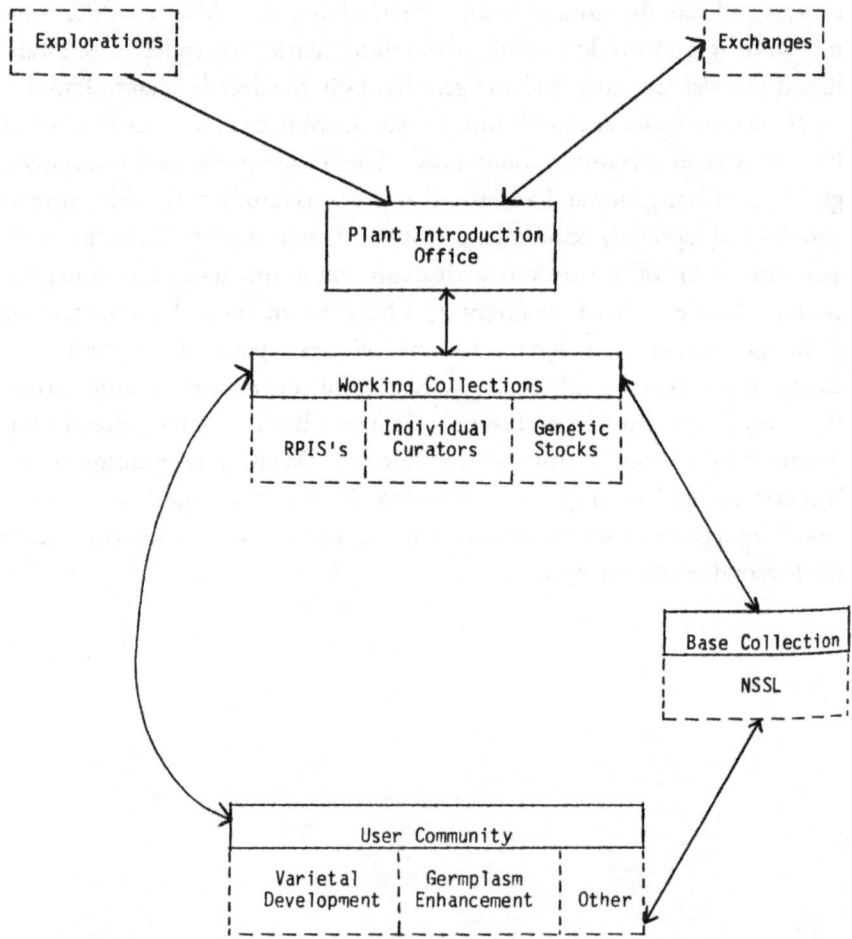

[16] Known duplication defined the effective functioning of the US National Plant Germplasm System, as shown in this 1981 diagram. The "User Community" of plant breeders and other researchers were to obtain seeds from official "Working Collections," such as those maintained by regional plant introduction stations. In principle these collections would have all material introduced into the United States through exploration and exchange. However, just in case, the working collections were also backed up in the national "Base Collection," the National Seed Storage Laboratory. From *National Plant Germplasm System*, I-6.

FOUR

COPY

IN THE 1960S FEARS MOUNTED among plant explorers and breeders that landraces of many crops, not just corn but also wheat, rice, sorghum, millet, cotton, and others, were disappearing as the world's subsistence farmers, encouraged by technical assistance programs, transitioned en masse to breeders' varieties. Researchers imagined a steady and dangerous destruction of genetic diversity, a process they now described as "genetic erosion." They also recognized that decades of collecting activities, in which landraces and farmers' varieties had been targeted precisely because of their perceived value, had for the most part produced ephemeral assemblages of poorly understood, badly cataloged material. Accumulated observations of mishandling led one prominent scientist to conclude that most collections of landraces and crop wild relatives were "distinctly less than effective."[1] Even the best managed seemed to be riddled with problems that ultimately resulted in the loss of samples. In other words, extant collections were as endangered as landraces in the field.

This recognition generated a new conservation impulse. Now the call came not just to gather farmers' varieties whose presence in fields was considered increasingly precarious, as the Committee on Preservation of Indigenous Strains of Maize had done in the 1950s, but also to collect existing collections into a global master collection. An "international germplasm bank" or "international seed bank" would provide long-term security for all crops and for all breeders, ideally for all time. As the institutions necessary for this international conservation enterprise took shape in the 1970s, the duplication of a complete collection and its storage in a second, maximally safe location came to be an unquestioned technical requirement for effective conservation. The demand for what would eventually be called "safety duplication" made seed collections and seed banks the objects of conservation, much as seeds themselves were.[2]

Today safety duplication is carried to its logical extreme at the Svalbard Global Seed Vault, a cold-storage facility dug into the arctic permafrost where duplicates of many world crop collections are warehoused.[3] Although its snowbound facade understandably invites analyses of what Joanna Radin and Emma Kowal call "cold optimism"—the faith that freezing living materials can indefinitely postpone decay, death, and even extinction—the vault's chief claim to security lies not in freezing but in duplicating.[4] The distinction is important. The vision of achieving security through the duplication of existing collections, most of which are also kept in cold storage, is not just a comment on the fragility of life or expression of fear about its loss or transformation. It is a statement about the fragility of institutions, the failures of technologies (including freezers!), and the disintegration of infrastructure.

It is ironic, then, that collection duplication, an acknowledgement of institutional failings, took center stage as a conservation strategy at a moment of unprecedented institution building in international agriculture, a time when today's infrastructures were first erected. This institutional proliferation was tightly bound up with other growth—that of the human population. Agricultural experts, and those who funded them, linked the need for research on food crops to the idea of a "population bomb" whose detonation presented an existential threat to all of the earth's inhabitants, rich and poor.[5] Although many advocated population control as the ultimate solution, the specter of teeming masses of hungry and potentially volatile people created opportunities to press for agricultural research too, especially for technologies to drive the production of staple grains and permanent institutions where scientists would be employed to develop these crops.[6]

Scientists who urged the safekeeping of crop diversity in internationally governed seed banks aligned themselves with dire portrayals of resource exhaustion and human misery in a decolonizing world. They piggybacked on urgent calls for international initiatives and especially research institutions to resolve impending food shortages. This strategy raised the stakes for the next phase of maize conservation as well as for the conservation of diversity in all crops. By the 1970s genetic diversity was no longer a local or even a hemispheric resource but a global one, required to tackle the "world food problem." So conceived, it inevitably invited conflict over which scientists, and which institutions, ought to control new international collections. Conflict bred distrust. And distrust provided additional impetus to recommendations of safety duplication, one conservation strategy that everyone could agree on.

PLANT INTRODUCTION AND EXCHANGE

Plant breeders and explorers had highlighted the abandonment of landraces and farmers' varieties as a potential international concern since the 1890s, but to little effect. The activities of the Maize Committee marked an important transition toward greater action. Still, it had considered the future of just one crop. It wasn't until the late 1950s that a coordinated international response began to take shape, driven forward by plant scientists convinced that "modern" and "improved" varieties were rapidly, and potentially disastrously, displacing the world's landraces.[7] These scientists worried in particular about Nikolai Vavilov's centers of diversity, historical regions of domestication or diversification. These were known to be rich in the populations prized by plant explorers and breeders, not only landraces but also crop wild relatives, the extant ancestors or close relatives of crop species.[8] Prized for decades as the source of interesting breeding materials, the centers of diversity now gained status as threatened habitats too.

One forum in which concerns about the loss of crop diversity achieved new visibility was the United Nations Food and Agriculture Organization (FAO). Founded in October 1945, the FAO was tasked with bettering human nutrition, food production and distribution, and rural living conditions.[9] The years leading up to FAO's founding had seen world leaders increasingly accept adequate diet as a human right and an issue deserving of international action. During World War II the Allies had gone so far as to erect "freedom from want" as one of the pillars they envisioned for the postwar world, and FAO took shape as a means to ensure this freedom.[10] The first FAO director general was the Scottish nutritionist John Boyd Orr, who initially advocated direct interventions to address hunger. Boyd Orr wanted to reorganize commodity markets, deliver emergency food relief, and devise credit schemes to foster agricultural development. These ambitions were quickly curtailed, not least by officials from the United States, who wanted to promote free markets, retain food aid as a means of influencing foreign governments, and disperse grain surpluses in ways that benefited US farmers. Thanks in part to US obstruction, and that of the United Kingdom too, FAO never received the funding or powers it needed to tackle hunger head-on, as Boyd Orr and others imagined. In its early decades its role was limited to offering technical assistance and compiling statistics.[11]

Among the technical tasks that the FAO took on were plant exploration and the exchange of breeding materials, essential elements of agricultural research that might benefit from international coordination. A Subcommittee on Plant and Animal Stocks, formed in 1947, set ambitious goals, including a global catalog of all "genetic stocks" available to breeders; exploration, testing, and propagation of useful lines; and international "stock centers" and "living collections in centers of origin."[12] This dazzling agenda ultimately fizzled, largely for lack of resources, but the will to play a managerial role in these activities remained—and for good reason too, as researchers demanded breeding materials. By the mid-1950s staff of the FAO Plant Production and Protection Division were receiving 100 to 150 inquiries from breeders annually. They were also responsible for delivering seeds to FAO fieldworkers and collaborators. The net effect was that, even without an overarching strategy, FAO operated as a clearinghouse for crop varieties.[13]

For some, this was insufficient. Throughout the 1950s delegates to FAO repeatedly requested further attention to international exchange of breeding materials. They asked for assistance with plant quarantine protocols, world germplasm catalogs, and an international bureau of plant introduction. These appeals pointed to FAO as the potential coordinator of a global system for sourcing diverse crop-breeding materials and made it clear that such a system would be welcomed.[14] Although there were few resources available at FAO to generate the institutional structure and activity that calls imagined, its staff made inroads through smaller initiatives. In 1961, in response to ongoing interest in these issues, FAO convened a Technical Meeting on Plant Introduction and Exploration. Experts from twenty-eight countries came together to discuss exploration and exchange activities, debate the finer points of collecting practice, and, above all, consider the possibilities for greater cross-border partnerships.[15]

While addressing these topics, participants in the 1961 meeting repeatedly returned to the conservation of crop diversity and asked FAO to take on leadership in this area too. "The pool of genetic variability which now occurs in the major 'gene centres' of the world" was "threatened with destruction," they reported. Perceived global agricultural change, amplified by the projected consequences of decolonization across Africa, Asia, and the Middle East, had converged to generate new anxieties about the survival of species and varieties in Vavilov's centers of origin of cultivated plants.

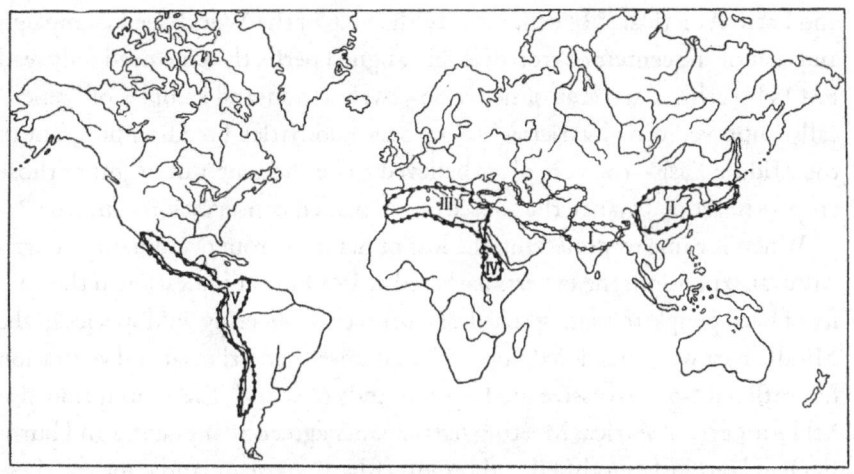

[17] In the 1920s Nikolai Vavilov sought the places of origin of the world's crop plants, sites often made apparent through the rich genetic diversity found there. Regions where many different crops had originated, such as the "basic centers of origin" identified on this map (nos. I–V), featured this richness both across crop species and within them, making them doubly important for explorers and breeders. From Vavilov, *Origin and Geography*, 127. Reproduced with permission of Cambridge University Press through PLSclear.

With one exception (the Mediterranean), the centers lay outside the industrialized West: in Central and South America, North and East Africa, the Middle East, South and Southeast Asia, and East Asia. Fixated on imminent transformations in these places, the conference participants warned that "agricultural development and general progress" would eliminate "primitive cultivated forms," while overgrazing, agricultural expansion, and industrial development would destroy the habitats that sheltered crop wild relatives.[16]

As several scholars have pointed out, the first of these predictions—elimination of farmers' varieties—relied more on extrapolation of past experiences in industrialized countries than on systematic field observations among farmers elsewhere in the world.[17] Reading FAO press releases surely sustained the prognostications. In 1957 FAO launched the World Seed Campaign. Premised on the idea that "one of the cheapest and most effective means of producing more and better food is the extensive use of high-quality seed of superior and well-adapted crop and tree varieties," the campaign helped establish seed certification bureaus, deliver new varieties and equipment, and set up distribution channels. Its ultimate target was the "millions of people in the economically underdeveloped countries of Asia, Africa, and Latin America."[18] In other words, the map of the World Seed Campaign and that of the centers of crop diversity aligned perfectly. One need only read FAO's breathless account of delivering twelve thousand samples of "genetically-improved cereal varieties" to a wheat and barley breeding program in the Middle East—the very place believed to be the center of origin of those crops—to see at a glance the source of intensified conservation concerns.[19]

When it came to projecting the loss of habitats from overgrazing or agricultural expansion, the experts gathered at FAO in 1961 questioned the ability of local people to manage natural resources. Collecting wild species in the Middle East was considered imperative because there "the natural vegetation has suffered from excessive use for thousands of years." The assumption also held for parts of Africa. Meeting participants agreed that grazing and burning had "seriously modified and impoverished" African landscapes.[20] These assumptions had roots in earlier imperial perspectives. Conservationists in the late nineteenth and early twentieth centuries had aligned with colonial administrations to establish reserves and restrictive regulations. They made local people into trespassers (rather than residents) in parks and into poachers (not hunters) of wildlife. By the late 1950s conservationists faced the prospect of independent governments setting their own agendas and began campaigning African leaders to keep established measures in place.[21] Echoing this call for postcolonial oversight, participants in the 1961 FAO meeting suggested the creation of "conservation areas" and "exploration stations" in areas boasting significant crop diversity, along with financial support for studying and distributing the plants found there. These would facilitate the salvage and circulation of endangered breeding materials, which they identified as a common global resource.[22]

Although the displacement of farmers' varieties was thought to be the case for most staple crops, in the 1960s it was maize that provided the go-to example of imminent peril. Scientists who sounded the alarm about endangered crops at FAO and elsewhere made frequent reference to early farmers' varieties of the United States, Latin American landraces, and Western European corn varieties now challenged by imported US hybrids.[23] The scenario of the rapid adoption of commercial hybrid corn, both observed and assumed, along with actions already underway to conserve corn varieties, clearly influenced analyses of the potential loss of crop diversity. The Maize Committee's work and other corn salvage programs also offered models for the shape that internationally coordinated collection and conservation should take.[24]

Almost from FAO's inception, political pressure from the United States, Britain, and other industrialized countries had hampered its ability to directly address hunger and malnutrition, forcing a focus on technical assistance instead. Yet hunger abroad remained a potent concern of these governments. The idea of a world food problem threatening global stability took on new valences in the 1950s and 1960s, becoming increasingly entangled with a perceived world population problem and Cold War strategizing. This created opportunities for individuals and organizations willing to see hunger as solvable through investments in science and technology alone.[25]

By the late 1950s "development" was entrenched in US academic and policy-making circles as a concept and global action plan. According to expert analyses, the transfer of knowledge and technology from the rich world to the poor world would accelerate economic growth, raise standards of living, and prevent civil unrest or revolution.[26] Unfortunately, challenges to the project of development seemed to be growing faster than the economies it purported to encourage. Advances in medicine and public health had stimulated population growth in the recently dubbed "underdeveloped" countries. This demographic increase was thought to be outstripping supplies of food, fuel, and other resources, with the effect of creating masses of hungry, unhappy peasants. Few visions could be more alarming to US policy makers obsessed with containing the spread of communism. The historian Nick Cullather describes how US experts "saw revolutions and social turmoil as political manifestations of mankind's persistent struggle to wrest sustenance from the soil." International affairs could potentially stabilize, save that "hunger remained an irrational element, turning citizens into mobs and giving demagogues license."[27] Population control, food aid, and efforts to bolster agricultural productivity therefore took center stage in US foreign policy during the Cold War.[28]

The Rockefeller Foundation took an active, if not leading, role in pursuing the US development agenda.[29] Having established a firm foothold in Latin America, in the mid-1950s it turned its attention to Asia, "that part of the world where there are the most hungry people." In Asia the foundation hoped to reproduce its apparent successes in Mexico and Colombia, again in places deemed to be both in need and of US strategic interest.[30] India, already receiving significant agricultural aid from the US government and the Ford Foundation, was a particularly promising site for intervention. An exploratory survey by the foundation described the "tragic situation" of the world's second most populous country. "Reduced to its simplest terms, the agricultural problem of India is that there are too many people on too little land," the study concluded.[31] It proposed that better crops, training, and agricultural extension might begin to address some of the problems created by these circumstances. In 1956 the foundation forged an agreement to pursue breeding and training in cooperation with the Indian Agricultural Research Institute at New Delhi.[32] Soon after it partnered with the Ford Foundation to launch the International Rice Research Institute (IRRI), which opened in 1962 in Los Baños, Philippines. IRRI's origins lay again in a dire assessment of population pressures, this time across Asia. "Rice is . . . the major food for those parts of the world which are underprivileged, and where the race between food and population is so grim that starvation is a constant threat and a not infrequent reality," declared an early foundation assessment. Guided by this assessment, IRRI's staff set out to transform rice production, especially through the development of higher-yielding varieties.[33]

As these Asian programs took shape, Rockefeller Foundation interventions in Latin America expanded along similar lines. In the early 1950s, with bilateral programs in place in Mexico and Colombia, foundation staff occasionally received requests for aid from other Latin American governments. They responded to these formally in 1954 with the Central American Corn Project, which targeted national corn-breeding programs in Costa Rica, El Salvador, Guatemala, Honduras, Nicaragua, and Panama. A few years later Edwin Wellhausen, now head of the foundation's Mexican program, convinced administrators to make the Central American program hemispheric. From 1959 he led the Inter-American Corn Improvement Program, the flagship enterprise of a larger Inter-American Food Crop Improvement Program, through which he sought to improve corn production at an international scale.[34]

Wellhausen placed the foundation's extensive collection of corn varieties at the center of his vision for better corn production. "The collections of corn being stored in the banks of Mexico, Colombia and Brazil are a gold mine in the further improvement of corn in Latin America," he declared in 1959. However, the "unlimited opportunities" that hypothetically existed in these collections were in practice limited by breeders' knowledge. "Without a doubt, there are hundreds of genes for yield, quality, insect and disease resistance, drought resistance, etc. scattered throughout the many different varieties stored in the corn banks about which we know little or nothing." If the varieties were only studied in greater depth, then "real super varieties could be developed for almost any set of conditions." Moving forward from this assessment, Wellhausen envisioned a program that would augment technical training and assistance with a program of "special basic research" on corn breeding conducted at a new international center, where "all the corn types of the world could be brought together." He imagined that this stockpile could be studied and deployed to the benefit of all corn-growing regions worldwide.[35]

Wellhausen was at liberty to imagine such an effort because the Office of Special Studies was, to quote a foundation report, "self-liquidating," mostly by turning over key positions to Mexican scientists.[36] Foundation reports described the program as having achieved its goals; later Mexican accounts stressed the inefficiency of maintaining two separate state agricultural research operations and the frustrations of Mexican researchers outside the Office of Special Studies.[37] Even if "success" wasn't the chief reason for its closure, however, it had contributed to appreciable changes in production. Corn yields per acre had nearly doubled since the 1930s. The gains were not sufficient to meet domestic demand every year, even with more land under cultivation, but it still looked like progress.[38] Wheat offered a more spectacular story. Under the direction of the agronomist Norman Borlaug, the wheat program had released high-yielding varieties that allowed well-to-do farmers in northwestern Mexico to increase their production. Wheat imports dropped about 90 percent, from an average of more than two hundred thousand tons per year in the 1940s to about twenty-three thousand by the late 1950s.[39]

In December 1960 an executive order of the Mexican president expedited closure of the Office of Special Studies. This directive created the National Agricultural Research Institute (INIA; Instituto Nacional de Investigaciones Agrícolas), into which the Office of Special Studies and the Institute for Agricultural Research (Instituto de Investigaciones Agrícolas), which had housed the state-run breeding program, were merged. Over the next few years, the Mexican research agency took over the various projects of its predecessors, including maintenance of the extensive corn collection of the Office of Special Studies, which by the early 1960s numbered around six thousand samples.[40]

An emerging complex for agricultural research at Chapingo, already the site of the National School of Agriculture and now home to the National Agricultural Research Institute, included the construction of a graduate school (Colegio de Postgraduados).[41] Wellhausen imagined this graduate center as the home of research on corn diversity that would underpin varietal development by the Inter-American Corn Improvement Program.[42] During its first years in operation, students and staff of the graduate school characterized the assembled maize landraces and refined their classifications. They tested crosses among different kinds to determine the most productive.[43] They also developed a second corn collection for use by the Inter-American Corn Improvement Program. Convinced that this collection ought to be a more manageable size, Wellhausen selected representative or "typical" samples of each race and otherwise combined similar samples into "composites."[44] In the end there were 670 composites and about 1,500 typical samples—a partially duplicate collection that was also kept on the Chapingo campus.[45]

Wellhausen's agenda for the future of corn breeding inched closer to reality in October 1963, when the Rockefeller Foundation and Mexican government agreed to establish an International Center for Maize and Wheat Research. This cooperative venture, which existed amid the resources of the graduate school and INIA in Chapingo, aimed to extend the successes of the Inter-American Food Crop Improvement Program to countries beyond the Americas. The cooperative arrangement proved unworkable, but the ambition for an international research center remained. In April 1966 a new agreement created an independent institution, the International Center for the Improvement of Maize and Wheat (CIMMYT; Centro Internacional de Mejoramiento de Maíz y Trigo), which would be run more along the lines of IRRI, the International Rice Research Institute in the Philippines.[46]

A year later two additional institutes took their places alongside IRRI and CIMMYT, the International Institute of Tropical Agriculture in Ibadan, Nigeria, and the International Center for Tropical Agriculture (Centro Internacional de Agricultura Tropical) in Cali, Colombia. Together with IRRI and CIMMYT, these foundation-led programs formed the backbone of a reorganized agricultural aid portfolio, which the foundation thought pointed "Toward the Conquest of Hunger."[47] In time these institutes did contribute to a transformation in agricultural aid and food production, even though the "conquest of hunger" remained a distant goal.

CIMMYT and its antecedents claimed the first apparent successes. After Norman Borlaug had developed higher-yielding varieties of wheat for Mexican farmers at what was then still the Office of Special Studies, he had looked for further routes to increased productivity. He began experimenting with short-statured or "semidwarf" wheat varieties. Many Mexican wheat varieties, unadapted to heavy fertilizer use, shot up when synthetic fertilizers

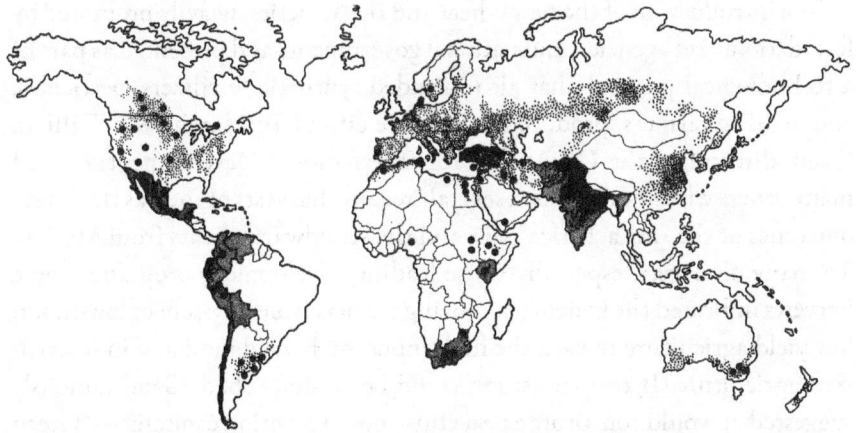

[18] By 1967 CIMMYT wheat varieties had traveled far afield, both as varieties tested for yield by agronomists (large black dots) and in cultivation by commercial farmers (shaded regions). The small dots indicate significant sites of wheat production worldwide, perhaps hoped-for destinations of future CIMMYT distribution. From CIMMYT, *1966–67 Report*, 56. Courtesy of CIMMYT.

were applied, leaving their tall stalks especially susceptible to falling over (or, in breeders' terms, lodging). Borlaug anticipated that genetically short-statured wheats would avoid this problem, benefiting Mexican farmers who could afford synthetic fertilizers. He was right. When his semidwarf varieties went into circulation in the early 1960s, production in Mexico rose even further—so significantly, in fact, that in some years wheat could even be exported. The pronounced effect inspired efforts to spread the varieties further. In 1966, 18,000 tons of semidwarf wheat seed, produced in Mexico, were shipped to India as part of a campaign to raise wheat productivity. The next year 42,000 and 22,500 tons went to Pakistan and Turkey, respectively.[48]

As the Mexican wheat varieties traveled across South Asia and made inroads in the Middle East, a new high-yielding rice variety was poised to follow a similar trajectory. Since 1962 breeders at IRRI had been trying to create a variety that would behave much like Borlaug's wheats. They wanted a short-statured plant that would not collapse under its own weight when fertilized, and they wanted it to be dramatically higher in yield than its predecessors. In 1966 they debuted a variety that met these requirements, IR8, to farmers in India and the Philippines. Thanks to tireless promotion, it quickly spread across South and Southeast Asia.[49]

The introduction of the new wheat and rice varieties, heavily promoted by foundations, aid agencies, and national governments and presented as part of a technological package that also included synthetic fertilizers, pesticides, and loans to farmers, produced immediate effects. In March 1968 William Gaud, director of the US Agency for International Development, echoed many others when he described several "record" harvests of 1967 as the direct outcomes of the "miracle rice" IR8 and the semidwarf wheats from Mexico. To many observers, especially those leading development programs, these harvests indicated the benefits of moving farmers from a system of low-input, low-yield agriculture toward the high-input, high-yield model of industrialized agriculture. If this transition could be accomplished, Gaud famously suggested, it would constitute a peaceful—not to mention capitalist—"Green Revolution."[50] It would also cement the pattern of agricultural aid established in the wake of World War II, in which hunger and poverty would not be addressed through the restructuring of global markets and the redistribution of agricultural goods, as had once been envisioned at FAO, but instead through the technological enhancement of commodity crop production.[51]

FAO AND THE INTERNATIONAL BIOLOGICAL PROGRAMME

Although the widespread circulation of IR8 and the semidwarf wheats inspired celebration in policy-making circles, elsewhere they provoked concern. One worry that surfaced almost immediately was the catastrophic effect it would have on crop diversity. Tellingly, by the early 1970s the rapid diffusion of hybrid corn was no longer the go-to example of the process that had only recently been dubbed genetic erosion. Instead, experts invoked the spread of IR8 and Mexican semidwarf wheat varieties.[52] Piecemeal regional change was supplanted by wholesale global change.

When agricultural experts gathered at FAO in 1967 for another meeting on plant exploration, introduction, and conservation, they again pinpointed "development" as a problem. They reflected on the cruel irony that aid efforts, intended to raise standards of living by ramping up agricultural production, might eventually undermine them instead by depriving plant breeders of "the very raw materials upon which they depend." And they deemed a crash program of collection and conservation essential.[53] Although the basic elements of this narrative remained as they had been in previous discussions, a heightened sense of global crisis pervaded the 1967 meeting. The loss of crop diversity was no longer just a threat with respect to specific crops or in localized places—as had been the case, for example, for corn in the Americas a decade earlier. Now the steady "genetic erosion" of crops was globalized. It was a "potential danger of the utmost gravity to all of man's crops" and a problem "sufficiently large to transcend the frontiers of nations." Meeting participants declared that there might be "no second chance" if the action plan they recommended was not followed. Emphasizing the need to save the world from this looming catastrophe, the assembled experts positioned themselves as humanitarians, taking action to sustain well-being everywhere, rather than self-interested researchers whose work would benefit if access to crop diversity were to be both facilitated and guaranteed.[54]

This presentation also aligned them with a growing environmentalist consensus in the late 1960s, one that blamed industrial societies with generating large-scale environmental problems that seemed to threaten the future of humanity.[55] The suffusion of this perspective at the 1967 FAO meeting can be traced in part to the participation of the International Biological Programme (IBP) as the meeting's cosponsor. This international research initiative, founded a few years prior, had been inspired by environmental concerns, including worries about population growth and the resulting strain on natural resources. The scientists behind IBP argued that well-financed biologists organized across national borders could generate knowledge crucial to facing these challenges. Organizers outlined two broad domains for contributions to IBP: assessing the present and future biological bases of global natural resources and understanding human adaptability in light of environmental change.[56]

Although some biologists and ecologists surely looked on IBP as little more than a route to research funds, its organizers embraced the idea that the program would preserve the earth's biological resources, ensuring that "samples of these wonderful creations of evolution . . . will continue to flourish for subsequent generations of humans to appreciate and use."[57] This preservation agenda—which tellingly promised no more than the survival of "samples"—was informed by assumptions about the peoples and places of the past and the sciences of the future. IBP-funded population geneticists imagined preserving human evolutionary history by extracting blood samples from men and women they identified as "primitive." Supposedly isolated from "modern" life, these individuals were thought to be free from exposure to things like radiation, chemicals, and population mixing, which would muddy their evolutionary story. Their blood was seen as a promising but endangered resource for future biological research, containing a wealth of information that tomorrow's researchers would unlock.[58]

The IBP's human genetic salvage operation shared key assumptions with an older vegetal counterpart. Missions to collect crop landraces had for years assumed a threat to "primitive" genetic assemblages posed by modernity, promised the extraction of value from these in the future, and insisted on the need for urgent action. These views were repeated in the IBP's agenda for crop genetic preservation, which would focus on "primitive races" and crop wild relatives. These were deemed to be of greatest scientific interest, containing unique and potentially valuable "genes and gene combinations" but unable to survive the "advance of civilization."[59]

IBP plans sparked new activity at FAO around crop diversity. An IBP working group led by the Australian wheat breeder and science administrator Otto Frankel brought FAO an itemized plan of action, beginning with expert assessments and extending through the organization of field expeditions, long-term seed preservation, and on-site conservation of wild populations.[60] Frankel's agenda nudged the FAO toward more decisive intervention. As plans unfolded, the idea of an international center for maintaining breeding materials was revived. In early planning IBP organizers had floated the idea of establishing "World Banks of germ plasm, both plant and animal." This was subsequently revised to using existing facilities to preserve whatever "genetic stocks" IBP projects assembled.[61] In collaboration FAO and IBP eventually settled on an "international gene bank." This was not to be a physical structure but instead a network of national and regional banks linked through international agreements and overseen by a central office. In the proposed model participating banks would agree to provide material to requestors from all countries and commit to specific routines for maintaining and renewing collections.[62]

The researchers behind this call for an international seed bank considered formal agreements and monitoring of maintenance regimes to be crucial. Their conversations circled around a concern that the world's existing collections of landraces and wild relatives—of which there were many hundreds worldwide, in the hands of breeders and botanists—were poorly kept. Regeneration of samples was needed too often, did not happen when it was needed, or was conducted haphazardly, all of which contributed to the loss of genetic diversity within collections. Frankel concluded that, of existing institutions, "few [could] be regarded as gene banks," places where long-term survival would be guaranteed. The plant explorer Jack Harlan agreed that losses from collections were "often rather alarming." He stoked fears among colleagues by describing the rapid disintegration of a world sorghum collection to half its original size within a few years. The upshot of observations like those of Frankel and Harlan—two of the most influential researchers who contributed to the joint FAO and IBP proceedings—was a recommendation that secure storage and ongoing maintenance needed to be considered even before gathering endangered varieties from farm fields.[63] Careful external oversight, such as that promised by an international seed bank, was essential to ensure the security of both existing and future collections.

Having agreed that the effort to coordinate existing and planned collections into a more effective global conservation system could "only come from [a] United Nations agency," participants in the 1967 meeting suggested that FAO would be the obvious center for this operation.[64] But the creation of an international gene bank, imagined as a collection of collections produced through intergovernmental oversight and management, also required money. In 1967 neither IBP nor FAO had much of that crucial ingredient to spare.

THE WORLD GERMPLASM PROJECT

Although funding was perpetually short at FAO, Frankel and his colleagues succeeded in attracting modest sums to follow up on their recommendations of 1967. Following the meeting FAO funded a new Crop Ecology and Genetics Resources Branch. A Panel of Experts on Plant Exploration and Introduction (hereafter the Panel of Experts), established several years earlier, was already in position to advise it. Working with scarce resources, staff of the new branch and members of the Panel of Experts did what they could to both generate and capitalize on concern about crop diversity.[65]

Among other activities they hammered out a more detailed vision for coordinating existing institutions into an international system for the exploration, introduction, and conservation of crop diversity. They formulated these tasks as distinct but related activities. Whereas exploration and introduction required trained collectors to travel to the field and breeders to assess, develop, and distribute collections, conservation simply required the sequestration of seeds in long-term storage under the best possible conditions. This led the Panel of Experts to distinguish between materials in frequent use by researchers and those slated for long-term preservation, referring to "working collections" versus "conservation collections." It also led them to the idea that these two kinds of collections could be managed by different institutions. Conservation collections would comprise duplicates of working collections, with samples set aside solely for conservation purposes and not used in research or exchange except as a last resort.[66]

Given that it had no money with which to create facilities for conservation, FAO looked for institutional collaborators. Its key targets included a handful of institutions already, or soon-to-be, equipped for long-term storage, such as the national seed storage facilities of the United States and Japan. Other potential collaborators were IRRI and CIMMYT, along with their patron, the Rockefeller Foundation. The agricultural research centers had dedicated seed storage facilities and boasted breathtakingly large collections.[67] IRRI scientists maintained a world collection of rice that by 1966 already had more than ten thousand accessions.[68] At CIMMYT the corn collection had grown to a similar size, despite Edwin Wellhausen's intention of keeping it manageable.[69] Meanwhile, the Rockefeller Foundation continued to see IRRI and CIMMYT as flagships of its aggressive charge "Towards the Conquest of Hunger." Its continued investment in collections of seeds through the centers led Otto Frankel to the foundation's doorstep in 1967, in hopes of convincing its administrators to support the initiatives taking shape at FAO.

Although they judged Frankel's proposals to be "of no possible interest" in 1967, foundation officers soon reversed course, at least with respect to considering crop diversity worthy of closer attention.[70] The CIMMYT wheat team helped change their minds. Norman Borlaug thought that the "spectacular success" of the semidwarf wheat varieties demanded that attention be given to preserving wheat landraces.[71] By 1969 he and another CIMMYT wheat breeder, Keith Finlay, were reportedly "pressing very hard . . . for some urgent action." Finlay, previously an IBP collaborator and associate of Frankel, thought that CIMMYT-produced wheats were displacing "the very material which will provide us with resistance and adaptability genes." He found further justification for conservation in his new employer's history. "CIMMYT, more than any other organization, has been responsible for the rapid expansion of new genetic material . . . and therefore we have a very firm moral obligation to do something," he urged.[72] Despite possessing significant collections of crop diversity, researchers at CIMMYT had not prioritized conservation. Their collections were resources for today's breeders, not tomorrow's.

In the case of corn, which had been sampled and curated far more extensively than wheat at CIMMYT, maintenance of the collections was mostly neglected—this despite the fact that its richness in diversity had been among the inspirations for CIMMYT in the first place. In the late 1960s the CIMMYT staff member Mario Gutiérrez Gutiérrez was assigned to create order from the chaotic mess that CIMMYT's Maize Germplasm Bank had become. As Gutiérrez later recalled, "What CIMMYT had in 1967 could not be classified as a bank but merely as a dump." Seeds from recent summer harvests intended for the bank had been placed in tins but not into cold storage. Two-hundred-pound sacks of corn filled the aisles of the cold chamber. The refrigerator was out of operation. Above all, "no true inventory existed," which meant that "staff members had to spend sometimes as much as a week trying to locate the seed they wished."[73]

While Gutiérrez labored in the chaotic CIMMYT corn collection, administrators at the Rockefeller Foundation's New York office took a rosier view of the situation. "During the past 25 years, the Rockefeller Foundation has contributed to science and to world agriculture by assisting in the collection, storage and evaluation of the world's basic food crops—corn, wheat, rice, sorghum, and the millets," boasted the foundation's director for agricultural sciences, Sterling Wortman. Even as he touted these achievements, Wortman acknowledged a "need to complete the assembly and evaluation of germplasm of these five basic crops" and proposed to do so through a foundation-led World Germplasm Project. Wortman's proposal carefully articulated the imperative to collect and conserve crop diversity as a product of the foundation's past successes in collection rather than its complicity in the destruction of landraces.[74]

The activities outlined for the World Germplasm Project looked similar to those being pursued by FAO. The project comprised a survey of experts in the field to determine the status of existing collections and further exploration needs, make collections, identify safe storage facilities, encourage evaluation of collections, and, finally, prepare a status report on the "collection of the world's germplasm." There were important differences, however. A first distinction was that, although the Rockefeller Foundation plans discussed "world germplasm resources," these were limited to "crop plants important to the conquest of hunger."[75] As the project took shape, this was taken to mean corn, wheat, rice, sorghum, and the millets. These were the staple grains at the center of the foundation's particular vision for feeding the world and staying the perceived population crisis. By comparison the plans laid by FAO and IBP cast a much wider net. For example, the preliminary world survey of urgent needs in field collection, begun in 1971, included these staples but also gave attention to other grains such as oats and barley, pulses like lentils and chickpeas, root vegetables including yams and cassava, and also tree fruits, coffee, and cacao. Rather than focus on a handful of commodity grains, it recognized the importance of subsistence crops, local or regional crops, and cash crops that provided a livelihood for peasant farmers.[76]

Another important distinction was that the Rockefeller Foundation had a clear time horizon. Wortman imagined that, within five years, seeds would be collected and "appropriate germplasm banks would be in operation serving scientists everywhere." This would not be an ongoing effort, like the proposed FAO oversight and coordination of a global network for collecting, exchange, and maintenance. Quite the opposite: with a short time horizon and bounded goals, the World Germplasm Project would be an effort that adhered to the foundation's requirements for its own grant recipients.[77]

Once in motion the World Germplasm Project operated through committees, one each for corn, rice, wheat, and sorghum and the millets. These were tasked with assessing the status of collections, identifying priorities for further collecting, and proposing a long-term preservation strategy for each crop. In this process a new Maize Committee was born. William Brown, former member of the Committee on Preservation of Indigenous Strains of Maize and now vice president of the Pioneer Hi-Bred Corn Company, accepted chairmanship. He immediately brought together the handful of scientists he considered most knowledgeable about the status of corn collections worldwide, which included those responsible for the large collections in Brazil, Colombia, and Mexico.[78] The summary assessments of this committee's initial report seemed rosy. There were few gaps in existing collections of corn diversity, thanks to earlier preservation efforts. In spite of difficulties, "including problems of maintenance, lack of manpower, neglect of collections at some storage centers, etc.," most of the corn samples gathered in the previous two decades survived. The vast majority were in the banks of CIMMYT; the National Agricultural Research Institute in Mexico, which had taken over the Office of Special Studies collection; the school of agriculture in Piracicaba, Brazil, which had some remnants of the materials collected by Friedrich Brieger and his colleagues; and the Colombian Agricultural Institute (Instituto Colombiano Agropecuario), which had inherited materials from the Rockefeller Foundation's program there.[79]

Yet even if the collections existed, their survival was by no means assured, and here the report was decidedly less rosy. The collection in Brazil no longer included any of the original individual samples collected in the 1950s, only bulked populations made by combining materials that looked similar. More worryingly, the university's facilities did not provide adequate conditions for long-term storage. In Colombia the collections originating in Bolivia had been declared "impossible to grow," while "lack of time" prevented scientists from renewing the collections from Ecuador. In addition, two races of corn from Colombia had been lost, despite the fact that the samples of Colombian origin were the best attended to of all the material in that bank. In light of these circumstances, the World Germplasm Project's Maize Committee arranged for CIMMYT to take over management of the entire Brazilian collection and, despite the evident reluctance of its lead scientist to cede control, some of the samples kept at the Colombian Agricultural Institute as well.[80]

The stopgap measure of transferring endangered collections to CIMMYT was just that, a stopgap. It was desperately clear that more supplies and personnel were needed to keep up with regeneration and distribution at all the sites. As one committee member pointed out, even when collections were said to exist in various seed banks, they were often impossible to obtain.[81] Committee members' uneasiness over the long-term prospects of maintenance at the Latin American sites was evident in their recommendation of "storage for insurance purposes." They thought that the collections of the "three major seed banks" (by which they meant CIMMYT, plus those maintained by the Mexican and Colombian state research institutions) ought to be duplicated in comparatively small quantities and sent for "permanent, long-time storage," where it would be accessed only "in case of emergency." The US National Seed Storage Laboratory in Fort Collins, Colorado, seemed the likeliest candidate to serve as an ultimate fail-safe.[82] In other words, the committee suggested that all the collections now housed in facilities in Latin America ought to be duplicated and sent for storage at a facility thought to be more secure, technically and politically. As had been the case at FAO a few years earlier, those assessing existing collections decided that, by and large, they could not be trusted to succeed at the challenging task of long-term conservation on their own. Additional measures—in this case both centralized management and duplication of collections—were deemed necessary.

Even if the US National Seed Storage Laboratory could be engaged to take on the backup function that the committee thought essential, it could not possibly serve as an institution for the ongoing management, assessment, regeneration, and dispersal of collections of corn from across Latin America. These tasks were necessary for diverse corn varieties to be used as well as to in the long term. An "insurance" collection at the National Seed Laboratory would eventually die in cold storage if never replaced esh seeds. CIMMYT therefore emerged as the lynchpin—like it or not—for the future of successful long-term conservation and "orderly distribution" of corn germplasm. While recommending that support be provided for the National Agricultural Research Institute in Mexico to enable better maintenance of its unique collections, the committee wanted CIMMYT to be expanded and not just supported. It could become, the committee thought, a repository and distributor for nearly all tropical corn-breeding material worldwide.[83] Years earlier Edwin Wellhausen had envisioned the CIMMYT Maize Germplasm Bank as a distribution center. Participants in the Rockefeller Foundation's World Germplasm Project now hoped to transform it into the central hub of a hemispheric, and perhaps world, conservation operation.

The recommendations of the World Germplasm Project's committee of corn experts dovetailed with the recommendations of the committees assessing wheat, rice, and sorghum and the millets. These also recommended that the relevant foundation-funded international agricultural research institutes become central hubs for storing and circulating breeding materials. The committees similarly converged in requesting duplication of the world's collections, for security's sake.[84] A consensus was clear. From the perspective of the expert committees assembled by the Rockefeller Foundation, preservation of crop diversity for the very long haul would be best organized by the international agricultural research centers and achieved through the strategic duplication of collections for storage in those facilities felt to be most secure. This plan was not the only one in circulation, however.

THE INTERNATIONAL BOARD FOR PLANT
GENETIC RESOURCES

As the Rockefeller Foundation's World Germplasm Project unfolded in the years around 1970, so too did a reorganization of international agricultural aid. The latter had far more lasting effects for the conservation of crop diversity. By 1969 the Rockefeller and Ford Foundations had initiated talks with the world's major providers of development aid, hoping to convince them that it was in everyone's best interest to agree on priorities for agricultural assistance and to coordinate funding. These negotiations precipitated the

1971 formation of the Consultative Group on International Agricultural Research (CGIAR).[85] Soon after, CGIAR became the venue for heated negotiations over the management of the world's crop diversity, a contest over the what, where, and by whom of conservation—but, crucially, not over the how.

In the discussions leading to the creation of CGIAR, the foundations made a case to other donors that only scientific research and training could resolve world hunger. In 1969, at a gathering of representatives of national and international aid agencies, foundations, and intergovernmental organizations, the vice president of the Rockefeller Foundation insisted that most development programs had, over the preceding twenty-five years, produced "disappointingly small" results. Meanwhile the "food gap"—by which he meant the difference between a population's caloric demands and the food available—had grown in developing countries thanks to "rapidly expanding numbers of people." He identified only one exception to this pattern, describing how, very recently, "in such critical countries as India, Pakistan and the Philippines," cereal production had undergone a marked increase. This was an obvious reference to the record harvests attributed to the Mexican wheat varieties and IR8 rice. Here it was offered as evidence that dire predictions about world-engulfing famine due to population growth need not come true. Research could raise yields, "buying time" for population programs to achieve their desired ends.[86]

A subsequent elaboration by the Rockefeller Foundation's Sterling Wortman of how these record-breaking figures had been attained was intended to lead listeners to the foundations' desired conclusion. Because scientific research centers in Mexico and the Philippines took credit for the miracle crops and subsequent miracle harvests, it stood to reason that world agricultural production could be boosted further by the "formation of a worldwide, interlocking complex of national and international scientific institutions, programs, and projects." These would produce the "scientific information, materials, and manpower required to intensify agricultural production." At the center of this "interlocking complex" would be several internationally funded agricultural research centers.[87] By 1969 there were four institutes already in place: CIMMYT, IRRI, the International Institute of Tropical Agriculture in Nigeria, and the International Center for Tropical Agriculture in Colombia. As of 1968, both the Rockefeller and Ford Foundations contributed $750,000 annually to each of these four centers, while expenditures ran to between $2.7 and $4 million per center. If the donor community could be convinced to channel aid through the institutes, then these could be placed on more secure financial ground—and the foundations would have funds freed up to support new endeavors.[88]

Chief among their imagined future projects, unsurprisingly, were further agricultural research institutes. According to Wortman, the world lacked adequate research sites for tropical vegetables and fruits, sorghums, millets, and oil crops, as well as agriculture in arid regions. At least one international center needed to take responsibility for "every commodity of any importance." Among other tasks the staff of these centers would conduct the "biological engineering" needed to turn local crops into high-yielding ones so that the "farmer can be emancipated from the restrictions on yield imposed by his native varieties."[89]

For crop varieties, as for other aspects of peasant cultivation, the principal goal of this scientific research would be "modernization." Foundation officers insisted that, with new seeds and cultivation methods, improved delivery of fertilizers and pesticides, and government support, "traditional farmers will modernize . . . as rapidly as their personal resources and inherent propensity to caution will permit." A vigorous appraisal of this view might question how a cash-strapped peasant farmer, stereotyped as stubborn and unteachable, could overcome both poverty and prejudice to buy into a modern agricultural production system. But confidence ruled the day. The intensification of research would lead to the intensification of agriculture, transforming "vast land areas" still "untouched by modern technologies" into efficient, effective farm operations. Only this could meet the needs generated by the "seemingly inexorable growth of world populations."[90]

With the launch of CGIAR, which held its first meeting in May 1971, the foundations succeeded in aligning international aid to their own vision of science-led agricultural intensification. The mission of CGIAR centered on the agricultural needs of developing countries. It would assess research needs, direct funds to existing agricultural institutes, create new institutes organized on the same model, and coordinate international research efforts. Sponsored jointly by the World Bank, FAO, and the United Nations Development Programme, the initial participants included nine donor countries of Europe and North America, five international organizations, three foundations, and one research center.[91] In assuming for itself the mantle of global coordinator for agricultural aid, CGIAR usurped a role thought by many to belong rightfully to FAO. Among other effects this shifted power away from aid-receiving nations, who tended, as a result of their greater numbers, to have a meaningful voice in UN agencies such as FAO. This was not the case within CGIAR. At the first meeting a decision was taken to invite "not more than five" representatives from developing world governments to serve as members.[92]

One of the first arenas in which CGIAR's new authority in international agricultural development was contested was the management of crop genetic diversity. At the suggestion of the renowned Indian agronomist Mankombu Sambasivan Swaminathan, the CGIAR's scientific advisory committee tackled the issue of plant genetic resources at the committee's first-ever meeting in the summer of 1971. Discussion of a paper prepared by Otto Frankel sparked demand for a concrete proposal of action. Committee members wanted to know the specifics. What already existed in collections? What exactly remained to be gathered? What would it cost?[93] This request spawned two years of revised proposals and counterproposals, working groups, and subcommittees. Both participants and later accounts presented these exchanges as a contest between FAO and the organizations behind CGIAR, especially the Rockefeller Foundation.[94] And indeed it was. It was also a contest between a broad conservation vision and a narrower aim of agricultural industrialization. In many respects it was a contest between the needs of the poor world and the interests of the rich.

In response to the demand for more details, a working group led by Frankel, who was by this time the FAO's chief consultant on plant genetic resources, laid out a plan for an "international network of genetic resources centres." At the center of this network would be ten "regional centres" located in places of high crop diversity, including sites in Brazil, China, Costa Rica, Ethiopia, India, Indonesia, Mexico, Nigeria, Peru, and Turkey. These regional centers, based at existing institutions as a cost-saving expediency, would manage working and conservation collections of regional landraces and wild relatives, which would be gathered, ideally, by trained local collectors employed at national institutions. Although the plan designated "priority crops," these included a wide range of items in addition to staple grains, such as legumes, fruits, tubers, oil and fiber plants, and forage grasses. To these were then appended lists of "other important crops" and still more crops of "secondary importance." The regional centers would be expected to work with national agricultural institutions in their region, genetic conservation programs in industrialized countries, and "crop-specific" centers such as CIMMYT and IRRI. All these activities would be coordinated by a small, independent committee of scientists "representative of the network," working in concert with several staff persons at FAO.[95]

Administrators at the Rockefeller Foundation chafed at this plan. It was too all-encompassing and provided little means of ensuring that collections, always difficult and expensive to maintain, would not be duplicated among sites or grow too large to manage.[96] Although known duplication, as in the duplication of working collections in conservation collections, was a good thing, they saw unrecorded or unintended duplication as a liability, entailing unwanted expenditure. Foundation officers also believed that the plan ignored the ongoing work of the international agricultural research institutes like IRRI and CIMMYT and risked duplicating this as well. Surely, they thought, well-resourced organizations with experience in managing global crop collections were better positioned at the center of any world network rather than at its periphery as the working group plan had indicated.[97] It could not have escaped their notice that placing the foundation's research institutes at the center would allow them to retain some control over collections. With support from the lead US delegate to CGIAR, they pushed a different vision. They wanted the focus of collection and preservation to be staple grains and for the international agricultural research institutes, existing and projected, to bear the greatest responsibility for these activities.[98]

The FAO and other participants in CGIAR questioned the narrowness of this vision.[99] In the end, however, their critiques made little difference. After two years of often-heated debate, a subcommittee succeeded in generating an operational plan for a new research organization, the International Board for Plant Genetic Resources. The broad conservation vision articulated in earlier proposals was curtailed, through not outright denial but rather a clear indication that activities would focus on important economic crops and agronomic characteristics. This in turn meant that the international agricultural research institutes, already focused on key crops like rice, corn, and wheat, featured centrally in subsequent activities rather than the regional centers tasked with capturing crop diversity across entire geographic regions.[100]

Many were unhappy with this arrangement, not only at FAO but also within CGIAR. At CIMMYT, which had been nominated by CGIAR-based planners to take a lead in conserving both corn and wheat, administrators outright objected. An observer for the new International Board for Plant Genetic Resources sent a worrying report from the first meeting of a committee tasked with overseeing corn conservation: "The director of CIMMYT's corn program made it quite clear that he did not want nor does he intend to have CIMMYT assume the international leadership role for maize germplasm conservation and utilization."[101]

As the effort to coordinate the collection and conservation of corn diversity as a truly international endeavor took shape, the institution finally agreed on to manage these activities balked. By 1975 there were signs that the Maize Germplasm Bank, once envisioned as the beating heart of CIMMYT's corn-breeding ambitions, was no longer felt to be so essential. A postdoc hired in 1973 to create a computerized catalog of the maize seed bank left the job disgruntled, claiming "a rather remarkable lack of interest in my work, or in any aspect of Bank operations."[102] Further evidence of disinterest came in 1976, when Mario Gutiérrez Gutiérrez left his position at the helm of the seed bank and was not replaced.[103] The leaders of CIMMYT's maize program judged that the collection and conservation of crop diversity far outpaced scientists' ability to use it—and they were determined to stay focused on the latter.[104] Within an enterprise driven by a commodity crop–centric and business-friendly vision of feeding the world, conserving crop diversity took a backseat.

This cold reception at CIMMYT left corn experts who were concerned with conservation to continue imagining the duplication of collections as their best option. William Brown, now at the helm of the committee advising the International Board for Plant Genetic Resources on corn, expressed skepticism about the capacities for long-term preservation anywhere but the US National Seed Storage Laboratory. Ever more familiar with global goings-on in seed collecting and storage, he was ever "more convinced that we face a serious risk of losing important germplasm if the foreign collections are not duplicated somewhere in the U.S.A."[105]

This fixation on duplication was also apparent within the International Board itself, as its members developed technical plans for the long-term conservation of crop genetic diversity. Preservation was not assured by collection alone, nor even by cold storage, but instead through duplication. With this view predominating, board members and consultants set out to identify the institutions capable of achieving the best possible conditions. Its goal was to enlist these institutions as "base collections" (previously called long-term or conservation collections) to house copies of all the world's collections of crop diversity. Unlike in "active collections" (previously called working collections), access to base collections would be limited, to minimize regeneration of seed samples and therefore unwanted genetic transformation of these. Whereas active collections might be kept in whatever conditions were achievable at a given facility, base collections would be housed under internationally recognized, optimal conditions for longevity and monitored to ensure their adherence to these standards.[106]

Although cold storage was considered important to prolonging the life span of seeds in base collections, the survival of the precious genetic material within "primitive" crops was not yoked to the freezer, as was the case, for example, with blood samples from "primitive" peoples collected during the International Biological Programme.[107] Preserved seed samples would slowly die in cold storage, so they needed to be continually renewed. This was a process that had to be handled with care. Scientists recognized that, with each generation, renewed seeds became less like the original samples and, for most researchers, retaining the original was the ultimate goal. Seeds had to be endlessly reborn and yet remain forever the same. Achieving this balance made concerns about maintenance paramount. In many assessments even the best human and institutional maintainers could not be trusted, making safety duplication essential.

The roots of today's backup culture, in which the risk of digital loss is managed through habitual redundancy so ubiquitous we hardly notice it, can be traced to the Cold War. In the United States missile defense demanded redundant generators and identical computers working in tandem, each backing up the other. As NASA undertook human space travel, its engineers provided backup versions of nearly every aspect of missions, even the astronauts, attempting to ensure reliability through repetition. Meanwhile, new businesses hawked the safe storage of carbon-copy documents in bombproof facilities. It's perhaps not surprising, then, to discover the origins of present-day seed backup in the same period. Nonetheless, as the media studies scholar Shane Brennan emphasizes, "backups are always constituted through, and contingent upon, the imagination of potential disaster."[108] Specifying that imagined disaster can be revealing. In the case of seed collections, the imagined failures were neither cataclysmic (as a nuclear strike would be) nor an expected cost of working on the cutting-edge. They were quotidian: funding shortfalls, broken refrigerators, poorly trained staff, and bad weather.

These quotidian catastrophes pointed toward certain kinds of places as ideal repositories for base collections. These should be sites where funds were secure, staff highly trained, power grids stable, and climates temperate. With these desired features in mind, a few research institutions in a handful of industrialized countries, and some centers under international control, topped the list of potential host organizations. This may have seemed both sensible and perhaps innocuous at a time when seeds were seen as a common good, but when the ownership of plant genetic resources became the subject of heated international debate in the 1980s, the decision would place the International Board for Plant Genetic Resources and its international conservation network in the hot seat.

[19] The Twenty-Third Conference of the United Nations Food and Agriculture Organization in November 1983 adopted resolution 8/83, an International Undertaking on Plant Genetic Resources. The agreement (about which there was considerable disagreement) declared diverse crop varieties a "heritage of mankind" and insisted they be made "available without restriction." ©FAO, image 3161.

FIVE

NEGOTIATE

AMID THE CLATTER AND HUM generated by several hundred delegates and observers to the 1981 Conference of FAO, a member of the Mexican delegation took the floor. Participants from 145 member nations had already reviewed the state of global agricultural production, assessed and commended ongoing FAO programs, agreed on budget appropriations, and wrestled over the wording of numerous conference resolutions. The Mexican representative opened discussion on yet another draft resolution, this one proposing "The Establishment of an International Plant Germplasm Bank." Two interlocked elements lie at the resolution's heart: a collection of duplicate samples of all the world's major seed collections under the control of the United Nations and a legally binding international agreement that recognized "plant genetic resources" as the "patrimony of humanity." Together the bank and agreement would ensure the "availability, utilization and non-discriminatory benefit to all nations" of plant varieties in storage and in cultivation across the globe.[1]

Today international treaties are integral to the conservation and use of crop genetic diversity. The 1992 Convention on Biological Diversity aims to ensure the sustainable and just use of the world's biodiversity, which includes plant genetic resources. Meanwhile, the 2001 International Treaty on Plant Genetic Resources for Food and Agriculture, also called the Seed Treaty, establishes protocols specific to crop diversity. Although it draws much of its power from the Convention on Biological Diversity, the roots of the Seed Treaty reach further back, to the 1981 resolution of the Mexican delegation and beyond.

Mexico's resolution, like today's Seed Treaty, offered conservation as a principal motivation. It told a story of farmers' varieties displaced by breeders' products, the attrition of genetic diversity, and the looming "extinction of material of incalculable value." Earlier calls for conservation had sketched the same picture. Yet those who prepared and promoted the Mexican proposal mobilized this narrative to different ends. They may well have wanted to protect crop diversity. Far more important, however, was the guarantee of access to this diversity, once conserved. They insisted that a seed bank governed by the United Nations and an international treaty were needed to prevent the "monopolization" of plant genetic materials.[2] This monopolization came in the form of control by national governments, the ultimate decision makers for most existing seed banks. It also resulted from possession by transnational corporations. By exercising intellectual property protections in crop varieties, seed companies could take ownership of these varieties, even if they were derived from seeds sourced abroad. In other words, the survival of a seed sample in a base collection, or its duplicate, did not mean this sample was available to breeders, let alone farmers, in its own place of origin. Binding international agreements were necessary to ensure access.

Mexico's intervention at the 1981 FAO Conference was just one volley in what would later be called the seed wars, a decades-long conflict over the granting of property rights in plant varieties and the physical control of seed banks.[3] Allusions to endangered crop diversity have been mostly rhetorical flourishes in this debate, deployed in defense of other things considered threatened by agricultural change—namely, peoples and governments across Africa, Asia, and Latin America in the later twentieth century. Seed treaties were meant to protect not seeds, but sovereignty.

Between the late 1960s and the early 1980s, in the midst of this struggle over seeds, consensus fractured about the loss of crop diversity—or, more specifically, about the meaning of this loss. When experts had gathered at FAO in the 1960s to discuss genetic erosion, most saw this as an inevitable consequence of a beneficial transition. Wherever farmers opted for breeders' lines over their own seeds, the value of these so-called improved lines was confirmed, and agricultural productivity inched forward. In the 1970s genetic erosion featured centrally in a very different narrative. It was offered as evidence of the misguided ideas and practices driving agricultural development, especially the Green Revolution, and of the dangers posed by powerful transnational seed companies. Corporate greed emerged as a new driver of crop diversity loss. The willingness of wealthy countries to sustain this greed through friendly regulations meant both were complicit in undermining the capacities of developing countries to feed themselves. The extinction of farmers' varieties and landraces was no longer an accepted byproduct of agricultural modernization. It was an argument against this development.[4]

This shift pitted scientists committed to saving crop diversity against activists ostensibly interested in the same thing. It brought competing visions of what agriculture could and should be head to head. Invocations of the imminent loss of crop diversity, the one element everyone seemed able to agree on, reached a fever pitch during the seed wars. This rhetorical barrage often obscured on-the-ground realities. While FAO delegates, government officials, NGO activists, and prominent scientists waged a war of words in meeting rooms and magazines, plant breeders and agronomists tended experimental plots, tested genetic combinations, and presented farmers with varieties they hoped would be improvements. In 1970s Mexico some of these researchers were newly resolved to use Mexican seeds and methods to address the needs of the country's poorest farmers. Keeping these individuals, their methods, and their corn collections in view grounds the seed wars in actual seeds. If the Mexican delegation's invocation of crop diversity at FAO in 1981 was a rhetorical flourish in a bid to defend national sovereignty, the concurrent use of crop diversity by some Mexican breeders was a practical strategy for getting Mexican agriculture out from under the thumb of the United States and transnational agribusinesses. On the ground, seeds were not ornaments in oratory but the very stuff of sovereignty.

REVOLUTION AND COUNTERREVOLUTION

In the late 1960s, while many government officials, foundation officers, policy makers and agricultural scientists celebrated the arrival of the Green Revolution, a counterrevolution was underway. One site of critique was Chapingo, Mexico, the place that many understood as the origin point of the Green Revolution model. Home first to the National School of Agriculture, and from 1943 the Rockefeller Foundation–sponsored Office of Special Studies, Chapingo had become a hub for the production of agricultural experts and the dissemination of agricultural knowledge. The inauguration of a graduate school (Colegio de Postgraduados) in 1959 and a subsequent reorganization and expansion of the facilities for agricultural research, education, and extension further consolidated the site's centrality in efforts to "modernize" agriculture in Mexico and beyond.[5] As Norman Borlaug's wheats traveled from Mexico to the world in the late 1960s, agricultural

[20] Students from across Mexico and Latin America came to Chapingo to pursue graduate studies in agricultural sciences at the Postgraduate College (Colegio de Postgraduados) after its founding in 1959. The Polish scientist Czesława Prywer Lidzbarska, who had emigrated to Mexico after World War II and was among the early teaching staff at the college, led instruction in cytology and botany. Courtesy of CIMMYT.

experts from around the world traveled to Mexico, hoping to learn firsthand how to drive development through agricultural research. This invariably meant a visit to Chapingo.

Although Mexican leaders and donors like the Rockefeller Foundation touted transformations at Chapingo as evidence of Mexico's agricultural development, not everyone thought the model espoused there served the best interests of Mexico's farmers. In the 1950s a few students had rebelled against the narrow technical focus of their education. At a time of rising anti-US sentiment across Latin America, they voiced concern about the intervention of US experts, including the Rockefeller Foundation. This unrest escalated after the 1963 announcement of Plan Chapingo, the proposed development on-site of a new headquarters for the National Agricultural Research Institute (INIA), as well as facilities for the national agricultural extension agency, a center for agricultural statistics, a library and teaching spaces, improved student housing, and other amenities. The additions and upgrades would be funded in part by the Ford and Rockefeller Foundations and the US Agency for International Development, patronage that some students saw as harnessing Mexico's agricultural future to foreign interests.[6]

Student dissatisfaction came to a head in 1967. In June a protest over poor education, inadequate facilities, and administrative corruption at a different agricultural college galvanized a strike among Chapingo students. Teaching, research, and extension activities came to a standstill for weeks, in a dispute that encompassed several colleges across Mexico. Although the quality of education was a core concern, student protesters linked poor agricultural education to the government's neglect of rural people more generally. Drawing on the ideals and rhetoric of the Mexican Revolution, they demanded training that served Mexican peoples' interests and not those of businesses or foreigners.[7] This revolutionary rhetoric also featured in the broader Mexican student moment, which rose to prominence the following year amid a global escalation in popular protests. In Mexico, as elsewhere, the student movement demanded social reforms and greater democracy. Student protestors also aligned with the wage laborers and campesinos exploited and abandoned during Mexico's rapid industrialization, contesting the inequitable distribution of benefits to economic growth.[8]

The strike of agricultural students in 1967 and the student movement of 1968 heightened political awareness among scientists and engineers who trained at Chapingo in the late 1960s.[9] Perhaps to their surprise, some went on to work in an environment a little more conducive to their politics. The marked inequality that encouraged student protest in Mexico in the late 1960s also generated agrarian uprisings and land invasions. A growing crisis in the countryside emerged from the impossibility of campesinos receiving a fair price for their grain, their losses of farmland to larger enterprises and potential jobs to agricultural mechanization, and their inability to find urban employment—all problems further exacerbated by population growth.[10] A growing number of researchers and administrators recognized the untenability of development that did not address the needs of these farmers too.[11]

At INIA maize researchers began to articulate new aims for their breeding programs. Early in the 1960s these efforts had focused on the development of hybrid varieties for the irrigated regions of the country, but by the end of the decade priorities had shifted. A 1968 report declared the hybrid programs a legacy of INIA's predecessor institutions, especially the Office of Special Studies. It called for the development of varieties to suit other regions

of the country, especially ones that would be resistant to drought. INIA's gaze was turning from better-off farmers to poorer ones and from global demands to local needs. One piece of evidence that this expansion was underway was the renewed study of maize landraces originating within different regions of Mexico. These studies aimed to find types with better resistance to diseases, cold, and drought and were accompanied by new collecting missions.[12]

Conditions for this and other state-funded agricultural research improved in Mexico after 1970. President Luis Echeverría Álvarez entered office that year during a crisis of legitimacy for Mexico's ruling political party, which stemmed in part from its violent repression of the student movement. The Echeverría administration channeled unprecedented resources into agricultural research and education in a bid to stimulate food production and address the ongoing rural crisis.[13] Both aims only increased in urgency during his six-year term, which saw a dramatic reversal of the self-sufficiency in maize production briefly achieved in the 1960s. Mexico went from meeting its own needs for maize in 1968 to importing two million tons in 1974, about one-fifth of its domestic consumption.[14] In its efforts to address this crisis, the government increased funding for agricultural research and training, which created opportunities for more Mexican agricultural scientists to direct their attention to the needs of campesinos. The shift was visible in Chapingo. At INIA, research activities expanded in the 1970s to respond better to the diverse needs of Mexican farmers.[15] Additional state experiment stations and staff covered more regions, aiming to provide solutions better tailored to the country's agricultural environments. The graduate school also saw an uptick in peasant-friendly research: studies of cultivation under adverse conditions, attention to regional ecological variations, investigations of soil conservation, and more.[16]

One prominent champion of this shift was the botany professor Efraím Hernández Xolocotzi. Hernández had been central to the collection and classification of Mexican races of maize at the Office of Special Studies in the 1940s. A subsequent Rockefeller Foundation fellowship enabled him to complete a master's degree with Paul Mangelsdorf and greased the wheels of his appointment at the National School of Agriculture in 1954.[17] In the 1960s Hernández began to formulate a critique of agricultural programs imposed by outsiders. He urged fellow researchers to refocus their energies on the "study of specific problems of the Mexican environment." He also encouraged them to show greater appreciation for the knowledge of peasants, suggesting that it was more appropriate for extension agents to learn from campesinos in the field than to dictate best practices to them.[18]

This was only the beginning. In 1968 Hernández left Chapingo to collect plants in Mexico and South America for the recently opened CIMMYT. After traveling among farming communities in Colombia, Ecuador, Mexico, and Peru, he was newly convinced that sound agricultural research started with farmers. Scientists needed first to understand the social, cultural, and environmental context of cultivation and to learn all that farmers had to teach, before attempting to impart other knowledge.[19] In the 1970s he extended this observation into an argument for ethnobotany as an essential component of agricultural development. By studying the relationships between plants and peoples, especially Indigenous peoples, ethnobotanists could recover knowledge and technologies that would inform agricultural interventions. This was an explicit challenge to the dominant model of modernizing agronomy—that is, to the vision of a Green Revolution based on foreign expertise and imported technologies.[20]

The distinction was evident in Hernández's evolving approach to locating genetic variation in crops to use and conserve. Although the missions he undertook for the Office of Special Studies, and later for CIMMYT, had been to locate materials for professional breeders, Hernández subsequently associated collecting with an appreciation of the knowledge of cultivators and the value of landraces as cultivated varieties. In the late 1960s he reported five new races of maize in Mexico, discoveries he attributed to his greater attention to the needs and practices of Indigenous farmers. Hernández celebrated this as a demonstration that his ethnobotanical approach would reveal new sources of genetic diversity. He also saw it as evidence of Indigenous farmers' role in selecting and maintaining varieties suited to particular ecological conditions and culinary preferences.[21] These were pointed observations. Farmers' intentions and ingenuity had been a source of contention in the *Races of Maize* studies completed in the preceding decade. And it was hardly the message espoused by CIMMYT scientists and others whose valuation of the same farmers' varieties relegated these to the status of raw inputs rather than well-adapted end products.

As he developed his ideas about the entanglements of plants, peoples, and environments, Hernández engaged with a group of scholars who similarly sought answers to Mexico's agrarian challenges in better knowledge of peasant and Indigenous culture, past and present. Together these thinkers advanced a view that it was the government's failure to attend to campesinos—in fact, its

[21] Efraím Hernández Xolocotzi ("Xolo" to many) and his students consider maize ears from Veracruz in 1981. From the 1970s onward Hernández espoused ethnobotany and agroecology, including engagement with local and especially Indigenous knowledge and practice, as routes to better agricultural outcomes in Mexico. Courtesy of Colegio de Postgraduados, Texcoco, Mexico.

collaboration in their exploitation in pursuit of profit and power—that had weakened the Mexican state. They created new accounts of peasant life in Mexico that identified Indigenous Mexican farmers, and not imported science and technology, as the key to restoring the country's agricultural production and therefore its cultural and economic stability.[22]

This resonated in 1970s Mexico. Political leaders embraced aspects of this narrative, if only to connect their leadership with the ideals of the Mexican Revolution. The Echeverría administration swept up the pharmaceutical promise of peasant-produced *Dioscorea* yams in its populist rhetoric, for example. According to the historian Gabriela Soto Laveaga, in this period several potent *Dioscorea* species harvested in the countryside for the international pharmaceutical production of steroids became a "public and symbolic Mexican commodity," capable of "liberating the nation" from its dependence on foreign companies and its perceived "technological backwardness." That the nation-healing tubers came from campesino labor was "tactically convenient," writes Soto. Echeverría could present the nationalization of pharmaceutical production, steroid manufacturing in particular, as a way to defend Mexican people and a plant that could be considered a "national patrimony."[23]

Similar nationalist narratives encompassed Mexican maize, stories that fostered, and were fostered by, a boom in the collection and study of Mexican maize landraces. This surge was sparked by the search for breeding materials to support INIA's development of drought-resistant varieties in the late 1960s and the field explorations of Hernández Xolocotzi and his students. Both led to the accumulation of new maize diversity at INIA's agricultural experiment station near Chapingo. In 1968 there were 4,000 samples of maize from Mexico in INIA's collection and another 1,300 from other countries, all maintained at the Maize Germplasm Bank (Banco de Germoplasma de Maíz) that INIA had inherited from the Office of Special Studies. Within the decade new collecting missions had increased the collection of Mexican samples to more than 8,100.[24]

These materials were resources for the expanded corn-breeding work at INIA in the 1970s and especially its efforts to address regions and peoples neglected in the research agenda of preceding decades. As two researchers responsible for INIA's Maize Germplasm Bank in the 1970s noted, collecting made researchers more sensible to the "ecological-social" dimensions of maize production, which in turn allowed for a "more appropriate focus for the regional and national breeding programs." When it came to use of the collections, they noted the contrast between their experience and that more typical for gene bank managers, in which banked samples were often ignored by breeders. In 1974 at least ten regional breeding efforts had asked INIA for seeds from the bank.[25]

The researchers working with the maize collections in Chapingo further set themselves apart from colleagues in the United States and Europe by insisting on the role of farmers in the creation and perpetuation of crop diversity. Through the centuries of persecution and upheaval that started with European conquest and continued through the twentieth century, "the poor campesinos and indigenous groups" had been the "most conscientious conservators" of diverse crop varieties. In the present day "millions of farmers" were as much evaluators of possible new introductions and combinations as professional researchers.[26] This close connection between the people of Mexico and the history of crop diversity meant that research in the collections was not just of agronomic or biological interest. For those working in the INIA collection, the taxonomic classifications of Mexican maize also generated insight into "the socio-cultural evolution of our nation, as well as the unity and the variation of the cultures that have developed in our current territory."[27]

Activities at the INIA maize gene bank and beyond indicate that many Mexican researchers saw in local farmers' lines—as opposed to more recently introduced types—the potential for addressing crises in agricultural production in the 1970s. This perspective valorized local varieties and landraces, lines stewarded by farmers from season to season, as the crops best adapted to Mexico's needs and as the product of Mexican people. It lent urgency to the projects of collecting, conserving, and classifying, and it gave these tasks a distinct new shape.

INROADS FOR AGRIBUSINESS

While scientists in Mexico searched for novel solutions to the country's rural crises, critical assessments of agricultural aid bolstered the case for these alternatives. By the mid-1970s studies by economists, sociologists, and other development experts indicated that the much-vaunted Green Revolution had done more harm than help, thanks especially to the input- and capital-intensive model of farming it espoused.

The first critiques of the Green Revolution followed close on the heels of its initial celebration. In 1973 the Oxford economist Keith Griffin joined a growing chorus when he cataloged the harms introduced with "high-yielding varieties," a phrase used to describe types bred to flourish with synthetic fertilizers. Their introduction had neither increased income per capita nor solved the problems of hunger and malnutrition, according to Griffin. They had produced effects, however: "The new technology . . . has accelerated the development of a market oriented, capitalist agriculture. It has hastened the demise of subsistence oriented, peasant farming. . . It has increased the power of landowners, especially the larger ones, and this in turn has been associated with a greater polarization of classes and intensified conflict." In 1973 Griffin thought that the ultimate outcome depended on how governments responded to these changes. Five years later he had come to a final determination. "The story of the green revolution is a story of a revolution that failed," he declared.[28]

Griffin was a researcher on the project "Social and Economic Implications of the Large-Scale Introduction of High-Yielding Varieties of Foodgrain." Carried out under the auspices of the United Nations Research Institute for Social Development, this project enlisted social scientists to document the uptake of new agricultural technologies—chiefly new crop varieties—and their social and economic effects across Asia and North Africa.[29] Mexico was also included among the project's case studies, since organizers pinpointed it as the historical site of the "first experiments in high-yielding seeds for modernizing nations."[30] An attempt to synthesize a single account from the case studies in the 1970s highlighted the problems arising from the integration of farmers into national and international markets. New varieties, chemical fertilizers, and mechanical equipment demanded that cultivators "become businessmen competent in market operations and small-scale financing and receptive to science-generated information." This was thought to be in marked contrast to their having once been "'artisan' cultivators" who drew on "tradition and locally valid practices" to sustain their families.[31] The fact that only a minority of better-off farmers could make such a transition meant that development programs benefited a few at the expense of the many. Drawing on her case study of Mexico, project contributor Cynthia Hewitt de Alcántara extended this observation about market integration into a reflection on the flow of economic resources around, and out of, the country— from laborers to landowners, from farms to industries, from national programs to foreign businesses. The reconfiguration of agriculture as what she labeled a "capitalist enterprise" had not brought more money to the countryside but instead robbed peasants of what little they had.[32]

This apparent contradiction in Mexico's agricultural development invited scrutiny from many besides Hewitt. The preceding three decades had been characterized by steady economic growth, thanks to increased international trade during World War II, government policies that encouraged national industry, and investments in infrastructure and education. This period of the so-called Mexican Miracle had also seen a transition from food dependency—needing to import grain to feed the nation—to self-sufficiency. At this level of abstraction, Mexico's prospects for sustaining adequate food and nutrition looked rosy. When sociologists and economists delved into specifics, however, the miracle revealed itself a mirage. Investments in agriculture had focused on supplying food to urban workers and developing new products for export. State food-aid programs, too, had been oriented to urban labor, with set prices that kept food affordable for consumers in the city but made its cultivation unprofitable for farmers in the countryside.[33] While well-off cultivators in the north of the country benefited from state-funded irrigation programs and guaranteed prices, poor farmers working small plots without access to state grain purchasers found that they could not sustain their families by selling surplus corn. Hewitt estimated that in 1969–70, one-third of the Mexican population experienced calorie deficiency. A 1974 national survey came to similar conclusions, calculating that 18.4 million Mexicans, over a quarter of the population, suffered from malnutrition.[34]

The persistence of poverty in Mexico, in spite of the country's celebrated economic growth, could be traced to the model of development embraced by national leaders since the 1940s. Politicians and policy makers had assumed that subsistence farmers could be made irrelevant, with their surplus labor absorbed into the growing industrial economy. Yet industry had not acted the sponge, with the result that this "irrelevant" segment of the population had grown while continuing to be neglected by the state.[35] The economist David Barkin linked faulty Mexican policies to a more fundamental problem of emulating the market capitalism of its northern neighbor. The apparently flourishing Mexican economy had invited the interest of foreign investors, in particular US corporations. Despite protectionist policies, these companies had moved in, and national industries had been sold off, leaving Mexicans vulnerable to the whims of private capital.[36]

Agriculture offered a prime example of this pattern. By the 1970s US firms dominated across the sector, from farm machinery (John Deere, International Harvester) to chemicals (Monsanto, DuPont, American Cyanamid) to production and processing (United Brands, Corn Products) to animal feed

(Ralston Purina).[37] Observing this trend, another economist pinpointed Mexican agriculture as the place of origin of a "new, world-wide modernization strategy." He traced a path from the interventions of the Rockefeller Foundation to the stimulus these gave to the importation of costly agricultural inputs to the management of Mexican farms by foreign firms. Foreign control and deepening ties to international markets affected food self-sufficiency. It made sense, from the perspective of increasing individual profits, for large and well-financed producers in Mexico to focus on the crops that would bring the best prices. These were more likely to be fruits and vegetables for US supermarkets or sorghum to feed cattle than corn or wheat to feed Mexican workers. Thanks to these patterns, it was possible to see much of Mexican agriculture as an extension of US agribusiness, operating chiefly "to exploit Mexican rural labour, Mexican land and water resources, and Mexican private and public capital for the principal benefit of US entrepreneurs."[38] The ultimate outcome of technical assistance to enhance agricultural production, ostensibly undertaken for the betterment of Mexican farmers and the Mexican economy, was the dominance of transnational companies in that very task, for their own aggrandizement.[39] This portended ill for Mexico and especially for the poorest Mexicans.

CORN BLIGHT

While agronomists and economists tackled crises of corn production in Mexico in the 1970s, US researchers confronted their own corn crisis. The outbreak of a virulent fungal blight generated new assessments of the costs and consequences of US dependence on hybrid corn and of the industrial calculations that motivated its mass production. Just as crises in Mexico had opened the door to critiques of agricultural development, so too did corn blight in the United States. Critics even found the same root cause: big corporations.

In the United States the crisis originated with the relentless abundance of corn production rather than shortfalls. Between 1935 and 1970 US farmers nearly tripled the amount of corn produced per acre. Yields rose from an average of twenty-four bushels per acre to seventy-two, with the effect that the volume of production increased even as acres planted to corn declined. These gains weren't solely a product of hybrid seeds, although they were often celebrated as such. In the same period use of synthetic fertilizers increased, new herbicides were introduced, planting and cultivation practices shifted, and a process of mechanization initiated earlier in the century now exploited the greater uniformity of hybrid varieties. With all these investments, corn, long king, retained its rank within the hierarchy of the agricultural economy. In 1969 the value of the US corn harvest topped $5.4 billion, while the next leading crop, wheat, trailed at $1.8 billion. Corn accounted for around 20 percent of the total estimated value of US crops that year.[40]

Bountiful harvests of corn and wheat were cause for celebration in the Cold War. US leaders routinely offered up cheap and abundant food as evidence that capitalism provided more and better for US citizens than socialism did for Soviets.[41] But agricultural productivity was also a political headache. When grain piled up, prices fell and farmers suffered. Politicians found themselves continually renegotiating schemes to prop up farms and dispense with stockpiled harvests. From the 1950s international aid featured prominently among preferred destinations for surpluses, an apparent win-win, in which the offer of grain as foreign assistance dispatched a domestic burden in the shape of a helping hand. In the 1960s presidents John F. Kennedy and Lyndon Johnson took this instrumental use of US grain in new directions. The historian Bryan McDonald describes a "new, offensive food-power doctrine" in place by the late 1960s, in which food aid "deployed under the guise of humanitarian assistance" was used instrumentally to pressure foreign leaders.[42]

The sovereignty of hybrid corn in US economics and politics entailed vulnerability. US Americans discovered this to their horror in 1970, when a virulent fungal infection, southern corn leaf blight, spread rapidly through the country's corn-growing regions. In 1969 scattered farmers in Iowa, Illinois, and Indiana reported spotted leaves and ears destroyed by a grayish-black rot. In January 1970, reports of the same disease wreaking havoc in cornfields began rolling in from Florida, then Alabama and Mississippi, as growing seasons progressed northward. By August it was apparent that the blight had reached epidemic proportions throughout the Corn Belt. One agricultural scientist dramatized the view from the cornfield: "It produced lesions on the leaves until they died and shriveled. It then attacked the ears, punching its way through the husks to the grain. Come fall, the mechanical corn pickers were enveloped in a black miasmic cloud of spores."[43] The USDA revised its projections downward for the corn harvest and then dropped them still further. Corn prices shot up, and forecasters warned that the price of chicken, pork, and beef would soon follow. The losses from the epidemic were estimated at more than $1 billion in 1972 (the equivalent of $6.25 billion in 2020)—at least 20 percent of the anticipated income from corn that year.[44]

As the damage unfolded, it became apparent that not all varieties of corn were equally susceptible to the blight. Hybrids were the problem and, more specifically, hybrids with a trait known as male sterility. This was a popular feature of commercial hybrids because it made them cheaper to produce. In the industry's first decades, one of the most labor-intensive and costly tasks of hybrid corn seed production had been detasseling. Farmworkers had to remove, by hand, the pollen-producing tassel from the top of every plant that was expected to produce seeds for later sale. The process ensured that one inbred line was pollinated by a different inbred line and not self-pollinated. It would therefore have the hoped-for hybrid genetic combination. Given the scale of corn seed production, the aggregated number of detasselers needed by seed companies was enormous. One common estimate held that on a single day at the height of the season, some 125,000 workers would be found out in the field detasseling.[45]

As early as the 1930s corn breeders had begun experimenting with methods for controlling the spread of pollen biologically, to eliminate detasseling. One promising route was to locate plants that had an inherited tendency to produce sterile pollen, also called male-sterile plants. If pollen were sterile, it wouldn't matter where it landed, and there would be no need to detassel. Not just any male-sterile plant would do, however. It would be impossible to maintain genetic male sterility, passed on through the chromosomes, as a consistent trait in an inbred line, because it could not be self-fertilized or crossed with other male-sterile plants. A promising alternative was male sterility conferred through cytoplasmic genes, which would be derived only from the female parent. This would enable a breeder to make a cytoplasmically male-sterile inbred line and maintain it through crosses with its normal cytoplasm counterpart. Although geneticists identified sources of cytoplasmic male sterility in the 1930s, more than a decade passed before their effective use was fully worked out. Paul Mangelsdorf was central to this development, as was Donald Jones of the Connecticut Agricultural Experiment Station and—most important—seeds of a corn plant from a colleague at the Texas Agricultural Experiment Station. Those seeds gave rise to a line known as "Texas male sterile," carrying "T" type cytoplasmic male sterility, or "Tcms." By 1970 the production of some 85 percent of hybrid varieties in the United States relied on the cytoplasmic male sterility derived from this one line and the methods worked out by Mangelsdorf and Jones.[46]

This apparent triumph of biological engineering was recast as a disaster when it became clear, even in the earliest days of the southern corn leaf blight outbreak, that hybrid corn varieties with Tcms were uniquely susceptible. Where farmers bucking the industry trend had planted open-pollinated varieties or hybrids that still required detasseling, the effects of the blight were nil. In other words, the new strain of the fungus devastated only one type of corn, which also happened to be the one type that practically every farmer was growing. The ubiquity of the cytoplasm from the Texas male-sterile strain was responsible for the intensity of the epidemic.[47]

It was only a small step from this realization to a further recognition that the problem lay as much with the entire system of corn production in the United States, and perhaps even industrial agriculture, as with the male-sterile plants. Modern breeding tended to produce genetically uniform varieties. Modern production standards encouraged monocrop fields of these varieties. Modern cultivation techniques increasingly enabled dense plantings. Together these provided ideal conditions for a plant epidemic. In subsequent assessments of the vulnerability exposed by the blight, experts struggled to lay blame on any one entity. Seed companies wanted to sell varieties over as wide a range as possible, to take advantage of economies of scale. Farmers preferred uniform crops, which ripened simultaneously and at a uniform height for mechanical harvesting. Consumers wanted uniformity too, casting aside misshapen corn at the supermarket. Who along this chain could rightly be said to have precipitated the disaster?[48]

PLANT PATENTS

Not every commenter on the corn blight struggled to pin down a culprit. "The proliferation of brand name maize varieties had disguised the fundamental genetic uniformity of corporate seed corn development," claimed the Canadian activist Pat Mooney in 1979. These circumstances had led farmers, unwittingly, to plant all the same plants, to their own peril.[49] When Mooney offered this diagnosis of the blight, nearly ten years after the fact, he was working with the International Coalition for Development Action, an association of NGOs he had cofounded. In 1977 members of this group began to wrestle with what they called the "seeds issue." This phrase encompassed a range of concerns, the most important being the increasing control exerted by private companies over the most basic component of agricultural production: seeds.[50]

In the 1970s activists increasingly identified transnational corporations, and private industry in general, as a chief source of problems in the global food system. US and European companies concerned with their bottom line did not care about equitable distribution, adequate nutrition, or environmental fallout. The consequences of corporate rapacity only ramified as these businesses made inroads into Asian and Latin America economies, in many cases having had their way paved by foreign aid agencies and agricultural development programs. To those advancing new critiques, the solution to hunger and poverty would be found in reforming not peasant agriculture but Western politics. They focused far less on checking population growth and much more on checking rapacious capitalism.

Consider Pat Mooney. In the late 1970s and early 1980s, he and his collaborators wove a complex narrative about social and political changes leading to ever-more-thorough dominance of industry in the supply of crop seeds and the social and environmental consequences of this new order. Their sprawling account dismissed the claimed benefits of the Green Revolution to peasant farmers and pointed instead to wealthy countries and, more specifically, transnational agribusinesses as the real beneficiaries of agricultural industrialization in the Global South. They observed that the seed companies that had stepped in to produce and distribute new high-yielding varieties stood especially to gain from this transition and that those same companies had recently convinced many governments to expand intellectual property protections for plant varieties. Seed activists linked the guarantees of this new intellectual property system to increasingly obvious industry consolidation, in which large companies were buying out the small seed firms of an earlier era, and also to the diminishment of crop diversity through enforcement of standards for uniformity and stability in registered varieties. Both processes were thought to destabilize food security. This destabilization was further compounded by a conservation system that centered on extracting seeds from farmers of poor countries for safekeeping and use in rich ones. With the enforcement of intellectual property, farmers who lost their seeds might get them back only by paying royalties.[51]

Mooney navigated the currents of these concerns in his influential 1979 book *Seeds of the Earth: A Private or Public Resource?* This account drew on the expertise of several associates. One was Cary Fowler, a Tennessean activist with a PhD in sociology. In 1975 Fowler had taken a research position at the Institute for Food and Development Policy. There he had been instrumental in preparing the organization's inaugural publication, *Food First:*

[22] Erna Bennett of the UN Food and Agriculture Organization meets with a group of women who live in the mountains outside Karpenisi, Greece, in July 1969, questioning them about the wheat varieties they grow. The accompanying FAO caption characterizes older varieties of wheat as falling out of cultivation and insists that "emergency measures are necessary to collect these old races before they disappear completely." ©FAO/Florita Botts, image 6411.

Beyond the Myth of Scarcity. The best-selling book showed that hunger was not the result of a shortage of food—a "world food problem," as many foreign aid agencies and philanthropies contended—but instead stemmed from ineffective and inappropriate policies.[52] Another important resource for Mooney was Erna Bennett, a plant geneticist at FAO and leader of its early crop conservation activities. Bennett had been angered by the change in FAO priorities after the creation of the International Board for Plant Genetic Resources. As she later remembered, the head of the board "led [it] into a wholly procorporate position" that she believed conflicted with United Nations policies.[53] *Seeds of the Earth* launched Mooney and colleagues like Fowler and Bennett into the media spotlight and energized an ambitious antiindustry campaign that eventually linked activists across organizations and continents.[54]

In its earliest years the International Coalition for Development Action's "Seeds Campaign" focused its energies on the issue of plant breeders' rights. This refers to the idea that professional plant breeders ought to be able to claim ownership over their creations, much as inventors do when they take out patents on new innovations. Plant breeders, and the companies that employ them, argue that without such protection the incentives to innovate are limited. Anyone could simply take seeds of a breeders' latest varietal innovation, grow them, and then harvest seeds from the resulting crop and sell these on as their own product. In the United States breeders and seed companies successfully demanded a system for patenting plant varieties in the 1920s. However, the scope of the initial US Plant Patent Act was limited to varieties reproduced as clones, such as from bulbs or cuttings.[55]

Breeders in a number of European countries were also successful in convincing their governments of the need to protect breeders' interests through certification schemes and patent systems. As Western Europe moved toward economic integration in the 1950s, diverging national approaches to intellectual property protection for breeders came under scrutiny. Seed companies in countries with fewer property protections were seen to have unfair advantages. In 1961 the International Union for the Protection of New Varieties of Plants (UPOV; Union Internationale pour la Protection des Obtentions Végétales) set a new coordinating system in motion, requiring that member countries offer intellectual property protection for at least fifteen years on all plant varieties shown to be distinctive (that is, new), uniform (genetically homogeneous), and stable (true breeding from one season to the next). Its entry into force in 1968 catalyzed changes elsewhere in the world. In the

United States a complementary system was established by the 1970 Plant Variety Protection Act.[56]

The fact that advocates of plant breeders' rights—chiefly, private companies—were able to successfully argue for UPOV and other systems indicated the extent to which the seed industry had changed in the preceding decades. At the start of the twentieth century, it consisted mostly in local and regional businesses. These had access to only a limited toolkit of genetic knowledge and breeding techniques with which to generate new varieties. In fact, rather than generate their own varieties, many agricultural seed companies simply multiplied the lines developed and disseminated freely by public breeders at state experiment stations or other research centers. By the 1970s more efficient breeding methods, greater research investments, and growing numbers of independently developed varieties meant that private companies had more at stake in the creation of enforceable intellectual property regimes.[57]

A still more dramatic change in the seed industry was only just beginning. During the 1970s the small seed companies that had typified the industry for several generations began to disappear. Unlike endangered species, or even threatened crop landraces, this extinction event was not due to neglect, undervaluation, or lack of attention. Instead the strengthened intellectual property protections put in place in the 1960s increased the perceived value of these companies. In the United States the decade following the passage of the Plant Variety Protection Act in 1970 saw dozens of independent seed companies bought up by larger firms. In most cases the buyers' core business lay elsewhere. They were oil companies (e.g., Royal Dutch Shell); chemical producers (Monsanto, Ciba-Geigy, Union Carbide); and pharmaceutical firms (Sandoz, Upjohn, Pfizer). These industry giants now saw potential for profit in seeds. In short, the ability to secure intellectual property rights precipitated a swift reorganization of the international seed industry.[58]

For many activists protesting the privatization of seeds, the history of hybrid corn in the United States illustrated its dangers. Producing and selling first-generation hybrid seed by crossing inbred lines had appealed to private companies because it enabled them to exercise control over the varieties they developed even in the absence of formal intellectual property protections. The genetic variability of plants raised from the seeds produced by a hybrid meant that growers were better served by returning to the company every season for a fresh supply of seeds rather than saving their own. This repeated business generated greater profits. A couple decades after the introduction of hybrids, a few hybrid corn seed companies had grown from modest family businesses into large corporations with significant economic clout. Within the seed industry as a whole, prior to consolidation, they represented the heavyweights.[59] These powerful economic actors also proved potent politically, arguing successfully against the release of new corn varieties by publicly funded programs. They insisted that government breeders, whose improved lines were freely available, represented unfair competition. As the involvement of public breeders in creating finished lines diminished, the handful of firms that dominated the hybrid seed industry had increasing discretion over what new varieties should look like and what models of agricultural production ought to be recommended.[60]

In one of the most famous accounts of this progression, the sociologist Jack Kloppenburg outlined two transformative outcomes of the invention of hybrid corn. When seed production by private companies displaced on-farm production, seeds themselves became a commodity. When these same firms shunted public programs to a supportive rather than competitive role, they established control over the form this commodity could take. The upshot, in Kloppenburg's estimation, was a system that afforded less autonomy to farmers, diminished government research, transformed public investments in agriculture into subsidies for private companies, and reinforced industrialized agriculture through the dissemination of varieties particularly suited to this style of farming.[61]

The rapid transition of corn seed companies to hybrid varieties and their vigorous promotion of these varieties testified as much—or, as some convincingly argued, more—to the appeal of the business model based on first-generation hybrids as to their superior agricultural performance.[62] Seed companies began to develop such hybrids in other crops, with an eye to achieving guaranteed return customers in these too. Not all were easily subjected to this approach, however. Wheat, a self-pollinating plant, was and remains notoriously difficult to develop into first-generation hybrid varieties on the model of corn. In the early 1980s, as debates over plant patenting intensified, 95 percent of the acres planted to corn in the United States relied on seed purchased that year, compared to only 10 percent of acres planted to wheat. In other words, the vast majority of the US wheat harvest came from seeds saved by farmers.[63] Without the ability to forestall seed saving, the possibilities of developing a vigorous private industry in wheat and similar crops seemed slim. Hence the appeal of plant breeders' rights: companies saw the case of hybrid corn as a positive one and wished to emulate it through other means.

Meanwhile, a growing cadre of activists like Mooney and academics like Kloppenburg saw hybrid corn as illustrative of the dangers that accompanied ownership of plant varieties. As their frequent references to the 1970 corn blight attested, these dangers included not only the aggrandizement of corporations to the detriment of some farmers and most consumers but also the vulnerability resulting from monocrop cultivation of nearly identical hybrid varieties. In response to the claim that only the strength of the public agricultural research system in the United States had saved the nation from worse ravages by the corn blight, Kloppenburg countered that in fact it was that system's "subordination to private enterprise" that had endangered the nation's agriculture in the first place.[64]

SELF-SUFFICIENCY

A similar account of the vulnerability in agricultural production created through the private seed industry—especially the increasing chokehold of large transnationals—was emerging among policy makers in Mexico. The continued failures of staple-crop production and ever-rising levels of hunger and malnutrition through the 1970s pushed government interventions to support peasant cultivation in new directions. Meanwhile, efforts to improve state seed production, in the interest of ramping up national crop production, led to close examination of the international seed industry. These intersecting lines of activity made clear to domestic policy makers and Mexican representatives abroad the connections between native maize and national security.

Just as maize breeding and research underwent significant changes in Mexico in the 1970s, so too did seed production. Since the early 1960s the seed industry in Mexico had been tightly regulated. In 1961, the year the government established INIA to lead agricultural research, it also created the National Seed Production Company (Productora Nacional de Semillas). Whereas INIA scientists were tasked with developing crop varieties, employees at the state-run seed producer oversaw the processes needed to make seeds of these varieties available. Unlike in the United States, where the varieties developed by state-employed breeders were public goods made freely available to any user, including seed companies, in Mexico the National Seed Production Company held a monopoly on the lines released by government breeders. In addition private companies were prohibited from conducting plant-breeding research.[65]

These restrictions entailed impressive national control over the development and distribution of crop varieties but did not eliminate seed companies from the landscape altogether. Nor could they, since the National Seed Production Company never had the capacity to meet all of Mexico's seed demand, and INIA lacked sufficient staff and funding to address the diverse needs of Mexican growers. The gap in supply was filled by government imports and private sales. During the 1960s a small number of US-based transnational seed companies, mostly specializing in hybrid corn and sorghum seed, set up shop in Mexico, while a larger number of Mexican businesses specialized in importing seeds for sale, especially of oil crops, fruits, and vegetables.[66]

The subsequent decade saw a sharp rise in the involvement of foreign companies in the Mexican seed market. This was partly encouraged by state policies, as the government slowly began granting seed companies permission to conduct research in Mexico. Of the first five companies to receive such permission, four specialized in corn and sorghum, and all were associated with parent companies in the United States. The number of foreign companies operating in Mexico increased steadily in the 1970s. By 1982 there were thirty-seven private seed companies producing (as opposed to importing) seed in Mexico, eighteen of which were affiliates of foreign-owned companies. These firms were, in general, larger operations than their nineteen Mexican-owned counterparts. They also included almost all of the largest transnational seed companies.[67] This growing presence meant that, even as state plant breeding and seed production efforts expanded, resulting in more varieties of a wider array of crops, the erosion of their dominance in Mexican crop development was well underway.

Foreign competition centered exclusively on lucrative markets. The new entrants into Mexico's seed industry, like many Mexican-owned firms, were not interested in providing seeds to peasant producers. They focused on high-value inputs, seeds that commercial farmers would pay a premium for. The involvement of foreign companies in Mexican agriculture therefore contributed to Mexico's food woes of the 1970s, emphasizing products that would fetch a good price, especially on international markets, rather than alleviate hunger at home. International companies provided seeds of winter vegetables ultimately destined for the United States, for example. Or they invested in hybrid sorghum, a feed grain that fattened the industrial beef, pork, and poultry consumed in ever-greater quantities by the growing Mexican middle class. The expansion of sorghum production—grown on more than ten times as many acres in 1980 as 1958—was particularly damaging because sorghum often substituted directly for corn. The two crops grew well on the same lands and soils, but sorghum was more profitable. As sorghum soared, corn sank. With it sank hopes for self-sufficiency, among both former subsistence farmers now tethered to the industrial commodity chain and policy analysts hoping to see Mexico resolve its food-grain supply problems. It was obvious to anyone that Mexico's inability to produce sufficient food crops contributed significantly to its growing trade deficit. As critics pointed out, seed importation and foreign-controlled seed production did too.[68]

By the end of the decade, the Mexican government could not afford to ignore its agricultural production problems. After a failed harvest in 1979, corn imports shot up, from an annual average of around 1.5 million tons between 1972 and 1978 to a staggering 3.8 million tons, an escalation in food dependence that was especially concerning in an era of food power.[69] The same year a government report estimated that 90 percent of rural Mexicans—some twenty-one million people—were malnourished. Later studies confirmed that almost half had a daily calorie intake that was 25 to 40 percent below the minimum standard. Urban Mexicans, though in general better fed, had also seen the quantity and quality of their diet decline.[70]

In March 1980 President José López Portillo authorized a new program, the Mexican Food System (SAM; Sistema Alimentario Mexicano). It aimed to achieve self-sufficiency in food production through a radical reorientation of agricultural science and policy. Rather than seek increased production solely through commercial farms, irrigated lands, expensive inputs, and the crop varieties these conditions demand—which had been the dominant government strategy from the 1940s—the architects of SAM wanted to provide incentives and support for peasant production on marginal lands. Mario Montanari, a Chilean economist in exile who helped devise SAM, saw the possibility of radically reshaping Mexico's global position. As he envisioned, by helping peasant farmers produce more and better crops, the Mexican state would achieve its goals of "withdrawing from external dependence, reaffirming national sovereignty, and redefining the terms of the relations between national and international systems."[71]

SAM was a complex bundle of interventions, many undertaken through existing state agencies. It included food subsides, state-run stores, transportation services, nutritional education, guaranteed prices for crops, credit and crop insurance for farmers, and the distribution of agricultural technologies. Within that last category seed supply was particularly crucial. To make Mexico self-sufficient in grain, SAM needed to rapidly expand corn production. To do this it had to ensure farmers' access to good seed, produced and

[23] A booklet produced by the Mexican government in the early 1980s explains important elements of the Sistema Alimentario Mexicano (Mexican Food System) program through the character of *el profe*, "the professor." *El profe* explains that, in the past, policies favored industrial development and supported farmers with access to irrigation. The new program would instead help *"los campesinos temporaleros,"* peasant farmers dependent on seasonal rains for cultivation who had been marginalized in previous state policies. From *Sistema Alimentario Mexicano.* Reproduced with permission of Secretaría de Agricultura y Desarrollo Rural, Mexico City.

certified according to careful protocols. SAM therefore needed to encourage domestic seed production. One approach it took was to provide subsidies to farmers who purchased certified seeds of corn and beans, which drove up demand. The greater number of buyers in turn incentivized the National Seed Production Company to scale up its efforts. In the first year of SAM, production of certified maize seeds quintupled.[72]

Because it was clear that even with subsidies many farmers would not be able to purchase certified seeds, a separate program set out to improve open-pollinated varieties and freely distribute these. Within two years this program managed to get seeds to an estimated one million farmers.[73] Still, these varieties lost rather than gained ground as a result of SAM. Expansion at the National Seed Production Company saw the share of hybrid corn, as opposed to breeders' open-pollinated lines, leap from 2 percent to 45 percent of its certified seed output. SAM policies prized rapid results and embraced the industrial approach even as they looked to devise a new vision of agricultural change.[74] One upshot was that private industry profited. Transnational seed companies stepped in to supply the hybrid varieties that many more farmers could purchase thanks to SAM credit and insurance schemes. One industry executive whooped that in Mexico the year 1981 "could not have been better for seed producing companies, regardless of who and where they are."[75]

SAM planners recognized seed production as an essential component of local food security and also national economic and political security. Within a program aimed at self-sufficiency, it was an area of clear dependency. Despite the leaps in domestic production in the program's first year, Mexico remained the world's largest importer of seeds. In 1981 SAM administrators commissioned a study of Mexico's seed industry, aiming to bring its development into alignment with their reform agenda. A small research team of economists and political scientists affiliated with American University in Washington, DC, charted the contours of the industry and conducted interviews with seed companies and government officials in the United States and Mexico over the ensuing year. Their final report recommended that the government restrict the operations of private seed companies. Allowing the large transnational companies that dominated the international seed industry to operate freely in Mexico created the "potential" that their actions would be determined by decision makers who didn't care about the government's priorities. Sidestepping the fact that a handful of transnationals were already operating in the Mexican market in precisely this way, the study encouraged actions that would keep private companies in check.[76]

The report also encouraged the Mexican government to adopt policies allowing the "freest possible international flow of germplasm."[77] This was not idle advice. The Congress of the Union was just then debating whether and how Mexico would pursue alignment with UPOV through its own plant variety–protection legislation. Efforts to extend compliance with UPOV beyond its founding members had started in countries with highly developed seed industries. By the early 1980s it reached into countries without these, including Mexico. This expansion aimed to make nascent seed markets lucrative for international corporations by limiting the opportunities for smaller, and often national, firms to reproduce and sell the products of corporate research investments. The Mexican government had participated in the 1978 meetings that created a revised UPOV agreement, but it had not yet become a signatory.[78] As the SAM analysts observed, any decision to do so would have implications for seed self-sufficiency—and therefore agricultural and economic self-sufficiency as well.

INTERNATIONAL UNDERTAKINGS

The American University team that prepared the seed industry report for SAM subcontracted part of their investigation to Pat Mooney and Cary Fowler, an association that sheds additional light on the report's dim view of Mexico's pending plant variety–protection legislation.[79] In 1981, buoyed by attention to *Seeds of the Earth*, seed activists were campaigning aggressively against such legislation and the growing power of transnational seed companies. As Mooney later recalled, the consulting work with American University gave him the chance to travel to Mexico City and to meet with government officials in hopes of enlisting them in a battle against seed ownership.[80]

The consultancy also positioned Mooney as a potential adviser to the Mexican delegation at FAO in Rome in 1981 and helped the International Coalition for Development Action's Seeds Campaign take center stage at successive FAO Conferences in 1981 and 1983.[81] In the lead up to the 1981 FAO Conference, the Mexican delegation circulated a proposal calling for an international seed bank and treaty. It outlined the case for "plant genetic resources" to be "conserved under the custody of FAO" in a location "legally recognized as property of the United Nations." This UN-governed seed bank was to be accompanied by a binding international agreement that secured the free availability of these resources to all nations. By the time the FAO Conference began, the proposal had the backing of countries in the Group of Latin America and the Caribbean Countries and the Group of 77, the coalition representing developing countries at the United Nations, supporters who ensured it was not swiftly dismissed.[82]

The Mexican proposal insisted that a seed bank and treaty were essential in light of the "rapid and profound erosion . . . that can lead to the extinction of material of incalculable value" to the future of food production. The

Mexican delegation and other supporters of the proposal understood this worry about total extinction to be secondary to more localized loss: the disappearance of diversity through loss of access. This was obvious in the initial draft, which insisted on measures that would recognize crop diversity as the "patrimony of humanity" and "ensure legally its free availability and non-discriminatory use, avoiding monopolies, economic speculation and political pressures on such vital resources." Although national seed banks and the international agricultural research centers possessed sizable collections, some of which were used by the International Board for Plant Genetic Resources as "base collections" for internationally coordinated long-term preservation, these fell, ultimately, "under the national sovereignties of the country in which they are located." They remained "without any legal guarantee" that seeds deposited with them would be available to requestors of any nation.[83]

The debate that unfolded around this proposal underscored that the loss of sovereignty, and not plant genetic resources, was the ultimate concern. Delegate after delegate from the Latin American and Caribbean countries and Group of 77 countries expressed their desire to safeguard seeds from becoming sources of political pressure. They saw a system that guaranteed access to all nations as the only way to achieve this. Meanwhile, industrialized countries, including Canada, Japan, Norway, the United Kingdom, and the United States, argued that such a system was already in place through the International Board for Plant Genetic Resources of the Consultative Group on International Agricultural Research (CGIAR). They also pointed to the CGIAR's international research centers as examples of effective, accessible international repositories. These objections made little headway with the proponents of a new system who insisted that the International Board and "autonomous" CGIAR centers could not be counted on to respond to all nations.[84]

The perceived shortcomings of existing conservation protocols were compounded by another issue, one that underlie the negotiations in 1981 but did not become a flashpoint until the 1983 FAO Conference. In 1981 the draft proposal put forward by Mexico led, after intense debate, to the adoption of a revised resolution. This called on the FAO director general to draft an international convention on the conservation of plant genetic resources and prepare a study on an international seed bank.[85] In the lead up to the resolution's adoption, several delegates referred to breeding materials rendered inaccessible through intellectual property protections. For example, the representative from Zambia observed that resources collected in developing countries were stored in banks in industrialized countries, from which they were made available to commercial breeders. As a result, access to some resources from one's own country could be achieved only by "paying heavily for them" in the form of proprietary varieties.[86]

When the FAO Conference two years later took up the director general's draft of an "International Undertaking on Plant Genetic Resources," these concerns about private ownership emerged as the central issue. The "plant genetic resources" whose conservation and free access was to be guaranteed by the undertaking was defined expansively in the draft agreement presented in 1983. The resources subject to the terms of the undertaking included cultivated varieties (lines created by professional breeders "in current use and newly developed"); "obsolete cultivars" (breeders' varieties now out of use); "primitive cultivars" or landraces; crop wild relatives; and "special genetic stocks" (such as those used in genetic and other research programs).[87] If approved, the agreement would therefore contradict legal structures such as UPOV, which granted exclusive rights to plant breeders and seed companies in the varieties they released. This strike against UPOV had been the aim of many agitating for FAO action on plant genetic resources from 1979 onward, but the elaboration of an agreement made it fully explicit—and unavoidably divisive.[88]

Issues of ownership were exacerbated by recent developments in biotechnology. In the 1970s novel techniques for manipulating genetic material, especially the creation of recombinant DNA molecules, had become the object of fear and financial speculation. The potential hazards of early gene-splicing experiments had provoked debate among scientists, which subsequently fueled public concern about the risks that genetic technologies posed to society, health, and environment. By the early 1980s activists had established anti-biotech organizations with an eye to disrupting the forward momentum of its development.[89] Meanwhile, biotechnology had become big business. Experimental outfits sustained by modest venture-capital investments transformed into headline-grabbing corporate enterprises. The pioneering biotech start-up Genentech went public in 1980, raising $36 million for a total valuation of more than $500 million. A few months later Cetus Corporation did the same; its initial public offering, which netted the company $120 million, was the largest ever to that date, in any industry.[90] Transnational capital contributed significantly to the biotech boom, especially investments by pharmaceutical companies anxious to prevent inroads into their established domains.[91]

As hype swirled and dollars flew, so too did patents, the coin of the realm in a high-value, and highly overvalued, industry that had yet to produce much in the way of commercial successes. In the 1980 case *Diamond v. Chakrabarty,* the US Supreme Court decided in favor of granting patents on organisms modified in the laboratory, opening a new vista for the exercise of patent rights. The microorganisms, and someday the plants and animals, produced through the new techniques of genetic engineering would be just as patentable as the tools and processes that produced them.[92] For the activists already aligned against plant patents and the movement of crop diversity from South to North, an unconstrained biotech industry portended new routes for exploitation. Farmers' varieties and crop wild relatives, long touted as sources of valuable genetic traits, seemed poised to become the target of precision extraction, thanks to the newfound ability to cut and move strands of DNA. With the long-standing biological barriers to their efficient use stripped away, they would be newly valuable, and vulnerable, to transnational seed companies. The strengthening of patent protections would both intensify the hunt for genetic materials and build further barriers to accessing breeders' varieties.[93]

As the debate opened at the FAO Conference in 1983—stoked by these concerns—the lines between developing nations and industrialized ones, between poor and rich, South and North, were clearly drawn. The Swedish delegation pointed out that certain varieties could not be included in an international agreement, as this contradicted national policies granting their ownership. The United States representative insisted that the proposed undertaking was in conflict with its legislation on plant variety protection and patents. Delegates from Canada, Japan, the United Kingdom, and other industrialized countries similarly refused the inclusion of breeders' lines. They argued that it was necessary to maintain plant breeders' rights, thereby excluding these lines from consideration for free exchange, to reward breeders for their labor and incentivize further breeding.[94]

Counterarguments took multiple forms. The Mexican representative insisted that "generations of peasants" were the ultimate authors of the world's rich crop genetic diversity. The Libyan representative demanded that the benefits arising from crop diversity—which was a "heritage for all mankind" but also "a national heritage and wealth"—ought to be "shared equally." Meanwhile, the delegation from Peru pointed out that, despite prizes and profits, breeders of rich countries still had not produced varieties that would address the needs of poorer ones.[95] For these countries it was essential that the outcome of the debates be a binding legal agreement that included "advanced breeder lines." A voluntary undertaking in which signatories would be able to exclude breeders' varieties would, as one observer noted, "probably do no more than legitimize the right of the industrialised countries to have access to the germplasm of the Third World" without providing access to anything created from this material.[96] It would exacerbate, or at a minimum reinscribe, inequalities in access rather than redress them.

Participants in these debates presented them then—and still do now—as a heroic battle between South and North, between the possessors of crop diversity and those who profited from its use. The conference passed a resolution adopting an International Undertaking on Plant Genetic Resources and encouraging member nations to agree to its provisions. The undertaking preserved intact the full list of materials constituting "plant genetic resources," including lines developed or in development by breeders, and the insistence on free exchange.[97] However, it was neither legally binding nor did it have the support of the countries whose policies had precipitated its preparation in the first place, those which allowed breeders and seed companies to own intellectual property rights in plant varieties. In practice little changed, and further contests lie ahead.

THE LOST DECADE

While Mexicans in Rome took the floor at FAO to champion peasant knowledge and the conservation of crop diversity and above all to assert national sovereignty and contest the power of industrialized nations and transnational companies over developing nations, economic crisis was brewing at home. Conservation advocates in Mexico saw campesino knowledge again pushed aside, seed banks stripped of resources, and free-market policies implemented at the insistence of the International Monetary Fund. As Mexico entered its so-called Lost Decade of stagnant growth, falling incomes, and growing unemployment, agricultural researchers entered their own lost decade.

In the early 1970s Mexico was heralded as an exemplar of economic development. Although its economic policies generated inequality, at the national level it looked like Mexico was achieving the transformation to an industrial economy that many developing nations sought. This trajectory took an abrupt turn mid-decade. Echeverría's policies of big spending and big borrowing, combined with the effects of the international oil crisis in 1973, precipitated an economic recession in Mexico in 1976. Although the discovery of a massive oil field in 1977 and its subsequent exploitation generated temporary stability, Echeverría's successor, José López Portillo, confidence bolstered by the oil boom, pursued policies of even higher government spending and borrowing. These led to a disastrous economic collapse a few years later, when oil prices dropped and international interest rates rose.[98] Faced with defaulting on its massive debts, the Mexican government agreed to economic restructuring dictated by the International Monetary Fund in late 1982. This meant cuts in public funding, deregulation of industries, privatization of some state-owned companies, and trade liberalization through the elimination of protectionist policies. The expensive Mexican Food System, SAM, was among government programs hit by the implementation of austerity measures. With the arrival of President Miguel de la Madrid Hurtado's new administration in December 1982, SAM came to an end.[99]

At INIA economic instability brought budget cuts and hiring freezes. Government spending on agriculture and forestry research had nearly quintupled from 1970 to 1980. In a rapid reversal it dropped by half between 1981 and 1986.[100] Research facilities deteriorated, administrators curtailed programs, and frustrated staff left in droves for more lucrative positions in industry. Researchers at Chapingo recalled the 1980s as a time when the campesino-oriented activities of the preceding decade essentially disappeared, a shift they attributed to renewed state emphasis on export-oriented production. Breeders went back to breeding lines developed in the 1940s and 1950s in the interest of increasing commercial production. The national corn- and wheat-breeding programs were turned into "simple extensions" of CIMMYT—at least according to some who worked through the transition.[101] In the midst of these shifts, INIA was folded into a new combined agency, the National Institute for Forestry, Farming, and Livestock Research (Instituto Nacional de Investigaciones Forestales, Agrícolas y Pecuarias). The restructuring left an agency that many felt was desk-worker heavy, with bureaucrats in place but experiments and extension work curtailed.[102]

For researchers working with collections of Mexican corn and other crops, the reorientation of the 1980s meant a slowed pace. The 1970s had seen an efflorescence of activities related to Mexican landraces and their recognition in new institutional structures. In maize the boom in collecting farmers' varieties extended from the late 1960s to 1978. Researchers in charge of maize had taken steps to reorganize their activities, including evaluations of landrace collections, to better reflect Mexico's diverse ecological conditions. INIA had created a separate unit to oversee the management of genetic resources.[103] The staff in charge of the collections still faced challenges, from occasional failures of the refrigeration system to defective monitoring equipment to flooding and rodent infestations. The economic downturn worsened these conditions. Resources to address these problems were not forthcoming from the Portillo government, in spite of requests for a new national gene bank from agricultural scientists and the endorsement of this need by outside observers.[104] In other words, Mexico's leadership at FAO in demanding an international gene bank and free exchange of seeds remained a distant spectacle for researchers in Mexico, individuals whose labor ensured that there would still be seeds to store and circulate. If they had thought that this attention to crop diversity as a national resource might generate new infrastructure and expanded budgets, the advent of austerity in 1982 surely extinguished such hopes.

Research persisted, however. A number of scientists retained a strong interest in maize varieties considered criollo or *nativo,* and farmers' varieties of other Mexican crops. These included Efraím Hernández Xolocotzi, his students, and a nascent community of Mexican agroecologists. Together with colleagues at other Mexican institutions, Hernández led an influential series of investigations into agroecosystems and "traditional agricultural technologies" in the early 1980s. As one participant remembered, this research program envisioned agroecology as a way for farmers' existing cultivation patterns, including their choice in crops, "to be studied, preserved, improved, and expanded."[105] Under Hernández's guidance scholars at Chapingo produced studies that interrogated the possibilities for conserving crop diversity, both in seed banks and on farms.[106] Meanwhile, the Maize Germplasm Bank of the new National Institute for Forestry, Farming, and Livestock Research continued to serve as a resource for scientists wanting to locate, classify, and conserve the races of maize of Mexico. The Chapingo graduate and Hernández student Rafael Ortega Paczka was one ardent champion of this work, continuing his studies of Mexican corn as a PhD student in Leningrad when labor politics made his position in Mexico untenable.[107] The interest of Mexican geneticists and agronomists in criollo maize, nurtured in the 1970s, kept collections extant through an otherwise bleak period.

An influx of money from an unexpected source also helped. Another devoted student of maize diversity in the 1980s was Major Goodman, a geneticist at North Carolina State University in the United States. He had studied maize diversity as a doctoral student in the late 1960s, work that had raised his awareness of the poor state of many collections, in the United

States and beyond. After discovering the collections of the Committee on Preservation of Indigenous Strains of Maize to be mostly neglected, he launched a crusade to save them.[108] In the early 1980s he remained incensed at continued failures in the conservation of crop diversity, especially when these concerned corn. He considered the fragile state of seed banks in Mexico and throughout Latin America a problem for the United States as well, and in 1981 he convinced USDA administrators to act.[109]

Thanks in part to Goodman's interventions, a "Plan for Preservation and Utilization of the Latin American Maize Accessions" took shape in the 1980s. A first ambition of the project was to develop a "complete inventory" of five major maize collections: those of CIMMYT and the Mexican national collections, additional collections in Colombia and Peru, and the corn kept at the US National Seed Storage Laboratory. A second was to evaluate the samples and renew the seeds. In keeping with technical standards in seed conservation and in light of continued bleak assessments from resource-short Latin American banks, the seed renewal would be large enough to ensure that part of each sample could be sent to the United States for long-term storage. In 1982 a USDA official and a plant breeder from Pioneer Hi-Bred dispatched as emissaries to Mexico reported that the head of the country's genetic resources program was "very receptive" to the proposal.[110]

This interest did not preclude administrative wrangling, however. In December 1983, as negotiations with Mexican officials dragged on, one of the project's organizers wondered whether Mexico's position in the recent FAO debates on plant genetic resources would affect scientists' and administrators' views on the US-led cooperative corn conservation project.[111] If there were concerns about partnering with the USDA, including arranging for Mexican collections to be duplicated in the United States, it is clear that they were not sufficient for administrators to refuse technical and financial support. Staff in the Mexican Ministry of Agriculture finally signed off on the USDA agreement in 1985, launching a multiyear collaboration that by 1988 had resulted in the shipment of at least 550 samples of Mexican maize to the United States for inclusion in its collections.[112] In the 1980s this bilateral agreement was far more consequential for Mexican maize collections than the better-known FAO resolution agreed to in Rome.

From the late 1970s onward, activists provided new perspectives on the loss of crop diversity, both its causes and its consequences. Pat Mooney thought genetic erosion was better described as "commerciogenic erosion," a result of private industry's steady accrual of power over seeds and crops—and ultimately over food and people as well. Far from being an unfortunate but inevitable side effect of a benevolent process of development, the loss of crop diversity was one among many destructive outcomes of mainstream agricultural science and its relentless privatization.[113]

This alternative narrative made many people angry. It's easy to see why. The prime mover of diversity loss was no longer the celebrated "improvement" of crops and farming systems. Instead, it was the imposition of socially and environmentally inappropriate agricultural technologies and especially corporate greed. Developing economies, purported to benefit from ex situ conservation strategies pursued in the early 1970s and the breeding programs these sustained, were now characterized as injured and exploited by the same efforts. Plant breeders and crop scientists, so recently heroes of a Green Revolution, were recast as culprits or, at best, accessories to a crime. And researchers who had labored, mostly thanklessly, to get policy makers and national agricultural institutions to care about maintaining collections of landraces and crop wild relatives in the 1960s and early 1970s were now positioned as enemies rather than allies in conservation.

The anger and frustration of these researchers was palpable. Longtime genetic resources advocate Otto Frankel castigated Mooney as a "self-seeking swindler." William Brown of Pioneer Hi-Bred despaired that Mooney and "others of similar ilk" were "solely responsible for stirring up discontent in the developing world" that would effectively shut down collecting in those countries. Another longtime maize conservation advocate, wounded deeply by what he considered the deliberate spread of lies about collection mismanagement at CIMMYT, speculated, privately and without evidence, that Mooney was funded by the KGB.[114] These scientists and their collaborators had been well intentioned in their efforts to preserve crop diversity, but they failed to appreciate how gross imbalances of power, between Global North and South and between private and public interests, would dictate who benefited from these efforts. They also afforded too much power to activists like Mooney, overlooking the history of grievances that fueled the demands of the developing country bloc at FAO and in similar forums. In the end individuals who ought to have been allied in conservation were in fact opposed. In the decades since, contestation has continued and, with it, competing political interpretations of loss, which alternately blame rapacious industry, neglectful governments, complicit scientists, angry activists, and poorly designed treaties for hindering the conservation of something that nearly everyone agrees is valuable.

As these debates raged, often in spaces distant from farm fields and experiment stations, farmers and researchers kept on with the work of growing crops. In the 1970s the study of Mexican maize diversity, and of the cultivators who had created and continued to care for this diversity, attained unprecedented visibility in Mexico. Converging currents in agronomy, anthropology, sociology, and even national policy making and international negotiating carried researchers toward new accounts of the roles that campesinos and criollo maize could and should play in the making of a stronger, more resilient nation. To support subsistence farmers in being self-sustaining was to contribute to making a self-sufficient Mexico, one freed from dependency on the "food power"-hungry United States and less prey to the incursions of profit-seeking transnationals. To study and conserve criollo maize was therefore to take part in this nationalist, largely anticapitalist, project.

By the mid-1980s, when Mexican scientists agreed to the offer of US assistance for its maize collections, the challenges to continuing in this vein could not have been clearer. Breeding research in Mexico had been reoriented again to serve commercial producers, as had agricultural policies more generally. President Miguel de la Madrid forged ahead with neoliberal policies that encouraged foreign investment and free trade. The recommendations produced during SAM for a national seed industry built around Mexico's abundant diversity in crop landraces and observant of farmers' varied cultural and ecological needs soon looked hopelessly idealistic in light of the continued foreign dominance of seed production. The assumed genetic erosion of Mexican crop diversity, and corn diversity in particular, seemed likely to continue unabated—unless, as some soon hoped, those in industry could be convinced that it was in their interest not just to stockpile diversity but to use it.

[24] Ears of corn from experimental crosses are set out for expert assessment at the 2008 Field Day of the Germplasm Enhancement of Maize project in Ames, Iowa. These plants, developed in part from tropical varieties, represent years of labor and especially data production and sharing on the part of US breeders and collaborators across the Americas. Photograph by US Department of Agriculture.

SIX

EVALUATE

THE GENETICIST MAJOR GOODMAN of North Carolina State University played a leading role in the US Department of Agriculture–sponsored effort to regenerate maize seed collections in Colombia, Mexico, and Peru in the mid-1980s. This had been conceived as a rescue mission, in which endangered seeds would be regenerated and copies placed in secure cold storage. Although frozen duplicate samples looked to many researchers like the endpoint of conservation, for Goodman it was simply the first step. Once the endangered Latin American collections were "safely in hand," they next needed to be incorporated into US breeding programs. This would diversify the genetic basis of commercial hybrids, which the 1970 outbreak of southern corn leaf blight had revealed as dangerously narrow. It might also convince breeders, and by extension government officials and taxpayers, that it was well worth the expense of keeping up collections.[1]

Persuading breeders of the value of seed bank holdings, especially landraces and wild relatives, was, as Goodman saw it, "'missionary' type work, rather mundane, but apparently necessary."[2] Other researchers involved in conservation and breeding agreed. Collections too often went unused, and disuse generated its own problems. As one breeder observed, "Even if the germplasm is adequately preserved in cold storage, under conditions one could depend on for many years, . . . someone could legitimately question the purpose of storing all that germplasm if it is not being utilized."[3]

A new phase in the conservation of maize genetic diversity took shape in the 1980s, one motivated especially by the idea of "genetic vulnerability" and the threat posed to commercial corn by its own uniformity. In the United States the southern corn leaf blight had dramatically illustrated the risks of genetic uniformity, which subsequent studies revealed as a problem across key economic crops. In its wake breeders considered whether and how to upend this pattern. For corn breeders using "exotic" materials from outside the United States stood out as a potential route to more genetically diverse, less genetically vulnerable lines. Varieties from the tropical regions of Latin America, relatively unexplored by US breeders, seemed especially promising. Their use was hampered, however, by the precarity of samples kept in seed banks and especially by the lack of knowledge about these samples. What traits lay within them? How would they grow in new environments? What types should they be hybridized with? Generating data, and making them readily available, would have to be the first step.

In wondering about the potential latent in tropical varieties, US maize breeders shared in long-standing assumptions about life in the world's middle latitudes. A pattern that the historian Paul Sutter describes as "tropical thinking" saw visitors to humid tropical lowlands and islands cast these places as distinct from temperate regions—in terms not just of geography and climate but also the qualities of the plants, animals, and peoples found there. Explorers, naturalists, and later biologists contributed to an enduring understanding of the tropics as a place of untamed, abundant, exuberant, and potentially wealth-generating life and of its peoples as incapable of harnessing this natural productivity to their own advantage. Tropical thinking justified imperialism, and with it extraction, exploitation, and racial discrimination.[4] In the 1980s these precedents fed into an emerging discourse of biodiversity conservation. Scientists, alarmed at the destruction of tropical habitats in countries eager for economic development, warned that their biological riches required care and preservation. Conservation campaigns advertised vulnerable tropical forests as the source of botanical cures whose potential could be revealed only through modern science. The yet-unrealized potential of biotechnologies heightened this valuation, galvanizing bioprospecting missions to identify and extract the "genetic resources" latent in tropical biota. Biologists, meanwhile, hoped that successful bioprospecting—and the promise of cold, hard cash—would make the benefits of biodiversity conservation apparent to all.[5]

Elements of tropical thinking circulated among researchers intent on salvaging corn diversity. They maintained that exotic tropical maize had enormous productive potential, needing only scientific research to be realized. They simultaneously saw this diversity as threatened by mismanagement and inattention within underfunded national agricultural research institutions. They therefore wanted to shore up seed banks to ensure that precious resources didn't go extinct. And they decided that one of the best ways to secure conservation facilities was to demonstrate the value of the seeds stored within them, for example, by using these to make better industrial corn. As a result, the hunt for genetic resources in tropical maize looked a lot like the better-known case of drug bioprospecting in the tropics, both in vision and in practice.

In untangling the complexities of late twentieth-century efforts to extract commercial pharmaceutical products from tropical biota, historians and anthropologists have focused on the routes through which knowledge travels, for example, among local plant collectors, market herbalists, ethnobotanists, and laboratory scientists.[6] The history of bioprospecting in corn reveals how scientists too were concerned with tracking these flows of information. Among maize researchers data generation and management were often identified as obstacles to realizing the value of stored diversity, roadblocks to the very knowledge of which seeds might resolve which woes. Their subsequent efforts to construct data infrastructures were never just about discovering useful "genetic resources," however. They instead addressed a widening web of political, social, and economic concerns, from national agricultural failures to career opportunities to corporate growth.[7] In the 1980s generating data about tropical maize would contribute to the ultimate salvation of corn diversity, and to many other goals besides.

EXOTIC POTENTIAL

The imagined potential of exotic plants and animals to enrich landscapes, and landowners, has a long history. In the eighteenth and nineteenth centuries, the introduction of species across ecological boundaries appealed to imperial administrators hoping to engineer colonial lands for economic gain. This science of "acclimatization" also promised profits closer to home, through the introduction of new game animals or garden flowers into Europe from far-flung places.[8] Acclimatization went out of vogue at the turn of the twentieth century, but interest in extracting value from exotic plants and animals persisted—and so too did the problems of adaptation to new environments that had first galvanized acclimatizers. Professional plant breeders responded to these concerns with the ideas and tools of their discipline, focusing on varieties, and later genes, as the objects of interest and experimentation, but their efforts to refashion crops and flowers traced an established pattern.[9] Corn breeders in the United States were no exception. From the 1950s onward a growing number thought exotic varieties, those originat-ing outside national borders, had potential riches locked within them. But these plants—unadapted, unimproved, unruly—needed expert coaxing for their value to be realized.

In the mid-1950s, while collecting and classifying corn with the Committee on Preservation of Indigenous Strains of Maize, Edwin Wellhausen of the Office of Special Studies in Mexico became one of the first vigorous pro-moters of tropical maize as a resource for US corn breeders. He and his col-leagues had discussed this possibility—and used it to motivate collection and conservation—but had done little to realize it.[10] Wellhausen wanted to turn the tide. With the corn blight yet to come, Wellhausen wasn't worried about genetic uniformity and the potential for plant epidemics. He simply thought that the spectacular gains in yield made by US corn breeders in the 1930s and 1940s were likely to plateau without an infusion of new genetic material.[11] Sitting at the Office of Special Studies in 1956, surrounded by collections recently enriched through the Maize Committee, Wellhausen had an obvious idea about where to find this new material. "Variation in corn in Mexico, Central America and South America is phenomenal," he declared. "Although we do not yet know its full extent, we know there exists a tremendous varia-tion in all characters that corn possesses." He urged breeders to begin the work of moving this variation into US hybrids.[12] Wellhausen knew this would take time, perhaps as long as twenty-five to thirty years. Breeders would be challenged by corn types poorly adapted to US conditions and likely to infuse as many undesirable qualities as potentially useful ones when incorporated into breeding lines. Wellhausen nonetheless considered the trouble worth-while. A decade later he was still convinced, even after struggling with some of the more immediate challenges himself.[13]

William Brown of the Pioneer Hi-Bred Corn Company agreed with Wellhausen. After exploring the corn of the Caribbean on behalf of the Maize Committee in the 1950s, he had continued experimenting with types he had first encountered there, crossing several Caribbean varieties with Corn Belt Dents. In 1958 he described this effort as "primarily a test of the value of the exotic germ plasm for use in the Corn Belt," from which he also hoped to acquire "some valuable breeding material."[14] Brown later recalled that experimentation with exotic lines at Pioneer had been influenced by the company's founder. Henry A. Wallace had "felt strongly that the program should be kept extremely diverse," Brown remembered. He was convinced that tropical maize could contribute to making better Corn Belt Dent and also interested in establishing Pioneer in Latin America, an expansion that would require experience with tropical types.[15] Whether following orders or instinct, or both, Brown explored the use of exotic materials in Pioneer's breeding stocks for decades. He did so from an increasingly influential position too, becoming the company's vice president and director of corporate research in 1965 and still later its president.[16]

The 1970 corn blight created many more advocates for introducing genetic diversity into US hybrids. For decades breeders and others had warned that landraces and wild relatives ought to be collected and conserved, judging these to be both threatened and potentially valuable. More recently, they had begun to worry about the samples of endangered crops that had been brought to seed banks for safekeeping, fearing that storage often meant neglect and therefore did little to ameliorate the looming danger of extinction. Now their attention swung from the vulnerability of landraces to the vulnerability of commercial crops. As the corn blight made clear, these were also endangered. For many observers the solution to this endangerment lay in diversification.

Although corn received the most attention in the aftermath of the blight, many crops came under scrutiny. A study undertaken at the US National Academy of Sciences identified the epidemic in corn as one instance of a more general phenomenon of "genetic vulnerability" caused by extreme genetic uniformity. This vulnerability was itself epidemic among US crops. Potato varieties had proliferated in the 1880s, but in 1970 a mere four dominated national cultivation. Across the country nearly all peas used in commercial processing traced back to a single 1914 introduction known as Perfection. Sorghum varieties had similarly narrow origins and, even more worryingly, shared cytoplasmic male sterility derived from a single source—an exact parallel to the circumstances that had produced the corn blight. The list went on. As the National Academy of Sciences study summarized, "Most major crops are impressively uniform genetically and impressively vulnerable."[17]

This vulnerability was tricky to document. No one had monitored the lines adopted by private seed companies, which became farmers' crops in the field. Still, specialists in many crops (potatoes and peas, for example) could examine the pedigrees of varieties released by public breeders to guess at the genetic makeup of lines now growing on farms. This approach was less helpful in corn. By 1970 corn breeding in the United States was characterized by a division of labor. Breeders at public institutions like land-grant colleges and state agricultural stations prepared new breeding populations by pooling together different varieties or races in an attempt to generate new genetic combinations. They sometimes created new inbred lines. Private companies then further worked these freely available materials, incorporating them into hybrid varieties that typically featured a mix of public and proprietary material. This division of labor meant that tracing the history of public lines provided only a partial account of potential genetic uniformity. The rest of the story was not forthcoming. Hybrid corn seed companies routinely insisted that industrial competitiveness required them to keep secret the exact pedigrees of their inbred lines and hybrid combinations.[18]

Researchers who wanted to assess the genetic background of the corn cultivated by US farmers therefore had to make do with limited knowledge. There were enough data available to indicate that private companies relied on the same handful of public lines out of hundreds released each year and that some of these remained popular for decades.[19] Given that the development of more vigorous inbred lines had recently enabled breeders to switch from double-cross hybrids to single-cross—that is, from using four parent lines to just two—this concentration was even more significant than it would have been a decade earlier.[20] It was also clear that a breeding strategy referred to as "recycling" compounded reliance on a small pool of public lines. Successful "elite" lines were often integrated back—that is, "recycled"—into breeding programs as the starting point for further development.[21] The incomplete but no doubt accurate picture of uniformity arising from these known practices was bolstered by a well-established genealogy. Both public lines and proprietary lines had originally been derived from a few open-pollinated varieties of Corn Belt Dent, and Corn Belt Dents had originated in the combination of just two of the hundred-plus known races of corn.[22]

If the continued reliance on a handful of public lines and the recycling of old proprietary breeding lines were sources of worrying uniformity, one obvious solution was to turn to new starting materials. In this context tropical lines, notoriously difficult to work with, looked newly promising, an almost untapped reserve from which to infuse Corn Belt Dent with fresh genetic traits. Breeders valued these as "genetic resources" in much the way that breeders had assessed landraces and wild relatives for many decades. Although experts agreed that "exotic collections" were often "poorly adapted and therefore low yielding" when grown in the United States, they could be "extremely valuable" as the sources of resistance to diseases and pests.[23]

Promising did not mean easy, however. One hurdle to using tropical corns in the temperate regions of the United States is that these tend to be sensitive to changes in day length. Adapted to short days, they often flower late when planted at higher and lower latitudes, only as the long days of summer come to an end. Unfortunately, in temperate climates plants that flower late are likely to be destroyed by cold long before they produce any seeds. So the first step to using tropical lines was figuring out how to make them available to US breeders, not just as seeds but as living plants whose qualities could be observed and evaluated in the field.

In the early 1970s Major Goodman's North Carolina State colleague Charles Stuber was among the first to make a go of incorporating tropical lines into a public breeding program. Drawing on Goodman's collection of maize seed, assiduously assembled from seed banks across the Americas, Stuber crossed a selection of "exotic" seeds from Latin America with Corn Belt Dent lines of the southern United States. Stuber anticipated that the resulting part-exotic lines would flower earlier, thanks to the infusion of genes from temperate corn.[24] Meanwhile, at Iowa State the geneticist and corn breeder Arnel Hallauer and his colleagues tried two methods, cross-breeding Corn Belt lines with varieties from Colombia and mass selection of the same Colombian varieties. The latter aimed to create a population insensitive to day length of wholly tropical origin by selecting seeds of the earliest flowering individuals each season as the parent plants of a subsequent generation.[25] Both efforts required several generations before evaluation could be initiated—and evaluation was still just a forerunner to breeding.

MISSING DATA

As attention focused on where and how to defend corn and other genetically vulnerable crops in the 1970s, agricultural experts agreed that fresh genetic material was only part of the solution. The 1972 National Academy of Sciences study called for improved plant quarantine, a national monitoring committee for crop vulnerability, and a coordinated system for the introduction, maintenance, and use of germplasm resources.[26] In 1973 an ad-hoc committee brought together by the USDA and the nation's land-grant colleges to evaluate genetic vulnerability arrived at basically the same conclusions. Its assessment of the existing US system for conserving crop diversity was dire. Despite "considerable effort," the assembly and maintenance of crop collections had "been too haphazard, unsystematic, and uncoordinated." Arriving at the stark conclusion that "the situation is serious, potentially dangerous to the welfare of the nation, and appears to be getting worse rather than better," the committee urged the creation of a national commission to guide plans for plant genetic resources.[27]

Policy makers and administrators made little headway on these recommendations in the 1970s. Calls for a monitoring body and national commission gave rise to a National Plant Germplasm Committee in 1974 and to the reconceptualization, if not the actual reorganization, of scattered institutions and programs for plant exploration, introduction, and maintenance into a National Plant Germplasm System. Postblight analyses precipitated the creation of a further oversight body, the National Plant Genetic Resources Board, tasked in 1975 with ensuring adequate attention to genetic conservation issues amid the competing demands of the national agricultural research infrastructure.[28] More meaningful changes remained on paper. When the US General Accounting Office conducted an audit of national management of "germplasm resources" in 1981, it concluded that genetic vulnerability remained a problem. Decentralized management forestalled an integrated and comprehensive program. The upshot, according to the audit, was that the National Plant Germplasm System neither routinely assessed the risks associated with vulnerability nor "adequately perform[ed] the housekeeping chores of collection, maintenance, and evaluation of germplasm stock."[29]

A quick look at the recommendations of the National Plant Genetic Resources Board in 1978 offers insight into the measures it thought necessary to reduce genetic vulnerability—and also why it was difficult, if not impossible, to achieve them. The board insisted that effective defense against vulnerability demanded a seven-phase program encompassing all aspects of crop research and development. This started with seed or plant acquisition, maintenance, and distribution and extended all the way through to breeding new varieties and producing improved seeds for farmers. This comprehensive trajectory from collection in the field to use by farmers also incorporated research into genetic variability, geographic distribution, and mechanisms of inheritance. The board, a group of crop scientists who surely would have loved to see more federal funding for these items irrespective of the need to address genetic vulnerability, urged attention to all phases, in a continuous process, to minimize the risks attendant on genetically uniform crops.[30]

This call to rescue crops, and an entire agricultural system, threatened by genetic vulnerability extended the activities seen as essential to conservation and management of plant genetic diversity. This work could not be considered finished when seeds of plants with potentially useful genetic material made it to the seed bank, even if that seed bank functioned effectively. The

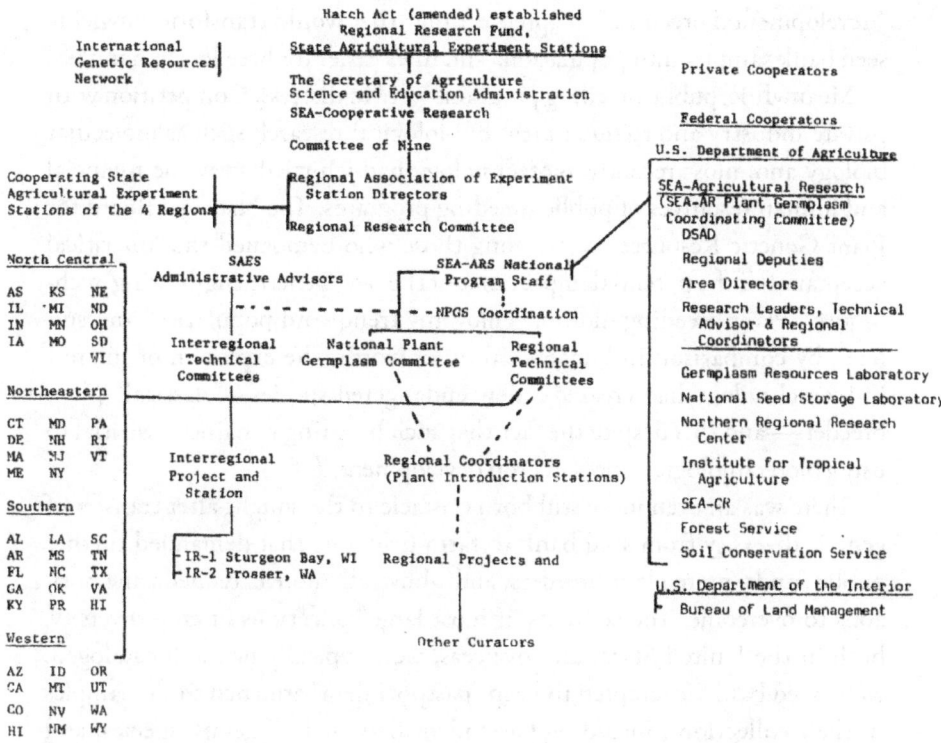

The diagram contains the following labels:

- Hatch Act (amended) established Regional Research Fund, State Agricultural Experiment Stations
- International Genetic Resources Network
- The Secretary of Agriculture Science and Education Administration SEA-Cooperative Research
- Private Cooperators
- Federal Cooperators
- Committee of Nine
- U.S. Department of Agriculture
- Cooperating State Agricultural Experiment Stations of the 4 Regions
- Regional Association of Experiment Station Directors
- SEA-Agricultural Research (SEA-AR Plant Germplasm Coordinating Committee) DSAD
- Regional Research Committee
- North Central
- SAES Administrative Advisors
- SEA-ARS National Program Staff
- Regional Deputies
- AS KS NE
- IL MI ND
- IN MN OH
- IA MO SD
- WI
- NPGS Coordination
- Area Directors
- Research Leaders, Technical Advisor & Regional Coordinators
- Northeastern
- Interregional Technical Committees
- National Plant Germplasm Committee
- Regional Technical Committees
- Germplasm Resources Laboratory
- CT MD PA
- DE NH RI
- MA NJ VT
- ME NY
- National Seed Storage Laboratory
- Northern Regional Research Center
- Interregional Project and Station
- Regional Coordinators (Plant Introduction Stations)
- Institute for Tropical Agriculture
- Southern
- SEA-CR
- AL LA SC
- AR MS TN
- FL NC TX
- GA OK VA
- KY PR HI
- Forest Service
- IR-1 Sturgeon Bay, WI
- IR-2 Prosser, WA
- Regional Projects and
- Soil Conservation Service
- U.S. Department of the Interior
- Bureau of Land Management
- Western
- AZ ID OR
- CA MT UT
- CO NV WA
- HI NM WY
- Other Curators

[25] The auditors at the General Accounting Office were not impressed by the National Plant Germplasm System, shown here in a 1981 diagram. They found its organization "almost impossible to decipher," an assemblage of independent units with nebulous links among them. "The net result is a set of components that is not really a system at all." From US Comptroller General, *Report to the Congress*, 15.

country could not afford for conservation to be what the National Plant Genetic Resources Board called a "museum type of activity." It needed to include crop improvement alongside collection and maintenance.[31] An effective insurance policy was one that moved landraces, wild relatives, and other potentially useful seeds and plant materials into the bank and also moved them back out to be used by US farmers in routine cultivation.

The obstacles to this vision were formidable. Those who promoted the collection of endangered landraces had long acknowledged the difficulties of working with materials that hadn't already been genetically cleaned up and improved through inbreeding and selection or weren't adapted to temperate climates. They and others assumed, correctly, that private companies, concerned with short-term profits, would balk at the costs and risks involved in underwriting such work. Public breeding programs were a likelier site for "developmental breeding" or "prebreeding" that would transform unwieldy seed bank samples into populations and lines easier for breeders to handle.

Meanwhile, public breeding programs were in distress. Competition with private industry and trendier areas of biological research such as molecular biology and, most recently, biotechnology had whittled away the financial and human resources of public breeding programs. The National Board for Plant Genetic Resources was among those who bemoaned the "uncritical acceptance of the potential importance of the new 'genetic engineering' techniques." Plant breeding did not "enjoy any trendword popularity" and suffered by comparison in budgets. In other words, the explosion of interest in biotechnology had created a new endangered species—"classical" plant breeders—and this despite the fact that such breeding remained essential to using biotechnological tools in crop development.[32]

There was an even more stubborn obstacle to the sought-after transfer of genetic diversity from seed bank to farm field, one that demanded further public funds, more plant breeders, and, above all, more extended time horizons to overcome. The contents of most large collections of crop diversity, both in the United States and overseas, were typically not well cataloged. Most seed banks attempted to keep "passport data" attached to the samples in their collections, including basic identifiers such as genus, species, and common name, as well as collection data such as latitude, longitude, elevation, and date of collection. But even if samples had been diligently labeled, and they often were not, passport data were not as helpful to breeders as information on plant height, flowering date, yield, disease resistance, and drought tolerance. The essential first step toward greater use of seed bank collections was therefore figuring out what exactly was in them. Unfortunately, evaluation—the generation of basic agronomic data about seed bank samples—was also laborious work, "a routine, non-glamorous activity that no one is particularly interested in doing." Even public plant breeders would shy away from a task so unlikely to produce career rewards.[33]

At North Carolina State University, where Charles Stuber's effort to make use of unadapted landraces in US corn lines was underway, this initial hurdle of data generation was plainly obvious. The task had started with the agronomic evaluation of some six hundred individual landrace collections, conducted over multiple years, before settling on the ten most promising of these as the source materials for populations to use in breeding. Only then did Stuber begin the further labor of developing these heterogenous breeding populations, which comprised mixed exotic and temperate materials, into improved populations that other breeders might be willing to work with.[34] The same initial surveying task would face almost any researcher in the United States wanting to work with landrace collections.

Stuber had obtained most of his starting materials from the collections of his colleague Major Goodman, who was in a better position than almost anyone to speak about the problems and possibilities of collections. Goodman knew firsthand both the recurring failures of collection maintenance and the frequent absence of basic information about samples in storage, thanks to more than a decade of dogged efforts to acquire the landraces of corn described in the *Races of Maize* series from seed banks in Mexico, Brazil, Colombia, Peru, and the United States. It is telling that, when he started assessing tropical lines for breeding in the mid-1970s, he avoided the little-known landrace samples and worked instead with commercial hybrids developed from these tropical races and marketed in Latin America. These already had been subject to genetic improvement in various breeding programs and analyzed by CIMMYT in the 1960s. Rather than start with preliminary screening and data generation, he could set to work on adapting the better-performing lines to conditions in North Carolina and eventually evaluating them in that environment, a process that would still take him ten years.[35]

REGENERATION

In its recommendations to the US secretary of agriculture in 1979, the National Plant Genetic Resources Board urged the creation of "crop advisory committees." These committees, which assembled experts knowledgeable about a specific commodity crop, would recommend actions to strengthen the genetic foundations of those crops.[36] By 1981 there were committees in place for alfalfa, corn, oats, peas, beans, potatoes, sorghum, soybeans, tomatoes, and wheat, among others. Their assignments ranged from providing crop-specific "descriptors" that would be systematically used to populate the databases associated with different collections to planning evaluation and breeding projects that would make use of exotic material to providing "early warning" of "impending... genetic vulnerability."[37] Clearly, securing genetic resources in the national interest was seen to require expert knowledge. Experts soon insisted that it also required basic information.

The new Maize Crop Advisory Committee, cochaired by William Brown and Major Goodman, held its first meeting in March 1981. Its members agreed on the fundamentals. There was no doubt that US hybrid corn had been created from a limited pool of genetic materials and that dominant breeding strategies were further restricting its genetic base. Meanwhile, *Zea mays* was acknowledged to be a wildly diverse species, which in theory meant many possibilities for discovering valuable characteristics and combinations. It had been collected and stored in seed banks around the world. Much of this diversity remained undocumented, however, at least for US-based breeders. Evaluation was essential to unlocking this potential and addressing the root cause of genetic vulnerability.

As a result, when asked to develop a "germplasm enhancement program" for corn in early 1982, as part of a larger national initiative, the Maize Crop Advisory Committee pushed back. "Germplasm enhancement," also referred to as prebreeding or developmental breeding, is the work of creating better breeding materials, especially by adding genetic diversity. Within the USDA it meant moving novel genetic material from landraces and crop wild relatives into "germplasm pools" for breeders.[38] The Maize Crop Advisory Committee insisted that, at least for corn, it was too soon to develop enhancement plans. As Brown explained to USDA administrators, the requisite data were missing. The world's largest corn collection was housed at CIMMYT in Mexico, with more than twelve thousand samples, but "with few exceptions these accessions carry essentially no data on traits that are of interest to the breeder." Brown insisted that without preliminary screening there was no way to incorporate these "intelligently" into an enhancement program.[39]

Other members of the Maize Crop Advisory Committee thought even screening was premature. Goodman stressed the need to acquire seeds first, and the committee ultimately agreed that US collections could be significantly improved.[40] Although it gave highest priority to acquiring samples from Latin American seed banks, it also identified issues that needed to be resolved domestically. Corn samples acquired from earlier exchanges with Latin American seed banks had been deposited at the US National Seed Storage Laboratory but not given official plant inventory numbers, meaning they could not be requested. Meanwhile, some samples registered in the US collection had either never been very large or had dwindled in number over time and were therefore unavailable.[41] In short, infrastructure and management in the United States also needed attention.

Convinced that stabilizing and evaluating seed bank collections needed to come before meaningful enhancement activities, the Maize Crop Advisory Committee forged ahead with its most urgent tasks. It arranged funding and oversight for a program of regenerating samples believed to be decaying rapidly in Latin American seed banks. The initial hope was to capture the four largest corn seed collections outside of the United States, including those from CIMMYT and INIA in Mexico, as well as from national collections in Colombia and Peru. It also moved to secure as many of these seeds as possible in the US National Plant Germplasm System.[42]

The initial purposes of this project (described in chapter 5) were multiple: to create a complete inventory with basic identifying (i.e., passport) information for the collections in Colombia, Mexico, Peru, and the US National Seed Storage Laboratory; to increase endangered seeds in sufficient quantities so duplicates could be placed in long-term storage; and to evaluate those endangered collections so that when rebanked they would be accompanied by information that would facilitate their future use. This would, it was hoped, remedy a continuing situation in which "germplasm stocks" of the country's "most important feed grain" were "difficult to obtain, poorly described, totally uncatalogued and often unavailable" from the seed banks charged with caring for these.[43]

The USDA-funded project soon narrowed to regenerating stocks, creating adequate catalogs, and entering duplicates into the US collections where possible—and this only for the national collections in Colombia, Mexico, and Peru. CIMMYT, reeling from recent budget cuts, declined to participate. The challenges faced in regenerating seeds, let alone conducting evaluation work, proved significant. Growing out the most endangered collections often meant working with the oldest and most poorly adapted seed stocks, which sometimes necessitated two or more regeneration attempts in successive seasons. The Latin American seed bank managers also experienced shortages of human resources, supplies, and money when earmarked funds were diverted to other needs. The lead Peruvian scientist involved in the regeneration project recalled still further interferences: "inflation, terrorism, the Niño, drought, etc."[44] It is little wonder that the scientists, US, Colombian, Mexican, and Peruvian alike, focused on the central task of regeneration and duplication rather than the additional step of evaluation.

Even if there had been more resources to support the manual labor of evaluation by the Latin American agronomists involved in the project, the transmission of their data would have posed another hurdle. As Goodman noted, "None of these programs has ready access to the mainframe, data entry, or microcomputer facilities needed to efficiently enter and transmit data." As a result, one of the most glaring shortcomings was in obtaining even basic information about the Latin American collections that could be used outside the participating seed banks. In 1987 only the team in Peru had come through with both the data generated by the regeneration project and complete passport data for the entire collection. Meanwhile, collaborators in Colombia and Mexico had as yet provided only their accession lists, with Colombia's list lacking passport data but Mexico's "reasonably complete."[45] At the end of the initial five-year period for the project, successes in data gathering focused on the basic identifiers included in passport data, considered to be "adequate" for Peru and Mexico and in need of a complete overhaul for Colombia.[46]

In 1983, with this regeneration program in progress and the never-ending failures of seed bank collections on his mind, Goodman presented a blistering critique of crop conservation efforts to fellow breeders and agronomists assembled at a Pioneer-sponsored research forum. Although his account of failings in maintenance and record keeping echoed earlier assessments, Goodman's personal experiences of USDA staff throwing away seeds and neglecting curation standards were surely more damning for the assembled crowd. He tethered these failures to a "lack of national leadership," without which there was "little incentive for a USDA scientist ... to attempt to assemble and maintain thousands of individual, unadapted collections." Anyone who did would face ruined budgets, a damaged career reputation, and, according to Goodman, "questions about one's sanity."[47]

Although Goodman emphasized the importance of securing endangered collections, especially through regeneration and duplication, the discussion that followed his presentation focused especially on problems of evaluation. The elite group gathered at Pioneer's research forum agreed that this needed to be a priority for all crops. A potato breeder, and former curator of the US potato collection, observed that he'd struggled to get any breeders interested in working with the many samples gathered under his care. A wheat breeder claimed that the USDA's otherwise impressive collection of this crop was useless precisely because it was so large that it had been assessed only in fragmentary ways. Their collective concerns pointed to the deeper worry that failures to maintain collections would never be resolved without concerted efforts to document and demonstrate their use.[48]

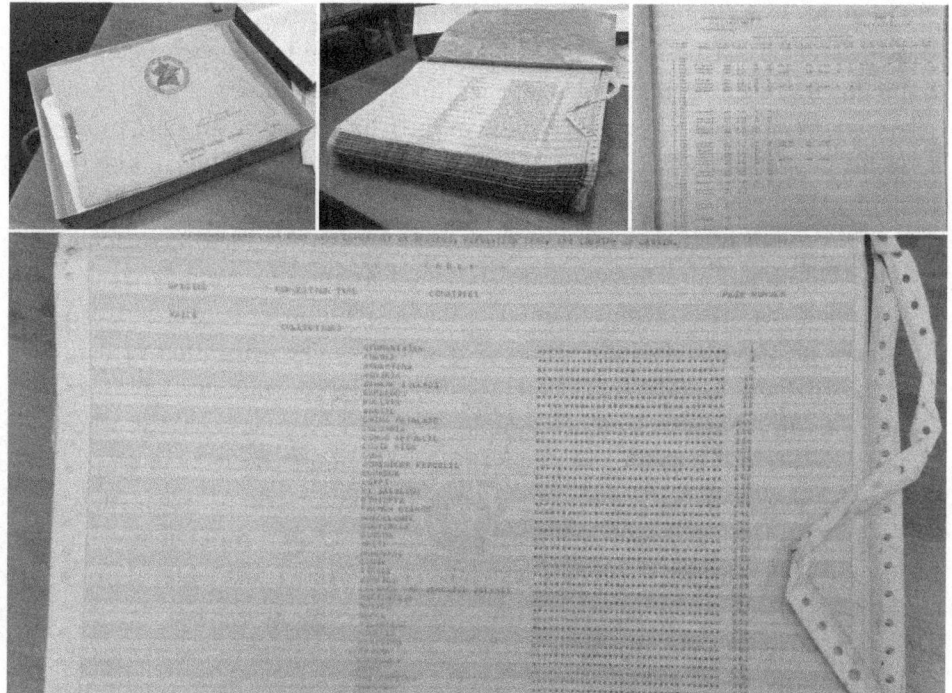

[26] At his office in North Carolina in the 1980s, Major Goodman relied on hardcopy printouts of the maize samples held in seed banks in the United States and abroad to monitor the collections and make requests. These hefty printouts from La Molina, Peru *(top left and right),* and CIMMYT *(top center and bottom)* have been preserved along with Goodman's papers, a reminder of how valuable he considered the data within them—even when the data at times consisted in little beyond a name and accession number. Photographs by author.

The timing of these conversations is revealing: they overlapped with the escalating negotiations over international management and control of plant genetic resources at FAO. While representatives in Rome in 1983 tussled for control of collections they touted as precious global resources and prized inputs into breeding programs, in less public venues scientists agreed that without more data these same collections were basically worthless.

EVALUATION

Many scientists agreed that evaluation could not be set aside for long, not if the entire National Plant Germplasm System or international conservation efforts were to be considered worth the time and expense they demanded.

For all his concerns over the physical security of collections, Goodman was acutely aware of this. In 1982 he assembled a proposal to discover the most promising breeding materials from within a diverse subset of the Latin American landrace collections already in the United States—namely, those in his possession at North Carolina State. The ultimate goal would be to develop these into lines that could be used by US breeders.[49]

The project, funded in part by a grant from Pioneer Hi-Bred, constituted what Goodman and his graduate student Fernando Castillo González described as a "first attempt to evaluate the broadest possible array of Latin American maize accessions." It was expedited by the utilization of data already in hand. Goodman had documented basic agronomic characteristics of his Latin American landrace samples whenever he grew them for other research. He and Castillo now used these data, which covered some 1,300 samples, to winnow out nearly 400 that they thought merited further study. Under Goodman's supervision Castillo began a multiyear project of growing out and evaluating these samples at sites in Florida and Texas, where shorter day lengths would enable easier cultivation of the tropical material.[50]

A far larger Latin American landrace-evaluation effort followed on the heels of this pilot project. In 1985 the USDA and Pioneer Hi-Bred announced a joint program for systematically evaluating the collections held in Latin American seed banks, which would be bankrolled by a $1.5 million donation from Pioneer. Describing seed banks as "pharmacies filled with wonder drugs that have no labels," William Brown, the impetus behind the Pioneer donation, reiterated the view that breeders were neglecting collections because they didn't know much about what was in them. Armed with more complete information, they could start using these materials to make better hybrid corn. He pointedly noted that Pioneer was supporting the effort only because government funding was not sufficient to meet this "critical need." The ongoing USDA-sponsored regeneration of endangered Latin American collections offered a useful illustration. Although it handled the very collections thought to hold rich possibilities for US breeders, it had not been allocated the resources required to assess these.[51]

The willingness of a private company—among the world's largest seed companies—to back a task generally agreed to fall within the purview of public institutions was evidence of Brown's continued influence at Pioneer. Although retired from day-to-day operations, he was still chair of the company board. Pioneer's donation to the USDA reflected his faith in the valuable qualities to be extracted from "exotic" corn. Although Brown often spoke at public events about epidemics and other risks that accompanied the genetic narrowing of hybrids, he was more concerned about reaching a plateau in corn breeding. "U.S. corn breeding is in the doldrums," he observed in 1987. "Breeders seem to spend just about all their time remixing the same old germplasm."[52] True innovation necessitated new materials. Brown also believed that private companies had an obligation to support the government seed banks and research programs that supplied the private sector with breeding materials. Brown and others at Pioneer hoped that the company's 1985 donation would spark similar donations from competitors and that this would in turn alert policy makers to the seed industry's concern over inadequate funding.[53] Although both seemed like long shots, Pioneer was in an exceptionally good position to make a risky investment. By the mid-1980s it controlled nearly 40 percent of the US hybrid corn seed market.[54]

Initial plans for the project incorporated yet another ambition, which Wilfredo Salhuana, a Pioneer Hi-Bred breeder and a chief architect of the developing project, described as a "change in the traditional viewpoint of germplasm collection agencies." The data that seed banks typically appended to samples in databases were those that accompanied collection. The plant characteristics included in these data tended to be basic morphological data such as height, flower color, seed size, and so on. The identification of further characteristics, including those relevant to performance in the field, was assumed to be the work of the collection user, not its maintainer. If this situation were reversed, such that the creation of agronomic data were an expected function of the seed bank, more breeders would draw on seed bank materials. This in turn would "maximize the usefullness of the collections and increase the world[']s genetic diversity [of] maize."[55]

The Pioneer-funded mass evaluation program therefore had multiple targets. Evaluating samples in the world's many corn collections and publicizing the resulting data could generate better corn varieties. This would be a service to farmers but, more crucially, to the seed industry. The project could also attract attention, and ideally funding, to evaluation and enhancement efforts across public and private institutions. Finally, it could encourage a reconceptualization of seed banks that would ultimately make their collections more user friendly and therefore more used. If all these were realized, the net result would be reduced vulnerability of national and international institutions, once their value could be observed rather than merely assumed or, worse, questioned.

The USDA-Pioneer evaluation effort planned to target the "major banks" in Mexico and Central and South America. It would ensure that both viable seed stocks and agronomic information generated through evaluation of the collections held by these banks would be made "available to plant breeders worldwide" through their deposition with the US National Plant Germplasm System.[56] Early lists of potential collaborators included the three state-run seed banks in Colombia, Mexico, and Peru already involved in regenerating seed in partnership with North Carolina State, plus seed banks in Argentina, Brazil, and Chile. As the project developed, it encompassed institutions in several more countries. The eventual list of participating Latin American scientists hailed from institutions in Argentina, Bolivia, Brazil, Colombia, Chile, Guatemala, Mexico, Paraguay, Peru, Uruguay, and Venezuela.[57]

This effort, dubbed the Latin American Maize Project (LAMP), got underway in 1987. Dozens of researchers representing Latin American partners and the USDA planted seeds from more than fourteen thousand seed bank accessions at seventy locations across North, Central, and South America. This was the first step in evaluation. The number was impressive, especially considering that each accession was represented by about a hundred plants in the field. Nonetheless, it represented only about half of the samples thought to be held across participating institutions.[58] In many cases the seeds listed in collection databases did not exist in the quantity or quality needed for evaluation—or did not exist at all. In addition not every sample could be coaxed into germination, let alone grow to be harvested. The eventual number of accessions evaluated was just over twelve thousand.[59]

Evaluations on the scale conceived for the Latin American Maize Project, which emphasized the creation of data useful to all participants, required a multisited, multistep process. A sample belonging to a particular region or country was first grown at a location roughly similar to that from which it had once been collected. There researchers conducted agronomic evaluations for characteristics such as height, days to flowering, number of ears, and yield. The best-performing 20 percent of these, determined for each country by the scientists in charge there, were grown again in the same country, this time in two locations. After another round of evaluation, these 3,000 or so best-performing lines were winnowed further, to the top 1 to 5 percent. The final 270 accessions—judged by local evaluation processes to be the most promising as future breeding materials—were then exchanged among countries for a further series of evaluations that would test their performance in these new locations.[60] The endpoint of these many evaluations and exchanges was a determination of the best candidates for further investigation by breeders in each country and the integration of this newly identified "elite germplasm" into breeding programs. Through this closely controlled data production, exotic accessions became elite breeding materials. As the Latin American Maize Project wound to conclusion in the mid-1990s, its organizers identified one of its key successes as having encouraged corn breeders to consider new breeding materials, in particular "superior foreign germplasm that was not looked at before." Seeds of corn landraces formerly "static in germplasm banks" had been mobilized, some for just a season and others potentially for the longer term.[61]

REGENERATION REDUX

The organizers of the Latin American Maize Project claimed one more achievement: a further cooperative international project centered solely on seed regeneration. In planning LAMP they had conducted a survey of Latin American seed banks. How many accessions did these banks have? How many seeds were available for each accession, and how viable were they?[62] It was this survey that had produced the sobering realization that only half of the nearly thirty thousand accessions on the books could be incorporated into the LAMP evaluation. That realization in turn motivated a broader Latin American seed regeneration effort along the lines of the North Carolina State–managed cooperative regeneration program, this time including many more countries and, crucially, CIMMYT.[63]

The Maize Germplasm Bank at CIMMYT had languished after the departure of its manager, Mario Gutiérrez Gutiérrez, in 1976. Throughout the 1970s and into the 1980s, the directors of the corn program at CIMMYT had shown little enthusiasm for the expansion and curation of the collection. They did not appoint a successor to Gutiérrez but instead relied on postdocs and staff members with many other responsibilities to keep up the bank. These arrangements did not allow for collections—especially the little-used landrace collections—to be regenerated at anywhere near the rate thought necessary to avoid losses. In 1980 Brown offered the assistance of Pioneer in essential maintenance activities, and the Pioneer employee Wilfredo Salhuana subsequently oversaw regeneration of about three hundred CIMMYT accessions per year between 1981 and 1987. Although portrayed as a rescue mission that would salvage endangered collections through renewal, restoration to CIMMYT, and duplication in the US National Seed Storage Laboratory, the arrangement also benefited Pioneer: the company kept seeds whenever an accession seemed potentially useful.[64]

The relationship between the international agricultural research center and the corporation was rocky, at least at the beginning. After two seasons of working with Pioneer, the associate director of CIMMYT's corn program complained about the company's difficulties in producing sufficient seeds to restock the collections. Pioneer had also approached CIMMYT with a proposed formal agreement that contained several "unacceptable clauses" concerning access to data on all the samples held in the seed bank and authority in choosing which seeds to regenerate. Even more infuriating was the circulation of stories in the US press that made CIMMYT out to be incompetent in the management of its collections. CIMMYT nonetheless needed the help. One of the chief complaints about the arrangement was that Pioneer was refusing to take on any more than three hundred accessions per year, when CIMMYT wanted them to accept one thousand.[65]

Attention to the Maize Germplasm Bank within CIMMYT shifted after the International Undertaking on Plant Genetic Resources was proposed at FAO in 1983. As battles over the ownership of seeds raged, CIMMYT repositioned itself as the logical central node for a "World Maize Network" dedicated to the conservation and free exchange of the world's extant corn diversity. The physical infrastructure of the bank was upgraded to create long-term storage capacity in 1984, thanks to an infusion of funds from the International Board for Plant Genetic Resources. The Consultative Group on International Agricultural Research delivered further funding that allowed for dedicated seed bank staff and much-needed computer capacity. Perhaps most significantly, the corn breeder Suketoshi Taba, formerly head of CIMMYT's Andean corn-breeding program, took over management of the bank.[66]

The quickly shifting sands at CIMMYT can be traced through its attitudes toward cooperation in seed bank activities. In 1982 the US Maize Crop Advisory Committee had failed to get CIMMYT involved in its initial seed regeneration program because the center had just entered a "'holding' phase on germplasm maintenance" as a result of budget cuts and subsequent relegation of the seed bank to a low priority.[67] By comparison, in 1985 efforts to enroll CIMMYT in what would become the Latin American Maize Project seemed to have stalled on account of renewed confidence at CIMMYT that the seed bank could look after itself.[68] The accounts of significant losses that emerged from the preliminary seed bank surveys of the Latin American Maize Project finally pulled CIMMYT into the expanding regeneration and evaluation activities. In fact, it offered an opportunity to regain a central place in international conservation of corn diversity by coordinating efforts to replenish stocks of endangered accessions from thirteen countries.[69]

This CIMMYT-led regeneration effort, like its predecessors, centered on the idea that a vast amount of irreplaceable genetic material housed in seed banks across Latin America was endangered by lapses in care. Reports from the thirteen participating countries submitted to a 1991 planning workshop estimated losses ranging from 5 to 100 percent for samples collected before the 1970s.[70] In Peru 575 of some 3,700 of these older samples had been lost, in Colombia 80 of 2100, in Mexico 500 out of about 9,400, and so on. The goal of the new program was to stem this tide by regenerating the "most endangered" of the surviving accessions. The seed bank managers estimated that there were, in total, a little over 10,000 of these. "Faced with the dire prospect of extinction for this maize diversity," CIMMYT staff and their many collaborators planned a "rescue mission."[71]

This effort was eventually funded by the US Agency for International Development, USDA, and CIMMYT and undertaken in the early 1990s. Although it had the renewal and safety duplication of endangered landrace collections as its core purpose, the organizers knew that this alone would not make collections, or the genetic diversity of corn, more secure. CIMMYT's director general characterized the situation bluntly. It would be difficult to defend spending money on the upkeep of maize collections without "more rapid progress" in using these to develop "new cultivars with high and stable yields."[72] The survival of seed banks depended on creating elite breeding lines from the materials in the bank. With this idea dominant the work of regeneration was accompanied by an emphasis on aggregating and entering data into a newly improved database system at CIMMYT and ensuring that all participants had access to it. Only with these elements in place could a seed bank collection be approached willingly, and effectively, by a breeder.[73]

In 1985, when he had just started the task of getting the long-neglected CIMMYT seed bank in order, Suketoshi Taba prioritized the problem of record keeping. The bank desperately needed a single exhaustive catalog, which necessitated gathering information, often incomplete, from scattered records and logs to first compile a full list of the bank's contents. Only then could this be populated with passport data and perhaps further documentation such as evaluation information. This in turn could be sent in theory to

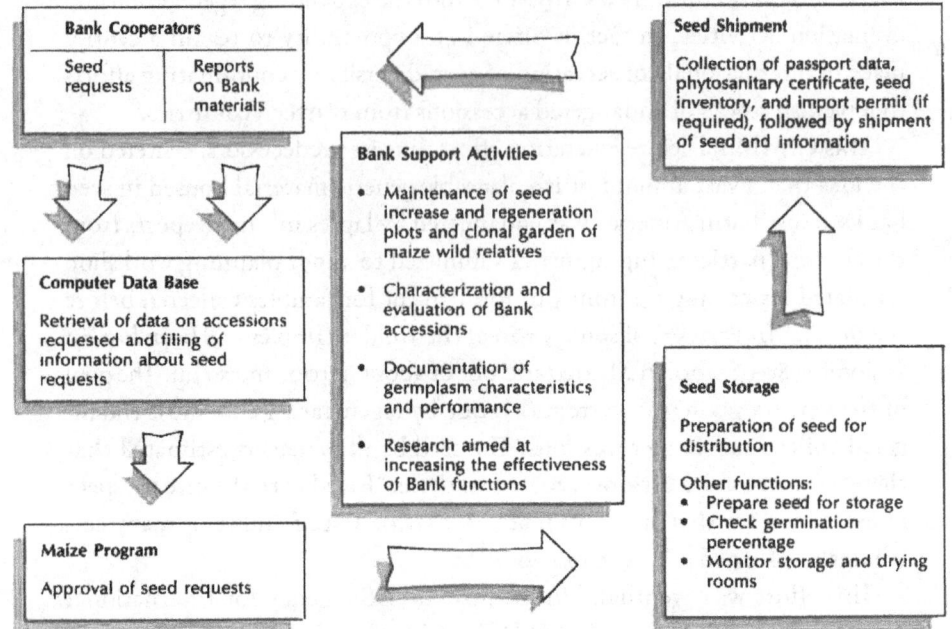

Bank Cooperators

Seed requests	Reports on Bank materials

Computer Data Base

Retrieval of data on accessions requested and filing of information about seed requests

Maize Program

Approval of seed requests

Bank Support Activities

- Maintenance of seed increase and regeneration plots and clonal garden of maize wild relatives

- Characterization and evaluation of Bank accessions

- Documentation of germplasm characteristics and performance

- Research aimed at increasing the effectiveness of Bank functions

Seed Shipment

Collection of passport data, phytosanitary certificate, seed inventory, and import permit (if required), followed by shipment of seed and information

Seed Storage

Preparation of seed for distribution

Other functions:
- Prepare seed for storage
- Check germination percentage
- Monitor storage and drying rooms

[27] A flowchart from 1986 representing the mechanics of seed requests from CIMMYT emphasizes the movement of information along with seeds: to and from cooperators, database managers, maize program managers, bank support staff, and shipment preparers. From CIMMYT, *Seed Conservation and Distribution.* Courtesy of CIMMYT.

any user who requested seeds. That individual would now not only be more likely to receive the correct seeds but also all the information generated in the past. Preparing the catalog required more secretarial assistance, a computer programmer and software development, updated practices of data gathering in the field, and even a reconnaissance mission to the Mexican national collection down the road at INIA to reacquire the collection information in its possession.[74]

Data management and circulation were laborious tasks. Yet, as Taba insisted, a "fundamental precondition" for an effective maize germplasm network of the kind envisioned was that all participants "possess a complete electronic database on materials they hold." CIMMYT would take the lead in aggregating, standardizing, and redisseminating this information.[75] Today most organism-centered research communities, such as geneticists who study fruit flies or plant scientists who work with the experimental favorite *Arabidopsis,* depend on community-wide electronic databases. Like the printed communications that preceded them, these databases simultaneously draw information from and discipline their users. They shape the language and standards of practice of geographically dispersed scientists and in the process create the very community they are purported to serve.[76] In proposing to centralize data on landraces at CIMMYT, Taba sought a similar transformation.

Of course, data were not the end point but rather an essential forerunner to use. As the CIMMYT-led regeneration effort got underway in the early 1990s and the earlier Latin American Maize Project claimed its first successes, Major Goodman viewed their ultimate payoff with some skepticism. US participants in the project had identified, through the exchange of accessions and evaluation data, collections that looked promising for their corn-breeding ambitions. These were, like all of the "elite" types identified by the Latin American Maize Project researchers, being officially accessioned into the US National Plant Germplasm System and assigned identifying plant inventory numbers, or PIs. "Unfortunately," Goodman surmised, "they will most likely suffer the same fate as most other PIs in that they will become part of our national germplasm retirement home." This stark assessment softened but conveyed the same sentiment as his other favorite metaphor for seed banks: "seed morgues."[77] Conservation advocates who agreed with Goodman had emphasized for more than a decade that if seed bank collections were not valued by breeders in the United States, both the corn in the field and that in the bank would continue to be vulnerable. Regeneration and evaluation, however urgent and essential, were still only initial hurdles to achieving the ultimate task of moving genetic material from seed bank accessions into breeding populations and eventually commercial varieties.

By the 1990s Goodman and his collaborators had two breeding projects underway that sought to adapt tropical exotic materials to the US Southeast. One relied on proven tropical hybrid varieties, obtained from commercial seed companies with operations in Latin America, and the second on landrace accessions. Both aimed at a clear demonstration that corn containing significant genetic material from tropical types could perform as well in North Carolina and elsewhere in the United States as Corn Belt Dents. The goals of this work extended beyond the creation of a few novel corn varieties, however. As one account summarized, "We feel that if the direct economic benefits of germplasm diversity cannot be conclusively demonstrated fairly soon, it is unlikely that the financial and political support for its conservation will endure."[78]

Although they may not have shared this concern about the fate of genetic conservation programs writ large, influential maize researchers, such as those on the Maize Crop Advisory Committee, did worry that the immense effort that had gone into the Latin American Maize Project would be wasted without some follow-up activity. The final component of the project had been directed toward "utilization" of the 268 lines that it identified as the most promising, agronomically, of the thousands of evaluated seed bank accessions. In the United States this meant getting the project's elite lines into the pool of materials used by commercial hybrid corn seed companies. Everyone knew that private companies would not invest significant resources in a long-term genetic enhancement program that offered only uncertain payoffs. Nor did the remaining public corn-breeding programs have anywhere near the capacity needed for such an effort. Arnel Hallauer at Iowa characterized ongoing public efforts to adapt tropical germplasm as being in a "sorry situation," thanks to competing demands on limited funds.[79]

The extent to which public corn-breeding programs were seen to be depleted, albeit to the detriment of the private industry more than the public good, was plainly evident in 1992. That December the corn- and sorghum-research subgroup of the American Seed Trade Association, the seed industry's powerful political lobbying organization, decided to press Congress for greater federal funding for corn breeding and research. The hybrid corn seed industry benefited significantly from public research, including the development of breeding populations, and industry representatives were keen to ensure they would not lose access to this windfall. When presented with the idea of a germplasm enhancement follow-up to the Latin American Maize Project, they embraced it as a vehicle for convincing Congress to direct more money toward corn breeding and research.[80]

Over the following year plans developed for the Germplasm Enhancement of Maize project (GEM), a joint public-private endeavor whose official mission was "to effectively increase the diversity of U.S. maize germplasm utilized by producers, global end-users and consumers."[81] It would aim to eventually provide the "maize industry," or private hybrid corn seed companies, with lines derived from "useful exotic germplasm"—that is, the elite landrace accessions identified by the Latin American Maize Project and some additional tropical hybrid lines. Armed with buy-ins from seventeen seed companies, each of which agreed to donate proprietary breeding lines, labor, land, and overhead costs to the effort, and the commitment of fourteen public breeders to serve as cooperators, the seed industry lobby succeeded in securing an annual commitment of $500,000 a year from Congress in 1995.[82]

Securing the support of private companies had been an essential, and tricky, component of the program. Their elite inbred lines, the best-performing corn varieties in the United States, were a prerequisite to creating competitive lines out of exotic materials through cross-pollination and subsequent selection. Those elite inbred lines also represented hybrid corn seed companies' greatest assets. In most cases knowledge of their genetic makeup was kept tightly controlled. Companies were reluctant to release their inbred lines into a publicly managed breeding program that would, presumably, make them part of the public domain again. This despite the fact that private companies had long depended on materials developed by public breeding programs and also would be on the receiving end of the various exotic materials gathered for GEM. The upshot was an agreement that made allowances for companies' proprietary interests in their inbred lines, such as a coding system that prevented competitors from knowing whose lines had been used in which crosses. Access to GEM lines was restricted to official cooperators, which had the dual effect of limiting the circulation of proprietary materials and encouraging companies to join the project rather than be left in the cold. That these conditions made the project appealing to private companies is evident in the ever-increasing number that signed on. By 2001 thirty-nine private companies had become involved.[83]

The protocol of GEM called for a subset of elite exotic accessions—just 51 of the 268 identified in the Latin American Maize Project—and 7 tropical hybrid lines donated by DeKalb Genetics to be divided among cooperating private companies. The private breeders would cross these exotic materials with proprietary inbred lines and return the seeds to the project coordinators, who would then redistribute them to other companies for further testing. This would lead to the selection of lines that would be further developed as breeding materials.[84] If achieved, the trajectory from regeneration to evaluation to enhancement would finally be complete—leaving only the daunting task of convincing companies to use and sell them.

It's hard not to be a little depressed by the idea that the Germplasm Enhancement of Maize project represented scientists' best bet for incorporating genetic diversity into an undiverse food system. As a result of the southern corn leaf blight, scientists had identified genetic vulnerability as a troubling byproduct of the predominant approach to plant breeding and industrial agriculture more generally. Some thought that crops and cultivation ought to change. In corn that instinct was channeled into a project that ultimately aimed to change very little. With big seed companies predominating production, their breeding lines had to be the target for introducing genetic novelty. But to be accepted even into the most basic of breeding projects, so-called exotic lines had to be confirmed as elite and to perform well according to the existing expectations of commercial seed companies. In other words, to change the genetic landscape, it was essential not to upset the agronomic or economic landscape.

Of course, there were additional motivations behind GEM. Broadening the genetic basis of corn grown in the United States might well have been an end in itself for some of its organizers. For others, including Goodman, success in this task would serve a larger agenda. In the process of adapting exotic corn varieties to US conditions, generating agronomic data about these, and demonstrating their potential value in breeding programs, researchers would stimulate greater interest in exotic materials typically considered impossible to work with. They would encourage more use of seed bank materials, which would in turn make it easier to sustain funding for these institutions in the future. Addressing the genetic vulnerability of US hybrid corn to environmental catastrophe would simultaneously address the vulnerability of seed banks and the genetic diversity within them to financial and political storms.

Private companies may not have been unduly concerned with some of the end goals touted for the project, which also included the "global sustainability of agricultural production, economic stability, and the nutrition and well being of society."[85] They nonetheless shared the view that their interests would be well served by bringing new genetic material into their breeding programs. If the government would subsidize a significant part of this effort, so much the better. To the extent that stimulating this subsidy was a larger aim of Pioneer Hi-Bred's initial investment in GEM, it reflected the company's interest in seeing the continuation of over a century of state investment in the importation of exotic plants for the enhancement of US agriculture and, ultimately, the benefit of private industry.

As it got underway, those organizing the GEM project celebrated its intention of moving genetic materials from the seed banks of Latin America into US hybrid corn. "Most of the exotic germplasm was being stored like museum pieces and helping no one," explained one GEM cooperator in

1997.[86] Now it was being used to generate a broader genetic base for US hybrid corn production through what Salhuana described as the "immigration of beneficial genes."[87] GEM celebrated the global circulation of genes, highlighting their movement from dormant seeds in Latin American storage centers into sprouting plants in US cornfields.

South of the US-Mexico border, however, conversations about the migration of corn genes had taken a distinctly different cast. By the 1990s, in the wake of the implementation of the North American Free Trade Agreement, US hybrid corn was poised to flood Mexican markets. This posed a threat to Mexican producers who couldn't compete, price-wise, with US grain, the production of which was heavily subsidized. As scientists, breeders, and peasant and Indigenous cultivators pointed out, it also posed a threat to native Mexican corn varieties. By the mid-1990s, thanks to the continued faith in biotechnological development that "classical" corn breeders so often decried, the first transgenically altered corn had hit the market. The importation of these varieties and their "unnatural" genes into Mexico risked their transfer through cross-pollination into criollo varieties. With this transfer, some contended, would come the degradation of Mexican corn and devastation of the communities that depended on it.

Southwest Traditional Crop Conservancy Garden and — Seed Bank

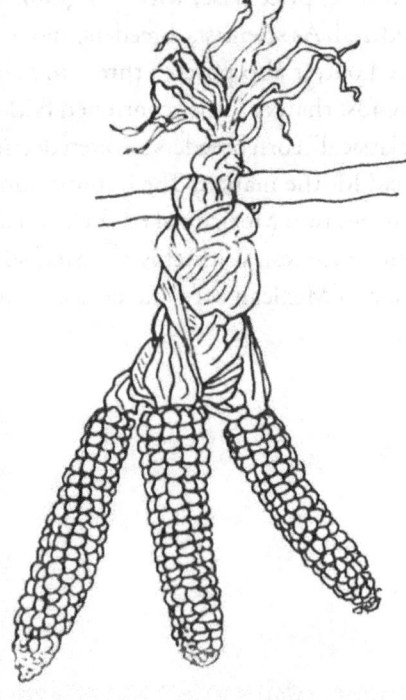

seed sources & resources 1982

meals for millions
P.O. Box 42622
tucson arizona   85733

[28] The Southwest Traditional Crop Conservancy Garden and Seed Bank, a forerunner to today's Native Seeds/SEARCH, sent its second annual seed list under this cover in 1982. The Native American man scattering seeds was a recurring motif, emphasizing that seeds circulated through the organization were intended for cultivation. Courtesy of Freedom from Hunger.

SEVEN

GROW

IN 2002 A GROUP OF INDIGENOUS revolutionaries in Chiapas, Mexico, accepted the donation of a freezer, which they hoped would help them address a pressing new concern: the "transgenic contamination" of their local maize. The previous year two US scientists had revealed the presence of a gene believed to have originated in transgenic (also called genetically modified, or GM) maize varieties from the United States in landraces collected in the Mexican state of Oaxaca.[1] This research, published in the journal *Nature*, proved controversial among scientists and especially industry observers in the United States. It also provoked unease for many Mexicans, whether dependent on maize as food, worried about the future of criollo varieties, or both.[2] In Chiapas it prompted leaders of the Zapatista Army of National Liberation to seek advice on the implications of these findings for their people, their cause, and their corn. A local ally linked them to an Arizonan ecologist, Martin Taylor, who advocated that they retain control over their maize by creating a seed bank. Taylor also bought them the freezer he considered essential to the bank's success.[3]

The first public action of the Zapatista Army in their fight for autonomy and self-governance within Mexico had taken place nearly a decade earlier. An armed uprising timed to coincide with the agreed start of the North American Free Trade Agreement (NAFTA) on 1 January 1994 signaled their rejection of Mexico's increasingly neoliberal policies. To the Zapatistas NAFTA represented the government's subordination of peasant interests to those of profit-seeking corporations. It constituted yet another instance of the exploitation of Indigenous peoples in Mexico by foreign powers.[4] When they became aware of the potential genetic mixing of transgenic maize with their local varieties in 2001, members of the Zapatista movement saw this as part

of the same larger, longer pattern. It was an unwanted and potentially dangerous imposition of foreign influence, in this case one that threatened the single
most important component of their autonomous food production.

In response to the specter of transgenic infiltration, the Zapatistas adopted
a familiar narrative of endangerment. They maintained that local farmers'
varieties were destined to be overwhelmed by industrial corn. So did other
Indigenous communities, peasant organizations, Mexican scientists, and
international activists who participated in a surging protest movement after
the 2001 *Nature* article. Declaring "sin maíz no hay país"—"without maize
there is no country"—new coalitions demanded that action be undertaken
to understand and prevent the flow of transgenic material into maize landraces of Mexico. Many insisted on policy changes and state sponsorship of
research as first lines of defense.[5]

Rather than wait for the government to respond, Zapatista rebels took
matters into their own hands. In developing locally managed preservation
measures, they shared in a trend toward *in situ,* or on-farm, conservation that
increasingly connected Indigenous peoples with scientists and activists in
many parts of the world at the turn of the twenty-first century. Initially
championed by critics and skeptics of the agricultural mainstream, by the
1990s efforts to promote local cultivation as a means of conserving genetic
diversity were themselves mainstream. Many factors converged to make this
shift possible. New research sustained more nuanced understandings of crop
diversity, especially in studies carried out by ecologically minded botanists,
agronomists, and social scientists. Legal frameworks like the 1992 Convention
on Biological Diversity fostered institutional support for community conservation, and novel proposals for in situ programs rendered them more politically and economically feasible. Even among seed bank advocates, the continued physical and political fragility of these facilities contributed to the
acknowledgement, sometimes grudging, that in situ conservation might usefully complement ex situ efforts.[6]

A revised narrative about the loss of crop diversity featured centrally in the
growing buzz around on-farm conservation. New field studies challenged the
anecdote-driven accounts of landrace extinction that had dominated for
nearly a century. In the late 1970s and 1980s, social scientists who fanned out
into farming communities to document changing cultivation patterns discovered that some farmers still planted local varieties even where breeders'
"improved" varieties were offered. This preference for, and persistence of,
landraces among small-scale producers belied their long-anticipated extinc-

tion. Although researchers rarely suggested that survival of these varieties in farm fields was guaranteed, they insisted that their elimination was not inevitable.[7]

Even with this counternarrative in play, extinction stories retained influence. The example of the seed bank in Chiapas points toward a different revision of the standard account of crop diversity, one that arguably did more to drive on-farm and community-based conservation forward. For the Zapatista rebels the swamping of farmers' varieties by industrial products was not chiefly a threat to posterity—that is, the livelihood of tomorrow's farmers—and not at all a problem for the future of breeders' work or state security. The loss of farmers' varieties endangered people of the present. The existential threat to local maize was an existential threat to its cultivators, and they pursued seed preservation as a component of self-preservation. Although this perspective did not, and does not, characterize all in situ projects, it represents the most important contribution of community-centered activities to the long history of crop conservation. In focusing on protecting culture and community, these activities reconfigured not only the means of conservation but also its ends.

SEEDING COMMUNITY

When Martin Taylor set off in his pickup truck for Chiapas to begin seed banking with the Zapatistas, his departure point was Tucson, Arizona, a city that might rightly claim to be the birthplace of contemporary community seed banking. Since 1983 Tucson has been home to Native Seeds/SEARCH, an organization dedicated to preserving the native crop plants of the arid lands of the southwestern United States and northwestern Mexico. The founders of Native Seeds/SEARCH saw a seed bank of those varieties as central to its aim of safeguarding them for future cultivators. Unlike its state-run counterparts in conservation—national or international research institutions staffed by professionals to serve the needs of plant breeders and other scientists—this grassroots collection would serve farmers, gardeners, and scientists alike. Put, very simply, it would be a "seed bank serving people."[8]

Native Seeds/SEARCH initially took shape through a nutrition-outreach program run by a local chapter of the hunger-relief charity Meals for Millions. In the late 1970s Meals for Millions staff in Tucson hoped to expand the options for healthy diets in nearby Tohono O'odham communities by distributing vegetable seeds for home cultivation. According to later accounts, their offers of broccoli and brussels sprouts underwhelmed the intended recipients, who instead kept "asking for the seeds their grandfathers grew." When Meals for Millions tried to respond to these requests, they discovered that sourcing seeds of traditional maize, bean, squash, and other vegetables was difficult, if not impossible. Since seed companies didn't sell them, Native American varieties could be had only from farmers still growing them. To increase access to these varieties, Meals for Millions staff would first have to find the farmers who had them, then purchase their seeds and multiply them for distribution.[9]

The seeds program that developed within Meals for Millions' Papago (Tohono O'odham) Nutrition Improvement Program soon blossomed into an independent initiative. In 1979 the staff moved from redistributing seeds grown by local farmers to producing seeds of about a dozen varieties in a "conservancy garden." The effort to grow and redistribute the seeds was managed by the ethnobotanist Gary Nabhan, his colleague Mahina Drees, and the regional director of Meals for Millions in the Southwest, Cynthia Anson. A year later, with a new space at the Tucson Botanical Garden and more varieties in hand, they distanced themselves from the nutrition program to pursue a broader agenda, as the Southwest Traditional Crop Conservancy Garden and Seed Bank.[10]

The goals of this project were wide-ranging. Its most immediate aim was to support Native American farmers of the arid Southwest. It sought to provide advice on seed storage, locate old varieties no longer available within a particular reservation or community, and make these and other varieties available to interested growers. The centrality of maize within long-held cultural and culinary traditions meant that sourcing and multiplying maize seeds was a central task, but there were plenty of other grain, dye, vegetable, and fiber crops that also merited attention. By helping farmers find and save seeds of desert-adapted crops like Hopi Blue corn, Wariho amaranth, Papago tepary beans, or Seminole Big Cheese squashes, the organization aspired to allow "traditional Southwestern communities ... the option of local food self-sufficiency."[11]

The conservancy garden and seed bank were central to this goal, places to renew and store the seeds that would flow back into farmers' fields. Meanwhile, these same fields and the harvests they produced were necessary to achieve a second objective, "to exemplify how regional seed banks can serve people, not just provide raw materials for esoteric genetic research." The critique accompanying this objective was explicit. Although professional plant collectors concerned about the loss of crop diversity often targeted "ethnic communities" and "their time-tried seeds," the resulting collections ended up in state-run repositories inaccessible to those same communities. The Southwest Traditional Crop Conservancy would be different.[12]

This perspective, which saw state-led conservation programs as fortresses accessible only to professional researchers, linked the seed savers in Tucson to another community-based seed organization. Seed Savers Exchange, founded as True Seed Exchange in 1975 by the Missouri homesteaders Kent and Diane Whealy, sought to preserve vegetable diversity by creating a network of gardeners and farmers with "vegetable seeds which have been passed down and improved for generations."[13] The Whealys were concerned that the increasing dominance of hybrids in seed catalogs and the dwindling number of gardeners committed to saving their own seeds spelled the extinction of earlier varieties. These might be prized for flavor, hardiness, or any number of qualities meaningful to a gardener but not a corporation. Kent Whealy first envisioned that these varieties could be preserved through a greater exchange of information and seeds. This was facilitated by the Seed Savers Exchange annual newsletter, a typescript compilation of all the seeds that members had on hand to share and those they wished to receive from other members. The ensuing swaps kept gardeners' favorite varieties alive and accessible and served, in Whealy's words, "as a supplement to government programs."[14]

Connections between the two organizations formed early and soon led to plans for a conference that would bring together other like-minded organizations and individuals. Whealy saw this as a way to extend the human connections essential to his model of conservation, creating a "network of networks that will accomplish much more."[15] The meeting would also create a space for advancing community seed banking, a concept already central to the Southwest Traditional Crop Conservancy but still only nascent within Seed Savers Exchange in 1980.

Held in October 1981 in Tucson, the Seed Banks Serving People workshop was, the local organizers declared, a first-ever gathering of people who wanted to save seeds for reasons other than plant breeding. The hundred-plus attendees conserved varieties "for their cultural and historic significance, their suitability to home gardens, their taste, texture and nutritive quality, and *for diversity's sake.*" They considered the conservation of crop diversity "an issue for them, not just for academicians and government administrators."[16] This implied that seed banking ought to be democratized, with all concerned citizens invited to decide on what to save and how to save it. The workshop itself went one step further, advocating that these individuals also become seed bankers. Rather than exhortations on the need for conservation, a perspective

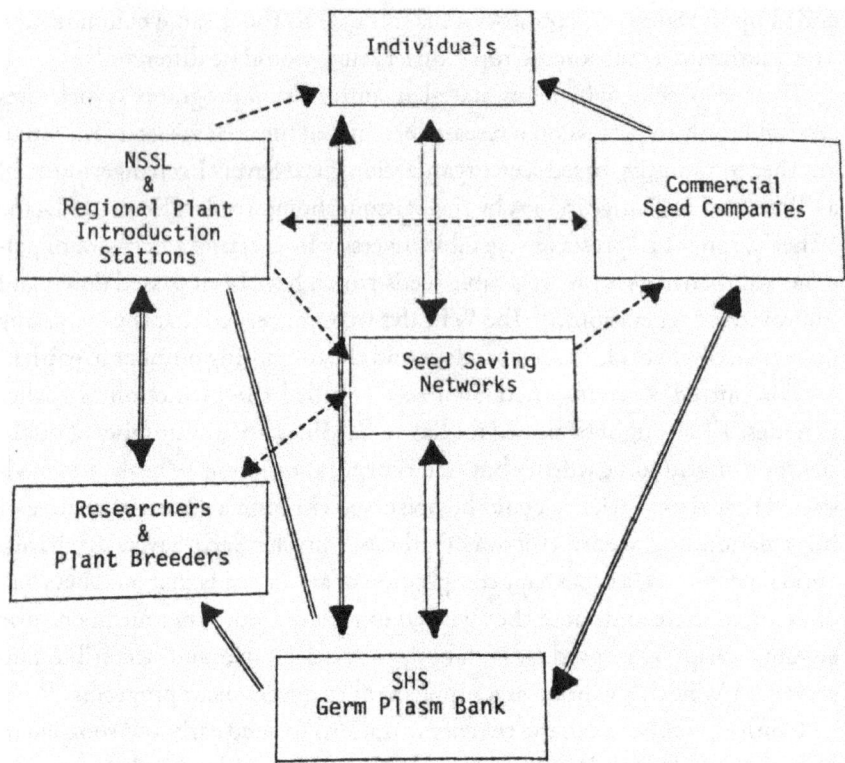

[29] A flowchart circulated by the not-for-profit Soil and Health Society as part of the 1981 Seed Banks Serving People workshop placed "Seed Saving Networks" at the center of seed conservation efforts. The society proposed a shared community seed bank (the "SHS Germ Plasm Bank," *at bottom*) that would serve seed savers' organizations, plant breeders, and seed companies and ultimately ensure that individuals would be able to grow treasured varieties. The concept of security through repeated known duplication was as integral to the conservation vision of grassroots organizations as it was to state-run seed banks. From Niethammer, *Seed Banks Serving People,* 64. Courtesy of Freedom from Hunger.

that participants presumably already shared, the presentations delivered practical advice on identifying conservation needs, documenting collections, safely storing seeds, increasing samples, cleaning and threshing seeds, running an exchange network, and more.

Many, if not most, attendees of Seed Banks Serving People were already experts in seed saving. Some were commercial suppliers of heirloom seeds or directors of seed and gardening organizations. Others were expert amateurs with extensive seed collections of their own. A handful were professional scientists. These included a few ethnobotanists, a teosinte researcher, a USDA bean collector, a seed physiologist from the US National Seed Storage

Laboratory, and the director of the US National Plant Germplasm System.[17] The activist Cary Fowler attended as cohost of the workshop and its keynote speaker. In 1981 he was still working with Pat Mooney and others on seed ownership and genetic conservation at the international level. Shortly after the workshop he would be part of the explosion of debate on these issues at the FAO in Rome. Occupying a very different place among crop conservationists in attendance, but no less passionate about the need to preserve genetic diversity, was William Brown of Pioneer Hi-Bred International. Former company president and now chair of the board, Brown represented the antithesis of small-scale and community-focused agriculture. He led the world's largest seed company—one that had made its mint in selling the same unsavable hybrid varieties that organizations like Seed Savers Exchange now decried. Yet with Brown at the helm, Pioneer invested in multiple approaches to crop conservation. In 1981 the company was the principal source of funding supporting the Seed Banks Serving People workshop.[18]

Participants declared the workshop transformative, an event that forged lasting connections among like-minded growers and seed savers.[19] Their interests in conserving diverse crop varieties had emerged from different experiences and political commitments. The Whealys shared the DIY ethos of the 1970s back-to-the-land movement. Meanwhile, Nabhan and his colleagues in Arizona were more closely linked with the aspirations of ethnobotany and agroecology, eager to show ecological underpinnings of Indigenous knowledge and influenced by a common assessment of Native American agricultural practices as inherently more environmental.[20] Others were connected to the older but now burgeoning US organic movement, which had for decades emphasized gardeners' decisions as socially and environmentally consequential acts.[21] No doubt most were cognizant of crop diversity as one issue among many pressing environmental concerns. None could have been ignorant of the profound industrial transformation in US agriculture that had unfolded in the preceding decades. Even those who were simply die-hard seed enthusiasts shared a view that their investments of time, energy, and money in saving and sowing certain kinds of seeds ramified well beyond their own farms and gardens.

EXTENDING THE MODEL

Strengthened by knowledge of their numbers, leaders in the community seed-saving movement forged ahead with new projects. Kent Whealy devoted energy to a next phase of Seed Savers Exchange by setting up an Heirloom Seed Bank.[22] Meanwhile, in Tucson the Southwest Traditional Crop Conservancy continued its expansion. To its existing projects of collecting, increasing, and distributing seeds of locally adapted crop varieties and maintaining its own seed bank and garden, the organization added seed bank building. Where it had initially advertised itself as a "model" bank, the organization was now "working with tribal personnel to develop regional seed banks and conservancy gardens" for the redistribution of "traditional seeds."[23]

Although the staff made some headway, their efforts to develop the central seed collection progressed far more rapidly than did their spread of seed banks.[24] Hemmed in by the boundaries of the regional Meals for Millions office, which stopped at the US-Mexico border even as the desert ecosystem stretched southward, and also by its narrow nutritional focus, Nabhan and Drees collaborated with two other enthusiastic seed collectors to create a new organization in December 1982. Together with Barney Burns, a graduate student in anthropology at the University of Arizona (and soon the husband of Drees), and Karen Reichhardt, a natural-resources specialist (and Nabhan's wife), they founded Native Seeds/SEARCH. This nonprofit was to be "devoted to the conservation and promotion of native, agriculturally valuable plants of the U.S. Southwest and northwest Mexico."[25] The quartet enthusiastically spearheaded new collecting missions and otherwise tracked down novel varieties. Their seed offerings rapidly expanded. The 47 varieties of 1983 leaped to 91 in 1984 and 216 in 1985.[26]

The seeds in turn dispersed farther afield. By the summer of 1984, after a year and a half in operation, Native Seeds/SEARCH had sold 1,500 seed packets at one dollar each and provided dozens of others to researchers. They had also distributed free seeds of varieties described as "formerly grown by these tribes" to farmers and programs associated with eight different Native communities. The latter achievement affirmed their sense both that "native seeds" were disappearing and that they were succeeding in restoring them.[27] The well-worn narrative of the fragmentation, assimilation, and disappearance of Indigenous Americans—and the capacity of outsiders to reverse undesirable outcomes of these changes—shaped and gave meaning to their efforts.

The centrality of maize to the Native cultures of the region and growers' continued use of many types in cooking and ceremonies contributed to an impressive assortment of maize varieties in the Native Seeds/SEARCH collection. The 1985 catalog listed sixty different kinds, most associated with a particular tribe, all characterized by color and kernel type (flour, flint, dent, sweet, or popcorn). Hopi Blue, Navajo Hominy, Papago 60 Day, Tarahumara Apachito—the aspiring cultivator could discover corn varieties suitable for a wide range of farm sites and culinary uses.[28] A few of these offerings derived from requests made of the USDA Plant Germplasm System. Mojave Flour Corn, for example, first listed by Native Seeds/SEARCH in 1987, traced back to a USDA accession collected in 1954 from the Fort Mojave Indian Reservation by Hugh Cutler as part of the collections he undertook for the Committee on Preservation of Indigenous Strains of Maize.[29]

Many more originated with farmers of the region, especially those connected to Native and Indigenous communities in the United States and Mexico. For example, Nabhan gathered a variety known locally as Onaveño from the farmer Casimiro Sánchez of San Ignacio, Magdalena, in Sonora, Mexico, who declared himself the only one still growing that particular type.[30] A sample of a sixty-day flour corn (so named for its ripening in sixty days) came from the farm of Patricia Smith, who knew it as huuñ and cultivated seeds that had been handed down through her family. As with Sánchez, the reported circumstances of Smith's donation dovetailed with the dominant narrative of disappearance. In the field notes Nabhan described her as potentially the only corn grower left on her reservation, the Ak-chin Papago Reservation (today the Ak-chin Indian Community).[31] Seeds of Onaveño and O'odham huuñ and hundreds of other collections went into the seed bank, which moved from an old refrigerator at Nabhan and Reichhardt's house to a chest freezer at the Tucson Botanical Garden, before landing in two freezers at the organization's new offices by 1991.[32]

Despite the rhetoric of loss, the collections of maize and many other crops only grew. The survival of these varieties in fields, and not just in the refrigerators and freezers of the Native Seeds/SEARCH seed bank, attested to their resilience and to that of the farmers who prized them. As the collections grew, so too did the distribution and cultivation of seeds on farms and in gardens. Many of these sites were not the Native communities targeted in the organization's early efforts. In 1988, after five years in operation, Native Seeds/SEARCH counted 1,700 members (at a cost of ten to twenty-five dollars per year but free to Native Americans); many were middle- or upper-class gardeners and presumably, like most of the board and staff of Native Seeds/SEARCH, White. The official roster included fewer than 125 Native American members.[33] Judging by sales totals, the distribution of seed packets numbered in the thousands; a tally for the first quarter of 1990 alone exceeded ten thousand. The same year seed donations were made to numerous Native American communities, in some cases reaching well beyond the region. According to the Native Seeds/SEARCH list, these included Pima (Akimel O'odham), Dakota, Lakota, Eastern Cherokee, Blackfeet, Havasupai, Tesuque, Tohono O'odham, Hia Ced O'odham, Yaqui, Navajo, Hopi, and Arapaho.[34]

The expanding annual distribution of seeds to farmers and gardeners, whether by sale or donation, was the yardstick by which Native Seeds/ SEARCH measured its successes. As its founders had emphasized, its ultimate goal for the "native American crop heritage" it sought to protect was not its conservation through off-site, or ex situ, storage in a seed bank but instead its continuation on farm, or in situ. This would ensure the ongoing evolution of varieties in particular ecological settings and secure the continuity of their "cultural-historical meanings." According to Nabhan and Reichhardt, these varieties contributed to the persistence of Indigenous agricultural systems, traditional diets, and cultural identities and, in part through these contributions, made possible better environmental stewardship and community sovereignty.[35]

Despite this ultimate goal of collecting to restore rather than simply retain, the idea of a fail-safe seed repository, embedded within the community, remained one of the most visible elements of Native Seeds/SEARCH's ongoing work. It was also an idea that the organization actively worked to spread. In 1991 staff members collaborated with the Zuni Archaeology Program and members of the Zuni tribe to establish a local seed bank. This would give Zuni farmers "direct access and control over traditional Zuni crops."[36] A year later Native Seeds/SEARCH hosted a workshop on seed banking for the Native American Farmers' Network. Participants from seven Native American groups learned techniques of collecting, cleaning, drying, storing, documenting, and testing seeds.[37]

A further contribution of Native Seeds/SEARCH to the promotion of community seed banking came through its advising on an influential how-to document published by the Rural Advancement Fund International (RAFI) in 1986. RAFI had formed in 1984 as an extension of the existing Rural Advancement Fund of the National Sharecroppers Fund, where Cary Fowler had been working since 1978. RAFI became for a time the institutional home of the group of activists who had made ownership and control of crop diversity an international concern, Fowler and Pat Mooney prominent among them.[38] As part of their continued campaigning, Mooney and Fowler

BUILDING THE BANK

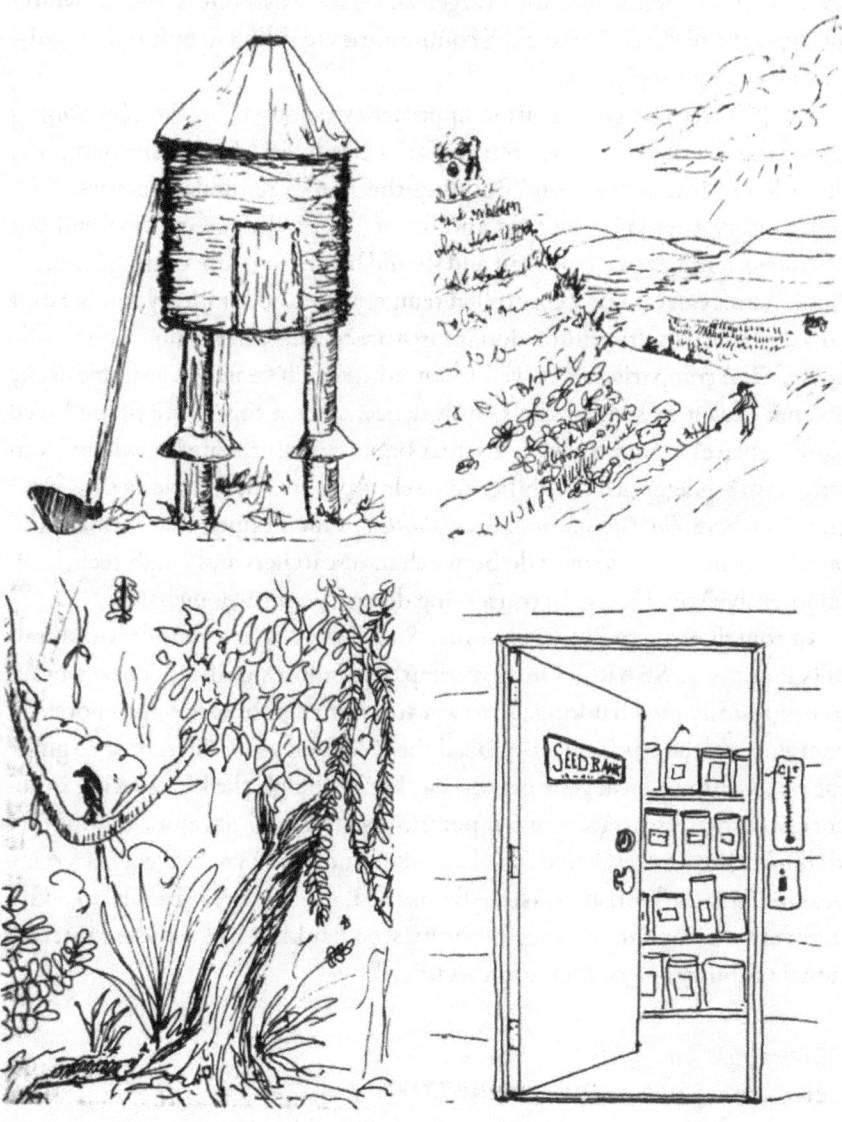

[30] Which is the best conservation facility? Although it provided instructions for setting up a seed bank *(lower right)*, the 1986 *Community Seed Bank Kit* reminded would-be crop conservators that the "best conservation" was carried out elsewhere. From Mooney and Fowler, *Community Seed Bank Kit*. Courtesy of ETC Group.

coauthored *The Community Seed Bank Kit* with input from Gary Nabhan and his colleague Kevin Dahl of Native Seeds/SEARCH. The kit comprised a series of short pamphlets instructing readers on community-level conservation and methods of seed banking; Nabhan and Dahl contributed chiefly to the how-to elements. The kit's target audience was NGOs, which would interpret the material for farming communities in different cultural, linguistic, and ecological contexts.[39]

The hierarchy of conservation approaches espoused in *The Community Seed Bank Kit* matched that seen at Native Seeds/SEARCH. The pamphlet that offered instructions on "Building the Bank" reminded readers, "The best conservation is in the field and forest." A further description laid out several options: "A seed bank can and should be a living, functioning farm. . . . A seed bank can also be a controlled temperature and humidity storage unit ranging from a multi-million dollar institute to a hole in the floor of a family home." The comparison of a well-financed research center to a simple cache was not meant to suggest their equivalence, since a community-based seed bank "replaces money with labour; text-books with intimate knowledge; and long-term storage with the ability to safely grow out the sample again." Here and elsewhere *The Community Seed Bank Kit* maintained that farmers and farming communities could do better than researchers and "high-tech, capital intensive" gene banks in conserving diversity for the long haul.[40]

In contributing to *The Community Seed Bank Kit,* the founders and staff of Native Seeds/SEARCH made perhaps their most significant contribution to community seed banking. Thirty years after its publication, promoters of community seed banking recognized the RAFI kit as the "first how-to guide for establishing a local gene or seed bank." Although the kit's specific influence is difficult to trace, contemporaneous efforts to promote community seed banking were effective. The late 1980s and 1990s saw projects targeting peasant farmers flourish in Asia, Africa, and Latin America. Thanks to NGO interventions, by the 1990s community seed banking had achieved international visibility as a conservation strategy.[41]

SURVIVAL

In the late 1960s and early 1970s, as scientists and administrators pushed for national and international responses to the loss of crop diversity, proposals for in situ conservation took a backseat to ex situ storage of seeds and other

plant materials in specialized facilities. To most scientists, attempting to preserve landraces and crop wild relatives on farms or reserves looked impractical and expensive when compared to placing seeds in cold storage. At a moment when the desirability of spreading breeders' varieties through economic development programs went almost unquestioned, in situ conservation also seemed to demand that some farmers be deprived the benefits of modern science. Most scientists agreed with the assessment of Otto Frankel, still a prominent international advocate for the conservation of crop genetic diversity, when he maintained that because in situ conservation required the "preservation of farming systems" amid rapid technological change, it was a "social impossibility" as well as an economic one.[42]

A few in situ advocates insisted that on-farm conservation was nonetheless essential, necessary for the continued evolution of crop species and to preserve genetic exchange with wild relatives. These concerns prompted the botanist Hugh Iltis to urge in 1974 that areas home to important crop diversity be placed off limits to development. He urged the preservation of designated areas through the "deliberate exclusion of agricultural 'improvements' as represented by the 'Green Revolution' and modern agricultural technology."[43]

Although they derived different conclusions, Iltis and Frankel based their ideas about conservation on a shared assumption about the inevitable trajectory of agricultural change. For decades plant explorers, botanists, breeders, and geneticists had warned that varieties created by professional breeders as a result of state, industry, or philanthropic investment would replace farmers' local varieties. It was not a question of whether it would happen but when. Why should it be otherwise, when in their view farmers stood only to gain from increased yields and therefore greater income? Evidence of this rampant genetic erosion nevertheless remained weak.[44] To the surprise of many scientists, when research focused on the loss of crop diversity finally got underway, some studies revealed the opposite of expectations. Even when offered "improved" varieties by state breeding programs or private companies, many farmers continued to grow landraces and to rely on farm-saved seeds and established networks of exchange.

Early investigations of farmers' decisions to keep landraces emerged from efforts to understand the shortcomings of aid efforts. In Mexico researchers associated with CIMMYT were tasked in 1967 with discovering how to deliver agricultural technologies to subsistence farmers who continued to use local varieties and older farming methods. Researchers' initial attempt to decide on the best "improved" maize type to promote revealed that none performed as well as farmers' own varieties.[45] A few years later two geographers returned to one village targeted by the CIMMYT effort to explain this "rejection of the Green Revolution." They reported that farmers considered the varieties promoted by researchers and extension agents the wrong size, unpalatable, susceptible to local pests, and unproductive of stalks needed for fodder production.[46] It was little wonder that many ignored state programs and relied instead on seeds that circulated locally.

Ethnobotanical and, increasingly, agroecological research produced similar accounts. When Efraím Hernández Xolocotzi invited his student Rafael Ortega Paczka to study maize samples from Chiapas in the early 1970s, Ortega Paczka confirmed the continued cultivation of the varieties that Hernández had found there three decades earlier—plus some newly introduced hybrids and other products of the state agricultural research system. The evidence suggested that crop diversity had actually increased thanks to development programs rather than decreased. As Ortega Paczka explained, farmers' varieties occupied different "niches" within the local ecology of agricultural production, being suitable for certain soils, climates, or cuisines. To the extent that these niches remained, so too would farmers' varieties. There was no reason to believe that the maize developed by professional breeders would rapidly vanquish the many diverse varieties that had preceded them, unless diverse cultural demands began to disappear.[47]

Thanks in part to insights like this, agroecology became an important resource for new models of in situ conservation in the 1980s. Agroecology, established earlier as the study of agricultural systems embedded in local ecologies, came to connote the study and application of ecologically and socially sensitive agriculture.[48] By the 1970s, especially with the influence of environmentalism, many agroecologists embraced this normative framing. They maintained that agricultural production informed by agroecological knowledge was better than "conventional" methods that relied on chemicals, irrigation, and other industrial interventions. Many also equated agroecological knowledge with peasant or Indigenous cultures. Thanks to more fine-grained investigations of Indigenous and subsistence cultivation practices, it was possible to see these as informed by intimate knowledge of land and climate, as well as more sensitive than conventional agriculture to ecological constraints and opportunities.[49] Although not all agroecologists were interested in this perspective, it found many ardent champions. In Mexico a growing community of agroecological researchers, many influenced by Hernández

Xolocotzi, pressed for the study of "traditional agricultural technology" across the agricultural curriculum. Knowledge of traditional agriculture would in turn point the way to methods and tools optimized for the resources, peoples, and national aspirations of Mexico.[50]

The increasingly well-documented importance of crop diversity in Indigenous farming systems led agroecologists to link the need for conservation of this diversity to the protection of Indigenous agricultural communities and practices. Here the aim of in situ conservation was not to conserve crop diversity as "plant genetic resources" accessible to future breeders. Instead, in situ conservation would target agricultural practices that sustained better social and environmental conditions for farming communities, which of necessity incorporated the maintenance of crop diversity. This represented a radical departure from the earlier understanding of in situ conservation as withholding desirable development from farmers. Instead, in situ conservation would benefit chiefly Indigenous or peasant communities, while incidentally serving wider regional, national, and international interests.

A model outlined by the University of California–Berkeley researchers Miguel Altieri and Laura Merrick demonstrated this different logic. Dismissing earlier proposals of conservation reserves as naive for failing to consider farmers' overriding concern with subsistence and survival, they proposed that development programs "incorporate native crops and wild/weedy relatives" into the "design of sustainable farming systems and appropriate technologies aimed at upgrading peasant food production for self-sufficiency." Peasant knowledge and local plants would underpin crop rotation, natural pest control, and soil conservation, helping farmers meet their subsistence goals and communities attain self-sufficiency while also making "important conservation gains."[51]

Agroecologists like Altieri and Merrick engaged with crop diversity as a substrate of socially and ecologically sound agriculture. A slightly different view of in situ conservation developed among social scientists whose chief  interest was crop diversity itself. Studies such as Ortega Paczka's analysis of maize in Chiapas confounded expectations of diversity loss and prompted closer analyses of how selection and exchange of varieties worked among peasant agriculturists. The anthropologist Stephen Brush produced an influential reassessment of crop diversity loss from just such circumstances. As a graduate student in the early 1970s, he had studied subsistence agriculture in Andean Peru. His experiences of potato diversity among the region's farmers led him to explore whether a new research institute, the International Potato Center, might diminish this diversity through the distribution of new varieties. His subsequent research revealed the opposite: diversity continued to flourish in Peruvian potato fields.[52] A follow-up study of maize farmers in Mexico produced similar findings. Farmers thought local varieties and recently introduced varieties each had distinct advantages and cultivated both kinds. They were knowledgeable and choosy about crop diversity, not blindly adherent to tradition, as often accused.[53]

These studies and continuing research led Brush to propose new roles for both peasant agricultural practices and the social sciences in crop conservation. The persistence of farmers' varieties in potato fields in Peru, cornfields in Mexico, and rice paddies in Thailand indicated that the Green Revolution had not produced the wholesale loss of genetic diversity so often claimed. Farmers in centers of crop diversity maintained a range of varieties, as a matter of preference, even where development programs had successfully introduced breeders' varieties. As Brush concluded, "Modernizing farmers also maintain diversity."[54] It followed that in situ advocates needn't stick to the idea of reserves protected from development (the "zoo mentality") but could instead take advantage of farmers' knowledge and dynamism to suggest conservation-friendly development. What was needed, besides supportive institutions and policies, was more information about the practices of peasant farmers. Brush suggested that "social science units join genetic ones in a common research effort" that would introduce new disciplinary perspectives into crop conservation—namely, views from anthropology, geography, and rural sociology.[55]

This call for change notwithstanding, Brush made a case for in situ conservation as, first, possible within the existing vision of agricultural development and, second, essential to shore up the shaky foundations of ex situ conservation. He further argued that it ought to be administered, at least in part, through the institutions that already had experience in studying and managing crop diversity. These included the international agricultural research centers of CGIAR, such as CIMMYT, IRRI, the International Potato Center in Peru, and others, supplemented by national agencies. Brush had collaborated with scientists at CIMMYT and the International Potato Center and saw these as potential homes for new in situ programs based around improved knowledge of crop evolution. They were also undeniable behemoths in agricultural development, institutions whose influence could not be written off and might therefore more usefully be harnessed, if possible.

This rosy view of the CGIAR centers provoked dismay among other advocates of on-farm conservation efforts. Responding to Brush's proposals, Miguel Altieri speculated that the agronomists and geneticists of the proposed interdisciplinary teams would not be swayed by their ecologist and sociologist colleagues and would continue to "discard farmers' knowledge, because they will think that it is primitive and the product of ignorance." Only when these farmers were recognized as the basis and beneficiaries of in situ efforts would conservation "protect sovereignty, ensure food self-sufficiency, and develop a sustainable agriculture."[56]

Despite their disagreement, the exchange between Brush and Altieri highlighted one thing clearly. Fostering in situ conservation of crop diversity by sustaining farmers' traditional agricultural practices had transitioned from an underdeveloped and often-dismissed concept to a persuasive proposal grounded in empirical research. The mechanisms through which this on-farm conservation could and should be achieved remained contested, but the notion that peasant farmers would be essential partners was increasingly uncontroversial.

FARMERS FIRST

As conservationists formulated new in situ models for protecting crop diversity, they drew inspiration from a nascent participatory paradigm in agricultural research. Social scientists tasked with addressing the shortcomings of the Green Revolution had pioneered a model in which researchers asked farmers their preferences among experiments, enlisted them in the management of test plots, and invited them to evaluate the results. Among conservationists this approach was embraced for valuing both farmers' and professional scientists' knowledge. Yet it also raised difficult-to-answer questions, from how to weigh competing demands of scientists and farmers to whether and how to extract labor from impoverished growers.

Social sciences were not a strength of the international agricultural centers in the early 1970s nor, for that matter, of many national equivalents. When critics first took aim at the Green Revolution, they had homed in on the inappropriateness of its preferred varieties for the poorest and most vulnerable of the world's farmers. Too often these required other inputs, like fertilizers or irrigation, and credit to secure loans. Many researchers insisted that creating interventions beneficial to peasant farmers would require greater engagement with those farmers, to understand better their social and economic constraints and to devise tools and programs that addressed their actual needs rather than their imagined ones.[57] The Rockefeller Foundation attempted to jumpstart this kind of research within the still-growing CGIAR network in 1974 by offering postdoctoral fellowships at the international agricultural research centers to social scientists. The influence of these researchers within and beyond CGIAR was visible in the flourishing of "farmer-centered" approaches to agricultural development from the mid-1970s onward.[58] These went by different names, such as Farming Systems Research, Farmer-Back-to-Farmer, and Farmer First, but shared in common social science methods and the desire to incorporate farmers' perceptions into research strategies and technology development.[59]

At CIMMYT one sign of the shifting terrain was a manual instructing agricultural researchers in the development of technologies suited to farmers' needs. Prepared by members of the CIMMYT Economics Program, the booklet *Planning Technologies Appropriate to Farmers* started from the premise that "effective research on agricultural technology starts and finishes with the farmer." It provided guidance on how to quickly and affordably gather information about the social and economic circumstances of a farming community and how to integrate this into research that could be tailored to the problems and opportunities actually experienced in that community. Such surveys, though important, were only the beginning. Many social scientists wanted to go further, inviting farmers into research as experimenters or planners. *Planning Technologies* therefore also described possibilities for these "on-farm experiments."[60]

On-farm experiments that flourished in the 1980s and 1990s drew inspiration from earlier anthropological work that characterized peasant farmers as practiced experimenters and inveterate innovators.[61] By casting farmers' annual selection of seeds as a deliberate process of adapting available diversity and their piecemeal adoption and modification of crops, tools, and techniques as an experimental enterprise, social scientists placed peasant farmers and professional scientists at two ends of a continuum of knowledgeable investigators. They differed in the type of knowledge they possessed rather than the amount.[62] This perspective also created room for farmers' expertise to be included in the making of agricultural innovations. Scientists took their experiments into farmers' fields and invited them to witness or help conduct trials of cultivation methods or crop varieties. Others invited farmers to help design experiments, soliciting their views on questions to resolve or crop varieties to test. In yet another form of collaboration, technical experts shared their knowledge and methods or created forums for farmer exchange and then left farmers to take action if, when, and how they chose.[63]

A recurring critique of breeders' high-yielding varieties of commodity crops as unsuitable for farmers working on the ecological or economic margin made varietal introductions promising targets for these participatory investigations. Researchers reasoned that if professionals had consistently failed to target the crop traits and types that would enhance poor farmers' livelihoods, perhaps those same farmers should be invited into crop-breeding programs as decision makers. They could advise professional breeders on traits and types of interest at an early stage, conduct trials of varieties in development on their farms, or accept instruction in selective breeding to carry this out on their own. Researchers tested these ideas among peasant farmers growing beans in Rwanda, potatoes in Peru, rice and sweet potatoes in the Philippines, and millet in northwestern India, among others.[64]

In Mexico participatory breeding programs that took shape in the 1990s focused on maize. Decades of development-driven crop-breeding programs had failed, notoriously, to generate varieties useful to, and used by, most maize farmers. In one of its earliest efforts, Project Sierra de Santa Marta (Proyecto Sierra de Santa Marta), an NGO founded in 1990 to foster participatory approaches to resource conservation in biodiversity-rich southern Veracruz, Mexico, tested the idea that farmers could help breeders do better. A product of collaboration between Mexican and Canadian researchers, Project Sierra de Santa Marta launched its first initiatives only after researchers had completed two years of multidisciplinary investigations of the local communities. Three biologists, two anthropologists, a geographer, an economist, an agronomist, a sociologist, an information scientist, and several students fanned out into the region in teams of three, conducting interviews with the Nahua and Zoque-Popoluca residents and creating maps of the region and charts of its resources. This experience laid the groundwork for a next set of activities, which included experimental programs on green manures, soil conservation, aquaculture, vanilla cultivation, wild-plant collection, ecotourism, and more, studies whose aims were decided in discussion with local farmers. It also led to tests of maize varieties and not surprisingly so, given an overriding concern of farmers in the region with regaining self-sufficiency in food production.[65]

Preliminary assessments identified a number of concerns in maize production among farmers in the region, including problems of stalks toppling over

[31] A researcher associated with Project MILPA (1990s) interviews a Mexican farmer. The project aimed to better integrate farmer knowledge and decision making into researcher-led programs of breeding and conservation. Photo by Eric Van Dusen; used with permission.

in high winds, lack of early maturing varieties, and the unsuitability of introduced breeders' varieties. Hoping to address these concerns, Project Sierra de Santa Marta partnered with CIMMYT to discover or develop varieties that would perform well in local conditions, relying on collaborations with local farmers. Organizers planned two experiments, which they hoped to repeat in four communities. The first consisted in a trial of about twenty different maize varieties, introduced and local, exposed to the same conditions in a single field, while the second saw one or two varieties tested in the fields of different farmers. A final hoped-for intervention was to train farmers to improve varieties on their own, using techniques of mass selection in the field. This aspect of the project aimed to counterbalance the possible loss of farmers' varieties that would result from new introductions, improving their productivity and, consequently, farmers' interest in cultivating them—generating in situ conservation.[66]

Obstacles to these plans unfolded one after another. A rumor that farmers had to pay to take part discouraged participation. Where farmers did sign on, many rejected the idea of placing the comparative trial of twenty varieties in a single participant's plot and insisted on their wide dispersal. Some took instructions on how to plant the experimental plot, then decided in the field that the plan involved too much labor and disregarded it. Depending on the volunteer labor of subsistence farmers proved untenable on the whole. In one extreme example of this, a young man named Andrés Gutiérrez took responsibility for the variety trial in the village of Mazumiapan Chico, only to be recruited into work with a government reforestation program shortly thereafter. The new job consumed most of his time. His mother then developed a hemorrhage that required expensive medical attention. His wife had a baby. These additional expenses forced him to look for paid work, which he found with the military at a barracks two and a half hours from his village. After two months of military training, he was sent a few hundred miles farther, to Chiapas, where the Zapatista uprising had reached a peak of intensity. A report of the project bluntly declared, "Andrés exchanged the machete and hoe for a submachine gun." It went without saying that his experimental plot never grew.[67]

Researchers' efforts to train farmers in the basics of breeding on their own farms also encountered successive obstacles. A key objective was to encourage farmers to select seeds for the next planting from healthy plants growing in the field rather than from a pile of harvested ears. Unfortunately, this neglected the role of women in choosing seeds from harvested ears as they processed these for family meals. Even where gendered work patterns were not a confounding factor, the basic physical labor of marking healthy, good-looking plants in the field for later harvest—the crucial component of the on-farm breeding procedure—proved too burdensome for most farmers. Their fields were often difficult-to-navigate slopes far from their homes. In some cases farmers' labor was absorbed with their cash crop, coffee, at precisely the times that the maize needed attention. Of the one hundred farmers who received seeds and training in 1994, sixteen still expressed some interest in 1996. Unfortunately, none were keeping up the selection process with their own local varieties—those that researchers had been most concerned with improving.[68]

On-the-ground realities challenged the lofty ideals of participatory research. The experiences of Andrés Gutiérrez pointedly captured how interest in the future possibilities of maize cultivation could not take precedence over present demands. Gutiérrez's circumstances, particularly with respect to the limited opportunities available to him and the necessity of his traveling far from the community to find work, were the product of decades of national policies that favored urban development over rural sustenance in Mexico. Given this, his deployment to Chiapas—to help suppress a peasant uprising that resisted the intensification of this inequality through the state's increasingly neoliberal policies—was a grimly incongruous outcome. The neglect of rural, and especially Indigenous, citizens by the Mexican state both inspired organizations like Project Sierra de Santa Marta and also limited the extent of their impact.

CORN AND NAFTA

Project Sierra de Santa Marta could hardly escape the confounding effects of shifting state policy on its earliest projects, which it launched just as the North American Free Trade Agreement between Canada, Mexico, and the United States came into effect. Signed in 1992 and implemented in 1994, NAFTA aimed to create a single North American market by eliminating protectionist policies like tariffs and import duties. In 1993, when the researchers of Project Sierra de Santa Marta were just developing their participatory maize projects, the government was paving the way for NAFTA by offering small subsidies to peasant farmers to offset their impending loss of income, a signal of changes to come. Subsequent years saw dramatic shifts in the conditions for maize production in Mexico, which affected campesinos across the country.

Experts diverged wildly in their predictions of NAFTA's effects in the years leading up to its implementation. Advocates argued that NAFTA would increase trade, create jobs, and reduce the cost of food and manufactured products. Critics worried about myriad consequences of this changed economic landscape, from the loss of blue-collar jobs in the United States to increased pollution in Mexico. The pros and cons for agricultural production loomed particularly large in debates over the desirability of free trade. US grain producers, with the advantages of a temperate climate, state-funded research, and access to credit, insurance, the latest technologies, and extravagant subsidies, outproduced Mexican farmers by huge margins. The US government was desperate to unload cheap grain into Mexican markets. In Mexico officials promoting NAFTA touted the anticipated flood of US grain as a boon for consumers who would see food prices fall, but many predicted the same influx would deliver a fatal blow to Mexico's poorest farmers. They would lose already-meager state protection and, with it, the ability to earn money selling their produce in Mexico. They would be forced to abandon their land in hopes of gaining employment elsewhere. The shift would therefore also end aspirations of achieving national food self-sufficiency.[69]

In the 1970s and early 1980s, the Mexican government had at times supported peasant cultivation of maize as a means of simultaneously providing welfare and boosting national productivity. As it adopted a free-trade outlook, the government had increasingly written off these producers as inefficient. Now with NAFTA this view finally dominated. If corn could be produced more cheaply in the United States, it ought to be produced there. Once this cheap corn was freely traded in Mexico, the price of corn to consumers would fall. Tortillas would be cheaper, and all Mexicans would benefit. Never mind that this assessment was complicated by US emphasis on producing corn for animal feed rather than human diets, a facet many pro-NAFTA accounts chose to ignore. Increased foreign investment in Mexico would create jobs that would in turn absorb the labor of peasants who could no longer survive by farming.[70] Even if these projections proved correct (which they did not), this rosy outlook on free trade did not ask whether rural Mexicans wanted to leave their land and their communities. The Zapatista Army of National Liberation's forceful rejection of NAFTA answered this unasked question. They did not want, and would not allow, their interests to be subordinated—not to the rich and powerful in Mexico, not to foreign powers, not to transnational corporations.[71]

Many Mexican voices lambasted the government for entering into a treaty likely to decimate the countryside. Among those invested in the outcomes of NAFTA on peasant farmers were scientists concerned about the conservation of maize diversity. Early environmental critiques of NAFTA cited the extinction of Mexican landraces as one among many potential negative outcomes of the treaty. The availability of cheap, homogeneous US yellow corn would eliminate the production of diverse Mexican varieties, a process that would be exacerbated by peasants abandoning rural areas to find paid labor. "If campesinos must move to the city to survive, the micro-habitats which have been sustained by their labor will be lost, and with them the maize varieties adapted to them," claimed one environmental economist, assessing the likely trajectory of post-NAFTA Mexico.[72]

Technological change in plant breeding heightened these baseline concerns. In 1994, the same year that NAFTA went into effect, the biotech company Calgene launched the Flavr Savr Tomato. This first attempt to market a transgenic crop to consumers failed, but it signaled the long-projected arrival of biotechnology in agriculture in the form of GM products. US regulatory agencies paved the way for the commercial sale of transgenic corn a year later, with their approval of varieties boasting two different engineered advantages.[73] This approval, coinciding with increased importation of corn as a result of NAFTA, generated additional concerns about the environmental consequences of free trade in Mexico.

One of the earliest adverse scenarios raised by anti-GM activists about the circulation of transgenic organisms outside the laboratory had been the possibility that their engineered genes might be transferred to other organisms, creating new weeds, pests, and diseases and other ecological disruption.[74] In the case of crops, an additional worry was the unwanted movement of novel genetic material from transgenic varieties into all varieties and potentially into wild relatives as well. Observers differed in their characterization of this process, some referring to transgene flow or transgenic introgression, where others saw genetic contamination or germplasm pollution. Although GM proponents tended to dismiss this scenario as improbable or harmless, or both, the possibility that transgenic material could spread outward from GM corn varieties was not easily denied. *Zea mays* is a prolific outcrosser, able to throw its pollen far afield and to be fertilized by any other kind of corn. It was also known to hybridize with its ancestor teosinte. It seemed eminently likely that if transgenic maize were grown in Mexico, its genetic novelties would make their way into other kinds of maize and into teosinte too.[75]

Concerns about these and other effects of transgenic maize haltingly influenced Mexico's regulatory policies. When the government received its first request for a field trial of a transgenic tomato in 1988, the Ministry of Agriculture initially approved it without special consideration. Subsequent years saw the approval of multiple field trials subject to containment procedures, the creation of a National Agricultural Biosafety Committee (Comité Nacional de Bioseguridad Agrícola), and ultimately a formal protocol to regulate transgenic products.[76] Those same years also saw increasing agitation against GM technologies outside Mexico, from political negotiations on the possibility of restricting these technologies through the international Convention on Biological Diversity to high-profile protests by Greenpeace against GM crops in the United States and Europe.[77] Mexican regulators approved laboratory and field trials of almost three dozen transgenic maize varieties, most requested by CIMMYT and three seed goliaths—Asgrow, Monsanto, and Pioneer Hi-Bred—before the government placed a de facto moratorium on them in 1998.[78] The ostensible reason for the moratorium was that the tested traits were unlikely to benefit Mexico, but mounting pressures from civil and scientific circles also played a part. In hopes of

[32] Beans grow up drying cornstalks in this field, a strategy for intercropping often associated with milpa cultivation in Mexico and across Mesoamerica. A group of farmers and/or researchers associated with Project MILPA
(1990s) is visible in the distance, perhaps assessing together the conservation possibilities of this plot. Photo by
Eric Van Dusen; used with permission.

addressing these anxieties, the government instituted a new Interagency
Commission on the Biosafety of Genetically Modified Organisms (Comisión
Intersecretarial de Bioseguridad de los Organismos Genéticamente
Modificados) in 1999.[79]

The projected effects of NAFTA and transgenic maize shaped conservation efforts as much as they did government policies. In 1995 US and Mexican
researchers launched a collaborative effort, Project MILPA (McKnight
Integrated Landrace Preservation Activity), which aimed to increase agricultural productivity and simultaneously promote in situ conservation. Placing
traditional Mexican crop associations and cultivation methods—often collectively referred to as the milpa system—at the center of their activities,
Project MILPA focused on diversity in maize, beans, squash, and leafy vegetables known as quelites.[80] The project was motivated by the demographic
and socioeconomic effects of government policies, such as the exodus of
farmers from the countryside in search of work and the increased pressure to
produce commercial crops rather than local varieties, which would lead to
the loss of diversity.[81] Another important inspiration was the desire to understand the exchange of genetic material between landraces and wild relatives.

Although pitched initially as a breeding strategy, in which the possible beneficial effects of this mixing would be explored, studies of this gene flow were quickly understood to deliver newly pressing information, answering questions about the consequences of introducing transgenic maize into Mexico.[82]

A substantial part of Project MILPA's funding came from the agroecology-inspired Collaborative Crop Research Program of the McKnight Foundation.[83] Although the project's on-farm trials and participatory breeding components looked a lot like those of the earlier Project Sierra de Santa Marta, there were nonetheless important differences between the two efforts. Project Sierra de Santa Marta incorporated the improvement of local varieties to offset the abandonment of these likely to be sparked by the project's introduction of new varieties. By contrast, in Project MILPA, whose contributing researchers included some of the most prominent US and Mexican experts in maize diversity, fostering the cultivation of landraces and crop wild relatives was a central aim. Its various subprojects sought better collection, knowledge, use, and conservation of these plants.[84]

With NAFTA-generated transformation of the countryside underway, and transgenic maize poised to assert its genetic influence over all other kinds, the mid-1990s were an inauspicious time for in situ conservation of Mexican maize diversity based around collaborations of scientists and farmers to finally take hold. As the anthropologist Elizabeth Fitting has shown, the "neoliberal corn regime" undermined the very fundamentals of small-scale maize farming. Farmers had to spend more of their time away from their farms. Often, their journeys were not to urban centers, where the employment promised by free-trade advocates had never materialized, but over the border as undocumented workers.[85] Even when farmers were able to stay put, their typical practices of seed exchange and experimentation entailed the likely planting of transgenic varieties imported from the United States as grain and subsequent genetic entanglement of GM corn with native Mexican maize varieties.[86]

Under these circumstances, ex situ conservation, rather than in situ, ought to have seemed more essential than ever before. Yet on-farm conservation such as that espoused in Project MILPA retained its new prominence. A tally of in situ conservation programs across Mexico in 2000 identified at least eleven, six of which involved participatory breeding work with farmers.[87] Depending on smallholder farmers to develop and preserve the varieties most useful to their own production—even as the conditions of possibility for this diminished—was of course consistent with the neoliberal corn regime. With reductions in government spending and the abdication of decision making to capital, who else but farmers could be asked to carry this burden?

SELF-PRESERVATION

Scientists were not the only ones in Mexico to champion on-farm conservation. In situ advocates had introduced the idea that saving crop diversity required saving peasant farmers. Meanwhile, peasant farmers were becoming more vocal about defending themselves. In 1993 farmers' organizations from around the world united to form La Via Campesina (The Peasants' Way), an international movement to sustain small-scale and Indigenous producers. Among its foremost concerns was defending the right of farmers to grow and share their own seeds, which La Via Campesina insisted was essential to the survival of rural communities. In Mexico protests against transgenic maize espoused this view as well and, in so doing, reversed the dominant logic of in situ conservation. Rather than protect people to save seeds, seeds would be protected to preserve people.

A turning point came in 2001, with publication of the *Nature* study that documented the biological effects of importing GM maize into Mexico. Its title claimed boldly, "Transgenic DNA Introgressed into Traditional Maize Landraces in Oaxaca, Mexico." Introgression, often used to indicate the movement of genetic material from one species into a second, here described the movement of genetic material from one kind of *Zea mays* to another. The study had detected the presence of "transgenic DNA constructs," presumably from imported transgenic maize, in criollo varieties growing in the region of Mexico where maize itself was thought to have originated.[88] This discovery of what the anthropologist Abby Kinchy dubs "genes out of place" sparked intense controversy. Biotechnology advocates moved quickly to dispute and discredit the study, even as anti-GM activists incorporated the evidence into their protest strategies.[89]

The finding also galvanized peasant and Indigenous organizations in Mexico. With the support of NGOs, they articulated a newly vehement stance against US maize imports. A Network in Defense of Maize (Red en Defensa del Maíz) took shape in early 2002 and declared the protection of Mexico's criollo varieties essential to the continuation of Indigenous cultures and peasant livelihoods. The network's concern over corn indexed a more sweeping discontent with the government's trade policies and welfare provisions.[90] NAFTA had been ruinous for rural Mexicans. In the intervening years since its implementation, the government had decided to phase out tariffs on corn imports more rapidly than the treaty demanded. Instead of instituting incremental change over fifteen years, it ended the tariffs in less than three. While yellow feed-corn imports from the United States shot upward, driving down the price that Mexican growers could ask for any maize crop, the promised markets for other farm products north of the border were never as extensive as advertised. The plummeting international prices of some key cash crops, such as coffee, caused further woe. Without government subsidies to help peasant farmers bridge the gap between the costs of production and their income from growing maize and other crops, they began migrating from the countryside in record numbers.[91]

A Forum in Defense of Maize was held in Mexico City shortly after the first account of transgenic introgression into Oaxacan maize. Participants included representatives of Mexican NGOs like the Environmental Research Group (Grupo de Estudios Ambientales), international organizations such as Food First and ETC Group (formerly RAFI), peasant and Indigenous organizations, and scientists. "The transgenic contamination of native maize varieties represents damage to the genetic memory of traditional Mexican agriculture," the official statement of this assembly declared. Although they had been brought together by news of transgenic introgression, participants had more than this all-but-invisible movement of genes on their minds. The forum's public declaration articulated the problem of transgenic maize as one among many affronts to rural Mexico and used its unwanted infiltration to demand broad reforms. These included declaring maize a strategic national resource, halting the import of transgenic crops, revising NAFTA policies on agriculture and grain imports, prioritizing the purchase of grain produced by Mexican farmers at fair prices, expelling transnational seed companies from Mexico, and recognizing the rights of Indigenous groups to self-determination.[92]

Just as mobilization against transgenic contamination of Mexican maize encompassed far more than a concern over "genes out of place," so too did the responses to that contamination. The Network in Defense of Maize offered a long list of potential actions to mitigate the threat of transgenic introgression.[93] These included education and outreach, promotion of criollo varieties, tests for genetic contamination, boycotts of certain tortilla producers, protests against the state-run maize supplier, and petitions for regulatory change. The network also encouraged the "registration, characterization, and duplication in situ of maize and other seeds culturally and biologically significant for communities and for the country" and declared their intention "to create

El maíz es el sustento de la vida, es sagrado. Es una relación vital que mantenemos cuando sembramos la tierra.

#NoTransgénicos #RedMaíz

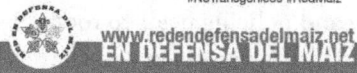
www.redendefensadelmaiz.net
EN DEFENSA DEL MAÍZ

Respeto a la autonomía de los pueblos campesinos e indígenas.

#NoTransgénicos #RedMaíz

www.redendefensadelmaiz.net
EN DEFENSA DEL MAÍZ

Un pueblo que produce sus semillas y comida, es un pueblo que no le pide permiso a nadie para existir.

#NoTransgénicos #RedMaíz

www.redendefensadelmaiz.net
EN DEFENSA DEL MAÍZ

Reconocer a todo México como centro de origen del maíz, prohibiendo las semillas transgénicas.

#NoTransgénicos #RedMaíz

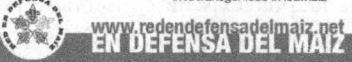
www.redendefensadelmaiz.net
EN DEFENSA DEL MAÍZ

[33] The predominantly Indigenous-led Network in Defense of Maize in 2015 equated transgenic maize and the structures that sustained it with genocide *(lower right)*. It insisted that "a people who can produce their own seeds and food are a people who do not have to ask anyone's permission in order to exist." Reproduced with permission of Red en Defensa del Maíz.

local seed banks in the hands of communities and organizations."[94] Articulated as part of a social movement centered on maize, the in situ conservation of Mexican landraces emerged as an obvious way for rural Mexicans to retain or regain some control over their own food and their own fates.

In this context in situ conservation did not mean preserving something "traditional" but instead investigating, and adopting or rejecting, new ideas and technologies.[95] This was certainly the case in Chiapas, where participants in the Zapatista movement invited the assistance of an outside expert, the ecologist Martin Taylor, to help them respond to the problem of transgenic contamination. When presented with the community's concerns, Taylor decided that a seed bank would provide the best means of protecting maize varieties central to Zapatista farming from unwanted cross-pollination with transgenic maize. Setting up a seed bank was not simply a matter of collecting, cleaning, and storing seeds, however. It required genetic testing of varieties to be banked and supplies from outside the community, like plastic bags and freezers. Above all, it required explanations of genetic technologies and seed banking for Zapatista leaders and community members who had been concerned by splashy headlines about the dangers posed to their maize but had limited education in genetics through which to parse them.[96]

The frozen Zapatista seed bank was abandoned after Taylor's departure, thanks to worries about the restricted number of samples in the freezer (just sixty-one maize varieties, of farmers' hundreds) and the capitalist implications of "banking" seed. Yet Zapatistas, working with collaborators in the United States, continued genetic testing and policing fields to root out contaminated plants. They also launched their own version of a seed exchange, international in scope, offering seeds of "Zapatista maize" to anyone who would promise to grow out their certified GM-free seeds and protect this heritage of Mayan peoples. The act of sowing Zapatista corn would be an act of conservation but also solidarity, a way of sustaining the Zapatista rebellion and, ultimately, the Indigenous peoples and communities allied to it.[97]

While peasants and their collaborators adapted existing conservation strategies to suit new contexts, escalating protests forced experts to adopt new perspectives on the nature of the problem posed by transgenic corn in Mexico and their response to it. Maize researchers had already noted the likelihood of gene flow from transgenic maize into teosinte and landraces.[98] In the wake of the 2001 study, CIMMYT called a temporary halt to adding samples to its gene bank, hoping to avoid inadvertent admixture. It also selectively screened accessions, to assure skeptics that its collections were transgene free.[99] Many institutions undertook further testing of maize samples in the field. Pro-GM allies, including CIMMYT, found nothing. Their negative results piled on to a steady flow of criticism of the original study and in some cases outright intimidation of its authors by those with vested interest in GM technologies.[100] Meanwhile, agroecology researchers, anti-GM activists like ETC Group, and environmental agencies such as Mexico's National Institute of Ecology (Instituto Nacional de Ecología) uncovered further evidence of transgenes in farmers' fields.[101]

Some Mexicans were unsatisfied with the narrow vision of these laboratory assessments of maize and especially with government stasis. In 2002 a collective representing twenty-one Indigenous communities from Oaxaca and three environmental organizations petitioned the Commission for Environmental Cooperation (CEC) to study the "potential direct and indirect environmental impacts on biodiversity" of transgenic maize in Mexico. The governments of North America had created the CEC as part of their agreement of NAFTA, authorizing this advisory body to study cross-border environmental concerns. The experts assembled by CEC in response to the 2002 petition, many of whom had long been engaged in studies of maize diversity, assessed the potential effects of transgenic corn on biodiversity, health, agriculture, ecosystems, and Mexican society and cultures.[102]

Thanks in part to the unexpected and passionate participation of activists, peasant farmers, and other community members, the final recommendations of the 2004 draft report were remarkably encompassing. The issue of whether or not transgenic material traveled was not the crux of the problem, the report reminded its readers. It had "become entwined with historical issues and grievances affecting rural Mexicans" that extended well beyond the care of maize landraces. Similarly, advocacy for GM maize needed to be seen as a product of scientific, political, and economic actors defending their interests in the larger agroindustrial system.[103] Taking this broad view of the issue, the authors of the CEC study recommended precautions: a continued moratorium on planting transgenic maize in Mexico, enhanced monitoring and conservation of genetic diversity, consultation of farmers on new agricultural practices, and support for campesinos pursuing "traditional farming." They also called for investments in on-farm conservation, including "community seed banks, farmer training and extension, registration and codification of local and traditional knowledge, and greater scientific research into landrace character and identity."[104]

Although the Mexican government largely ignored the CEC recommendations, the study did have a lasting effect: it forged links between anti-GM activists and scientists whose inclination was to stop the use and importation of transgenic varieties in Mexico.[105] In subsequent years, as regulatory policies remained in flux and as government agencies worked to bring the country into compliance with various treaties, especially the Convention on Biological Diversity, many Mexican scientists remained forcefully aligned against transgenic maize. Often they advocated its ban as a way of protecting Mexican people as much as Mexican maize. A 2006 "Manifesto for the Protection of Mexican Maize," drafted by scientists and NGO representatives in response to a new, weak law regulating genetically modified organisms (Ley de Bioseguridad de Organismos Genéticamente Modificados), called on the government to include the perspectives of rural and Indigenous peoples when revising this inadequate regulatory tool.[106] Scientists who participated in a comprehensive investigation of native maize (Proyecto Global de Maíces Nativos), which was created to inform a regulatory regime for maize required by the biosafety law, strongly advocated a revised law and continued moratorium to protect maize diversity. They also urged further collection of and experimentation with landraces and, above all, protection of "more than two million small scale or marginalized farmers" who they characterized as the "guardians of the native germplasm of maize."[107]

International agreements further shaped the possibilities for maize conservation in the late 1990s and early 2000s. The Mexican government's efforts to develop policies on biosecurity, mandated by the Cartagena Protocol on Biosafety of the 1992 Convention on Biological Diversity, created opportunities to contest transgenic maize and neoliberal policies and to make conservation in seed banks and on farms a national policy issue.[108] Meanwhile, the reformation of state policies in light of the 2001 International Treaty on Plant Genetic Resources for Food and Agriculture led to the creation of a National System of Plant Genetic Resources for Food and Agriculture (Sistema Nacional de Recursos Fitogenéticos para la Alimentación y la Agricultura).[109] In situ conservation was included in the planning and execution of this new national system, reflecting, though likely not resulting from, its prioritization in the Convention on Biological Diversity. Plans unfolded for twenty-one community seed banks, which were categorized in the plan as ex situ conservation—highlighting the ambiguity of these open-access, community-based institutions. The community seed banks would complement the eight long-term seed banks and eighteen working collections managed principally by researchers. Further in situ activities included identifying "seed guardians" (custodios) for native maize, farmers already cultivating local varieties "of the highest purity" who could be enlisted into the formal conservation system. In exchange for cultivating a hectare of their local variety, the seed guardians received training, technical assistance, farm inputs, and a salary. According to one tally, 350 farmers were successfully recruited to this scheme, whose harvests included fifty-two of fifty-nine Mexican landraces.[110]

The geneticist Flavio Aragón Cuevas, an employee of the Ministry of Agriculture, oversaw the state-funded community seed banks in Oaxaca, where the controversy over transgenic contamination had originated. Aragón was a specialist in genetic resources at a government agricultural station in the region's central valley. Already familiar with in situ activities thanks to an earlier a participatory breeding program, in 2005 he began developing seed storage facilities and management procedures, soliciting donations and training local Oaxacan farmers in the basics of seed banking.[111] Aragón was also a collaborator of David Cleveland and Daniela Soleri, former associates of Native Seeds/SEARCH. Having engaged in Native seed keeping and exchange through their work with Native Seeds/SEARCH, Cleveland and Soleri had begun to study farmer plant breeding in Oaxaca in the mid-1990s.[112]

Following the controversies over transgenic contamination, Soleri and Cleveland joined Aragón and other researchers for a survey of farmers in "traditionally-based agricultural systems" to understand the effects of introducing transgenic varieties. The results—launched into a still contentious space—indicated a "high potential for transgene flow" and called attention to the significant variability in farmers' assessments of GM varieties, which depended in part on their experiences of past technologies and outsiders' interventions. The authors' takeaway message was that any introduction of transgenic varieties needed to be tailored to local needs and carried out in consultation with farmers.[113] They posed an unusual question: Could GM crops support diversity in agricultural crops and sustain traditional agricultural systems? Their answer was yes—if GM varieties were created to specifically to support those systems, made free to use, and closely monitored for adverse social and ecological effects. Given the expense associated with achieving any of these, not to mention industry's indifference to the first and outright opposition to the second and third, it surely made more sense to invest in peasant farmers and their communities in other ways.

In other words, the real problem with transgenic maize was its embeddedness in an extractive production system that threatened farmers. Limiting its spread was a way of defending these producers. Like many campesinos, Indigenous people, and activists in Mexico who had aligned with the Network in Defense of Maize, these researchers saw saving seeds—either by storing them in community seed banks or keeping them safe from transgenic incursions—as a route to protecting people and communities.

Whether they were linked to subsistence and resistance at the grassroots or integrated into formal conservation agendas, in situ approaches to saving crop diversity such as community seed banks and participatory plant breeding brought the task of perpetuating crop diversity full circle. Decades of ex situ efforts had been premised on the idea that it was possible to disaggregate crop diversity and human diversity. The turn to farm- and community-oriented conservation signaled a recognition that crop varieties do not survive, or do not survive well, without cultivators. By the turn of the twenty-first century, the imagined future of maize diversity in Mexico, the United States, and beyond was no longer one in which farmers' diverse varieties lie dormant in cold storage, called up only when needed. Researchers and growers now imagined not just the possibility but the need for these to continue in cultivation.

Placing growers at the center of conservation efforts rather than cold storage, electronic databases, or formal agreements did not mean tossing away these tools. Instead, these were adapted to suit new contexts, in the form of freezers full of maize seeds in remote Chiapas, community mailing lists informing gardeners about the availability of Hopi Blue Corn in the Southwest, and enrollment forms for guardians of criollo maize to receive stipends from the Mexican government. Nor did it mean abandoning certain kinds of expertise, such as knowledge of genetics or plant breeding, which were shared and adapted by researchers and sometimes adopted by farmers and gardeners. Native Seeds/SEARCH reformulated technical seed storage advice to suit a suburban family home.[114] Scientists in Mexico produced pared-down guides to plant genetics to use in participatory breeding projects.[115] Zapatista activists developed a form of seed banking and exchange that would sustain their political agenda.

Backyard gardeners and Indigenous farmers no doubt seem unlikely guarantors of global food security to many experts on plant genetic resources stationed at research institutions around the world. Yet the case to consider these cultivators essential partners in conserving genetic diversity has only grown stronger, such that research institutions today routinely incorporate in situ activities into their slate of conservation interventions. This co-option of on-farm and in-garden conservation is in many ways a welcome change. In situ advocates resist, or create the possibility of resisting, the crisis talk of inexorable crop extinction. They highlight instead the possibilities for persistence, for adaptation and survival, and for repatriation and restoration—and for social and technical solutions premised on these.[116]

This shift has been bound up with a second transformation in the framework of crop conservation, one that emerged as peasants, Indigenous peoples, and other farmers became increasingly involved in these efforts. As agroecologists pointed out in the 1980s and as peasant organizations like Vía Campesina make clear today, those early champions of in situ conservation who had called for the protection of subsistence farmers as a means of ensuring the future of crop diversity had their story backward. The problem was not that plants were endangered but that people were.

This account of human endangerment is not the endangerment story of old—that is, the racist rationalization of inevitable Indigenous extinction. Nor is it the neocolonial view that peasant cultures must disappear as they are "developed" into "modern" ones. It is an endangerment story that is instead a calling to account. The infrastructures of global governance and finance profoundly disadvantage the poor, especially in the Global South. In agriculture international grain markets and free-trade mandates deliver rewards to aggressive transnational players and diminish farmers' possibilities for sustaining themselves and their families, communities, and cultures in the ways they would like. Demanding recognition and redress of this endangerment is a route to empowerment. The reason to sustain diverse farming practices and crop varieties is to foster human diversity and human flourishing. The future of crop diversity hinges on the survival of this worldview.

CODA

MANY SAMPLES OF MAIZE COLLECTED as exemplars of endangered types in the past century survive. My own experiences attest to this. I've seen Mandan Yellow Flour and several other kinds once offered in the catalogs of the Oscar H. Will seed company safely stored at the North Central Regional Plant Introduction Station in Ames, Iowa. According to a USDA database, descendants of a farmers' variety collected by Hugh Cutler in Cochabamba, Bolivia, in 1942 and sent to Paul Mangelsdorf were also in the introduction station's cold room that day, with others that can be traced back to cobs he and Marian gathered on their South American adventures.

I've encountered the seeds collected by the US National Research Council's Maize Committee, both original samples and later renewals of these, across multiple sites. Some remain as part of the Mexican national maize collection near Chapingo, where many of the early collections even remain in their original jars, with later generations of renewed stocks resting on nearby shelves. Further copies sit in the massive seed vault at CIMMYT, just a short drive down the highway, and that means they exist in off-site safety duplication collections as well. The duplicate samples demanded from the Latin American seed centers by the Maize Committee as its project came to an end are still kept in Fort Collins, Colorado, at what is today the US National Laboratory for Genetic Resources Preservation. Major Goodman has personal copies of much of the Maize Committee's world collection in North Carolina, and presumably many other researchers do, too—though surely none has quite so comprehensive a selection.

In Tucson, Arizona, I've walked through the seed bank at Native Seeds/SEARCH, which houses maize seeds gathered from Indigenous American farmers since the 1980s. Today the organization emphasizes more resolutely than ever before that cultivation, especially by Native communities, is the ultimate conservation measure and works to get its collections back into the hands and fields of these growers. Although I've spent time with the people responsible for all the collections just named and peered into many a jar of cold corn seed, it's not been my privilege to learn the extent of the samples maintained by Pioneer Hi-Bred International. In 2021 Pioneer is a subsidiary of Corteva Agriscience, itself a spinoff of the chemical giant DowDuPont (now just DuPont) and reportedly the world's largest agricultural organization. I'm guessing its corn seed collections are extensive indeed. I wouldn't be surprised if a relative of the lines made by William Brown from crosses of Iowa inbreds and Caribbean landraces in the 1950s survive somewhere within this industrial behemoth.

The projects that perpetuate maize seed samples today remain as varied, politically and socially, as those that first secured them. Seed samples survive as industrial inputs, scientific objects, and national patrimony. They are deployed as instruments of economic development and nurtured as the substrate of community life. They represent the foundation of future industrial lines and continued dominance of transnational agribusiness over global grain and human diets, even as they hold out the possibility of alternatives to this very fate.

In many ways conservators of maize diversity, and crop diversity more generally, have seen great successes. The German breeders who first wrung their hands over the possibility of farmers' varieties disappearing in the late nineteenth century would surely be delighted to discover not only that their understanding of the threat posed by the loss of varieties is widely shared but also that so many means of thwarting this loss have been pursued. The experts who assembled at FAO in the 1960s and insisted on the need for the international coordination of seed collections would probably see many, if not all, elements of the infrastructure they sought in today's Seed Treaty, in the networking of seed banks through transnational crop databases, and in the funding drives and disbursements of a Global Crop Diversity Trust founded in 2004. Millions of seed samples have been collected, classified, stored, duplicated, shared, and evaluated. Some have even found their way back into cultivation.

There are problems with interpreting this as success, however. First, claiming victory in saving endangered crop varieties depends on their actually have been in danger of wholly disappearing in the first place. While this is true of many farmers' varieties, it is not true of all, and, as others have pointed out, researchers are still working out where, how, and why diversity thrives.[1] Second, survival in cold storage provides cold comfort at best. Genetic diversity believed safe in freezer vaults encourages the view that monocropped fields of nearly identical plants are not so risky, even if the evidence suggests otherwise. As the history I've traced here shows, diversity in the bank and uniformity in the field have gone hand in hand. The perversity of seeing freezer storage as a triumph of conservation is further brought home by occasional accounts of duplication as a threat to seed banks. Since the 1990s the possibility that seed banks are, unknowingly, keeping identical samples has troubled some bank managers, who are both chronically short of funds and continually asked to take on more collections. To some, genetically similar samples within and across collections are dangerous. Spending money and time to perpetuate these contributes to the possibility that both resources will be stretched to the breaking point, causing banks to collapse and all samples to be abandoned.[2] The worry is undeniably a sensible one. Still, it is a strange form of biodiversity conservation that doesn't celebrate the proliferation of endangered diversity but instead identifies it as a problem.

Given these objections, it might be more accurate to say—as a significant number of people have said, especially since the struggles over seeds at the FAO in the early 1980s—that, in becoming the objects of conservation, farmers' varieties have simply been placed more conveniently within the reach of professional breeders and other scientists rather than rescued from oblivion. They have been saved as genetic resources, yes, but not as crop varieties.

There is a further issue with any putative success story that emerges from the history of crop conservation as it has most often been pursued. The narrative of endangered seeds has been, with rare exceptions, a distraction from—and in some cases an active deflection of—other endangerments. Farmers' varieties have typically been considered at risk of disappearing when farmers themselves have faced existential hardship. Professional breeders and the organizations that employed them, whether private industry, national agencies, or philanthropies like the Rockefeller Foundation, may have celebrated the replacement of farmers' varieties with breeders' lines as "improvement" and "modernization"—that is, as transitions that would benefit growers. But the most vulnerable farmers were not, and are not, the beneficiaries of modernization, and many have had their livelihoods destroyed and communities decimated in its pursuit. At too many junctures farmers' seeds were considered valuable but not their knowledge and their ways of life.

Peasant and Indigenous organizations, and the activists and researchers who align with them, have brought a sea change to this perspective in recent decades, insisting on the centrality of seed security, and seed sovereignty, to the communities they represent. Many articulate a worldview in which maintaining certain seeds is integral to the survival of peoples and cultures, while the survival of peoples and cultures is integral to the future of certain seeds. Within this perspective long-term storage of seeds can never be the ultimate guarantor of security, because the promise of such storage has always been in its minimizing the need for the continued cultivation of diverse crops by diverse peoples. Farmers' in situ conservation instead points the way toward a richer agricultural future.

Yet even the power of this view is sometimes missed when the seeds in question are fitted to a standard endangerment narrative. Many accounts of crop diversity, regardless of their politics, tell a single story in which this is both finite and rapidly disappearing. As I've described in this book, the success of this narrative has meant a focus within many strands of conservation work, whether grassroots or state-run, on salvaging endangered lines, stemming the tide of losses, and especially on fixing types as they are or, more often, as they are thought to have been. This work often depends on an understanding of varieties or landraces as discrete, settled entities, able to be identified in the field and extracted and to be carried to new locations, or even forward in time, and still retain a singular identity. It suggests that all the diversity we will ever have is already in place and keeps many would-be conservationists fixated on an idealized past even as they strive to secure the future.

There is an alternative to this line of thinking. As the historian Courtney Fullilove reminds us, "Seeds are always in a state of becoming, never fixable."[3] Social scientists who investigate seed maintenance and exchange among farmers emphasize that varieties are not stable objects, locatable as "pure strains" in the way that, for example, the Maize Committee once imagined. These are instead moving targets, continually altered by farmers who need to be flexible and adaptive to survive.[4] The insights of these researchers, although often hard-won in a world where knowledge about farmers' seed habits is not as prized as the means of changing those habits, reinforce a view that many farmers have of their own labor.[5] If there is a hope for an agricultural future with more alternatives than those apparent today, at least with respect to crop diversity, it lies in developing more accounts that reflect this understanding.

The history of a recent seed sensation provides a case in point. In 2012 a photograph of Glass Gem corn, a variety with glittering, multicolored kernels, went viral. The variety went viral too. Native Seeds/SEARCH, which had posted the pic, received thousands of orders for it. Demand later crashed the website of Seeds Trust, another purveyor of heirloom varieties. In 2020 Glass Gem had a Facebook group with more than twenty thousand followers. A quick online search turns up hundreds of proud photos of sparkling, home-harvested cobs.

Although it has been described as a "poster child for the return to heirloom seeds," Glass Gem is not an old variety but a new one. Its creator, the Oklahoman Carl Barnes, started collecting corn varieties in the 1940s, inspired by memories of the corn grown by his Cherokee grandfather and other members of his family. He especially prized varieties associated with Native American communities, which he gathered from across the country. By the 1980s he and a small collective of enthusiasts called CORNS were maintaining about a hundred of these and other open-pollinated varieties. Barnes was interested in preserving history, but for him this didn't mean keeping varieties as static as museum samples. It meant cultivating. It meant mixing. It meant . . . inventing.[6]

Barnes allowed different kinds of corn to cross-pollinate in the fields and selected new types from the subsequent mosaic, letting environment and interest dictate his choices. "I think most Indians did this too," he reflected in a 1982 interview. "In different generations and under different conditions [the corn] eventually kicks out what all's there," he described, referring to the different traits that his replanting and remixing made evident.[7] In the 1990s a small, rainbow-kerneled line that he developed from a mix of a Pawnee popcorn and the landraces Osage Red Flour and Osage Greyhorse caught the eye of another corn enthusiast, who started growing the novel line in New Mexico. There it cross-pollinated with larger, local flour corns, before making its way into the hands of the executive director of Native Seeds/SEARCH— and eventually into internet fame.[8]

The story of Glass Gem is an outlier. Accounts of nearly vanished varieties, recovered intact as they were once grown, often from an isolated farmer or an aged gardener, are far more common. Recovery, revival, and restoration feature centrally in these stories. But Glass Gem reminds us that there is potential for conservation in motion as well as in stasis, in mixing as much as purity, and in reinvention alongside restoration. The story of Glass Gem is one that foregrounds change and change makers. In contrast to the more typical conservation variety, frozen as-is, or the more typical grower of a prized landrace, too often essentialized out of time and history as "native," "primitive," or an "old-timer," Glass Gem and its growers offer a story saturated with transformation. It reminds us that crops and people travel together and transition together in an ever-changing world. It suggests that attempting to situate conservation in this flow rather than outside it might be possible.

Plant scientists and professional breeders recognize varieties as entities in constant motion but nonetheless look to accounts of fixity to define their contributions. When gene bank managers accept a seed sample as an accession into a collection, they register it as belonging to a specific type and coming from a particular time and place. This allows them, and the gene banks' users too, to imagine a precisely bounded entity, frozen in time—even as they acknowledge problems with the identification of samples, confusion about the relationships within and among collections, and the likelihood that conditions of storage and processes of regeneration over time alter the lines kept in the bank.[9] Likewise, when a breeder patents a new variety, that line ossifies in law. This is a convenient fiction to enable commerce and research. Varieties placed on the market as stable, identifiable, singular kinds are also recycled back into breeding programs. Today's inbreds are an input into tomorrow's inbreds. When a variety becomes "new" enough to warrant its own name or brand is open to debate.[10]

The makers of hybrid corn and other contemporary industrial crops are confident remixers. They have many tools at their disposal for keeping diverse varieties in motion by improving their fit to different places and peoples and preferences. Unfortunately, most are chained to a system of agriculture and food production that is not particularly interested in catering to differences, preferring instead to scale up. Industrial seed dynamism is not linked to industrial farm dynamism, at least in the sense of responding to needs and concerns beyond the corporate bottom line. As a result, professional breeders are denied opportunities to produce the kind of diversity in food crops that might be expected from a group that collectively strives to supply seeds to farmers in wildly divergent cultures and climates.[11]

Even grassroots seed savers are sometimes subject to the fiction of stasis. Certainly most recognize and celebrate that their contribution to conservation, keeping varieties in cultivation and circulation, enables crops to evolve and adapt to changing conditions. Yet many still stop short of self-describing as breeders—that is, as individuals whose choices shape plants to their own interests, abilities, and climates—preferring instead to speak of stewarding the "heirlooms" and "heritage varieties" of the past.

The problem is not that crop diversity is being lost. It is that so many people have abdicated the duty to create it, not least by inevitablizing loss in the stories they tell. Broadcasting extinction as the problem invites the solutions of salvage, storage, and defense against change. Advertising the need to reengage creation—to reimagine, revamp, and retool crops—might prompt a different approach. When I suggest that it's time to start thinking outside stories about diversity's loss, I have in mind our telling stories of its coming into being. I have in mind Glass Gem corn, a celebratory tale that is not premised in crisis and does not end in heroic salvage of a treasured relic—and which nonetheless is an example of the successful perpetuation of crop diversity. What would a science and practice of conservation rooted in this narrative look like? What if our efforts were directed at creative remixing rather than forestalling extinction?

This alternate narrative already has adherents. Many advocates of participatory breeding think that conservation is best achieved by continuing to develop farmers' varieties. There are also breeders, amateur and professional, who, like Carl Barnes, tinker on, proliferating types for pleasure or profit heedless of the trajectories of industrial production.[12] Native and Indigenous collectives reject the declaration of sweeping, inevitable loss of crop varieties— and other foods and lifeways—and champion instead accounts of resilience, resistance, and adaptiveness.[13] In a truly impressive embrace of conservation through genetic creativity, the Land Institute of Salina, Kansas, a nonprofit research organization, is domesticating new perennial crops that it believes could restore prairie ecosystems of the US Midwest while also feeding people.[14]

It's no stretch to suggest that the Svalbard Global Seed Vault, the "failsafe" storage facility for the world's gene banks, ranks as the most iconic crop conservation project today. The so-called Doomsday Vault, often depicted as a lonely pillar in an endless sea of snow, is a potent reminder of where narratives of loss and crisis end up. Diverse seeds are documented and duplicated, frozen in time and place, and secured against every disaster except the mundane reality most likely to beset them: death in storage. Although the vault's striking, sometimes sparkling, facade, is unlikely to be unseated as the face of global crop conservation anytime soon, it's surely worth considering alternatives. Perhaps Glass Gem corn will do.

ACKNOWLEDGMENTS

This book would not have been written without the support of the Riksbankens Jubileumsfond, which funded my Pro Futura Scientia fellowship from 2017 to 2020, as well as the Swedish Collegium for Advanced Study (SCAS) and the Centre for Research in the Arts, Social Sciences and Humanities (CRASSH) at the University of Cambridge. The time and resources offered through the Pro Futura program and the support of staff and fellow scholars transformed the possibilities of the project. I am grateful to Björn Wittrock, Christina Garsten, Simon Goldhill, and Steven Connor for their roles in sustaining vibrant intellectual communities in Uppsala and Cambridge. I am further indebted to the staff of SCAS and Michelle Maciejewska at CRASSH, who kept everything ticking along.

Several other organizations sponsored research underpinning different chapters of this book. I thank the Rockefeller Archive Center (Grant in Aid, 2015); the Wellcome Trust (Seed Award, 2016); the University of Cambridge (Humanities Research Grant, 2015–16); and the Humanities Research Centre of the Australian National University (Visiting Fellowship, 2018). Churchill College and the Department of History and Philosophy of Science (HPS) at Cambridge also provided resources for archival trips and other research expenses. I am grateful to both institutions for continued support.

This book benefited from the expertise, insight, and patience of many people. Since I started pursuing the history of maize conservation in earnest, I have spoken with dozens of individuals who have spent years and decades studying and conserving crop genetic diversity. Without their guidance I would still be hopelessly lost. A few people were singularly generous with their time and energy, no one more so than Denise Costich, who welcomed a total stranger on a cross-country road trip and graciously forged connections that enriched the project in ways I'd not realized possible. Others who went out of their way to teach me about conservation from their perspective include Candice Gardner, Major Goodman, Colin Khoury, Denisse McLean, and Rafael Ortega Paczka. I cannot thank these individuals enough.

My heartfelt thanks extend to many other insiders who did their best to keep me on the right path and who, in the process, made research more fun and friendly than it has ever been. In roughly the order that I met them, this group includes Lisa Burke, Mark Millard, Arnel Hallauer, Jianming Yu, Goran Srnic, John Torgrimson, Tor Janson, Tim Johnson, Zach Row-Heyveld, Sara Straate, Mike Ambrose, Neil Munro, Catrina Fenton, Christina Walters, Lana Wheeler, Patricia Conine, William Prange, Stephanie Greene, Eric Roos, Henry Shands, Peter Bretting, Karen Williams, Tom Payne, Carolina Camacho Villa, Cristian Zavala, Ernesto Preciado, Rosalinda González Santos, Mellissa Wood, Nicholas Garber, Sheryl Joy, Joy Hought, Lois Ellen Frank, Walter Whitewater, Mauricio Bellon, Juan Manuel Hernández Casillas, Jesús Axayacatl Cuevas Sánchez, Louis Goodman, James Holland, Matt Krakowsky, John Dickie, and Charlotte Allender. A handful of other experts and eyewitnesses took the time to share their experiences at a distance. I'm grateful to David Barkin, Steve Brush, Mahina Drees, José Esquinas-Alcázar, Pat Mooney, Gary Nabhan, Karen Riechhardt, and Daniela Soleri for the time they spent sharing insights by phone, Skype, and Zoom.

A special thanks is due to an extraordinary couple, Bill and Elisabeth Cutler. In late 2017 I directed an email to an account I thought might belong to the son of Hugh and Marian Cutler, with a small but absurd hope not just that my message would land in the inbox of the right Bill Cutler but that he would have some records documenting his mother's experiences traveling across South America with Hugh in search of maize. What a crazy idea! And crazier still that the end result would be the opportunity to see reams of his parents' letters, journals, and photos. Bill and Elisabeth's generosity resolved questions lingering over this book and has provided the story and source material for another.

The staff at numerous archives and libraries were indispensable. Particular thanks go to José Juan Caballero at the International Center for the Improvement of Maize and Wheat (CIMMYT); Andrew Colligan at the Missouri Botanical Garden; Robyn Diamond at the Australian Academy of Science; Janice Goldblum at the National Academy of Sciences; Lee Hiltzik and Michelle Beckerman at the Rockefeller Archive Centre; Dan Huff at the Research Corporation; Bill Musser at Seed Savers Exchange; and Jorge Gustavo Ocampo Ledesma and colleagues at the Universidad Autónoma Chapingo. Thanks also go to those who volunteered time to source images, especially Eric Van Dusen and Heike Vibrans, and to Jessica Lee and Daniela Sclavo for helping me to sort out image permissions.

Many scholars nurtured this manuscript. Rebecca Earle was the first to read any of the chapters, and her enthusiasm for those drafty first drafts helped me to believe I had something worth saying—and also to say it better. I'm grateful to Jonathan Harwood and Jim Secord, who read the first version of the manuscript and helped me see how it could be much better. Staffan Müller-Wille and Andrew Buskell also read every page, and their reflections and especially their encouragement were invaluable. A number of people read and commented on chapters at various stages in their development: Mary Brazelton, Tad Brown, Xan Chacko, Courtney Fullilove, Nick Hopwood, Ann Kakaliouras, Colin Khoury, Tim Lorek, Diana Alejandra Méndez Rojas, Rafael Ortega Paczka, Jonnie Penn, Helen Zoe Veit, and Emily Wanderer. Their friendship and collegiality have been as essential to my writing as their red pens and sharp-eyed critiques. I also had the luxury of whole seminars devoted to two of the chapters. Many thanks go to Sabina Leonelli and her colleagues at Exeter and to Jim Scott and the Agrarian Studies seminar at Yale for their substantive feedback.

I'd be remiss not to mention a wider community of scholars who shaped my ideas about the history of science, agriculture, and environment in the years that I've spent on this book. A few warrant very specific mention. Sara Peres and I worked together in the months of research and reading that preceded my Pro Futura fellowship; my thinking about crop conservation was profoundly shaped by our conversations. Ana Barahona, long an admired colleague, proved a true friend in a moment of need. Thank you, Ana. Dan Kevles remains a trusted mentor in all things. This project also profited from exchanges with Leah Aronowsky, Alison Bentley, Dominic Berry, Christophe Bonneuil, Raf de Bont, Derek Byerlee, Pia Campeggiani, David Cannadine, Lino Camprubí, Berris Charnley, Deborah Coen, Linda Colley, Henry Cowles, Angela Creager, Deborah Fitzgerald, Vivette García Deister, Meira Gold, Matthew Holmes, Hazem Kandil, Kim Kleinman, Emma Kowal, Prakash Kumar, Jung Lee, Kay Lewis-Jones, Aryo Makko, Kärin Nickelsen, Charles Pigott, Sadiah Qureshi, Greg Radick, Joanna Radin, Jenny Reardon, Harriet Ritvo, Libby Robin, Joëlle Rollo-Koster, Robin Scheffler, Simone Schleper, Sigrid Schmalzer, Susanne Schmidt, David Sepkoski, Gabriela Soto Laveaga, David Spanagel, Alistair Sponsel, Jan Stegner, Emilia Terracciano, and Keith Tribe, among others.

At Cambridge I work with tremendously talented people. My colleagues in the HPS department were generous enough not just to let me take three years away from teaching and admin but to encourage it. What can I possibly say in thanks? You guys are the best. Thank you especially to those who also took (or feigned!) an interest in learning more about corn and seeds as well, especially Anna Alexandrova, Jenny Bangham, Hasok Chang, Theo di Castri, Marta Halina, Boris Jardine, Lauren Kassell, Tim Lewens, Dániel Margócsy, Josh Nall, and Richard Staley. Beyond HPS I'm grateful to Christopher Clarke, Emma Gilby, Julia Guarneri, Inanna Hamati-Ataya, Martin Jones, Chris Sandbrook, Emma Spary, and Paul White, individuals who contributed to this project in different ways.

My editors at the University of California Press, Kate Marshall and Enrique Ochoa-Kaup, have shown great care, patience, enthusiasm, and firmness, always at the right time and in the right measure. They also guided the editorial team through the publishing process and located the anonymous reviewers, whose generous feedback helped shape the book at crucial junctures. Many thanks go to both and to all those at UC Press who contributed to this final product with their diverse expertise. Despite all this input, advice, and support, I'm sure there are places where I've erred and times when I've arrived at conclusions that might be contested. Any mistakes and misinterpretations in these pages are mine alone.

I dedicate this book to Andrew, who helped my interest in crop diversity flourish on the table and in our garden, as well as in these pages. There are others who sustain me, though perhaps not as literally. Candy and Dan have been unparalleled providers of corn curios. And I could not ask for a more supportive family. No day passes without my mind wandering to NoVA, wondering what everybody is up to there. Mom and Dad, Tom and Michelle, Bridgid and Amos, Hugh, Charlie, Elizabeth, Anna, and Thomas, I'm nowhere without you. Thank you all.

NOTES

INTRODUCTION

1. Global maize production first exceeded one billion tons in 2014. Shiferaw et al., "Crops That Feed"; OECD, *Agricultural Outlook,* ch. 3; FAOSTAT (Food and Agriculture Data), FAO, last accessed 2021, www.fao.org/faostat.

2. Denise Costich, quoted in Matthew O'Leary, "Maize: From Mexico to the World," CIMMYT, 20 May 2016, www.cimmyt.org/blogs/maize-from-mexico-to-the-world.

3. Hellin and Bellon, "Manejo de semillas," 11. On Mexican farmers' continuing contributions to maize evolution, see Bellon et al., "Food Supply Implications."

4. "Seed Bank," Native Seeds/SEARCH, last accessed 2021, www.nativeseeds.org/pages/seed-bank.

5. Vidal and Dias, *Endangerment, Biodiversity, and Culture,* 16.

6. On the history of wildlife conservation, see W. Adams, *Against Extinction;* and Barrow, *Nature's Ghosts.* The historical literature on conservation is extensive. Useful starting points include (on parks) Gissibl, Höhler, and Kupper, *Civilizing Nature;* (on technologies of restoration) Friese, *Cloning Wild Life;* and (on regulation) Cioc, *Game of Conservation.*

7. On the role of narrative in endangered species conservation, see Heise, *Imagining Extinction.* One premise of the emerging field of extinction studies is to develop new stories about species extinction and, in doing so, arrive at new understandings of its significance. See Van Dooren, Rose, and Chrulew, *Extinction Studies,* 1–18; and Van Dooren, *Flight Ways.*

8. Montenegro de Wit, "Are We Losing Diversity?," 627.

9. Brush, *Genes in the Field;* Brush, *Farmers' Bounty;* Nazarea, *Heirloom Seeds;* Galluzzi, Eyzaguirre, and Negri, "Home Gardens."

10. Van de Wouw et al., "Genetic Erosion in Crops." See also Bellon and Risopoulos, "Small-Scale Farmers"; and Chapman and Heald, "Agrobiodiversity Loss."

11. A useful review of published studies is included in Thormann and Engels, "Genetic Diversity and Erosion." On the challenges of this research, see Van de

Wouw et al., "Genetic Diversity Trends." For a range of perspectives on why conserving crop genetic diversity matters, see Hoisington et al., "Plant Genetic Resources"; McCouch et al., "Feeding the Future"; and Guarino and Lobell, "Walk on the Wild Side." On agricultural biodiversity more broadly, see Massawe, Mayes, and Cheng, "Crop Diversity."

12. For example, see Thormann and Engels, "Genetic Diversity and Erosion," 1:283–84.

13. Examples directly relevant to the history told here include Barrow, *Nature's Ghosts;* Matheka, "Decolonisation"; Qureshi, "Dying Americans"; De Bont, "'Primitives' and Protected Areas"; De Bont, "World Laboratory"; Raby, *American Tropics;* Sepkoski, *Catastrophic Thinking;* and Schleper, *Planning for the Planet.* See also contributions to Vidal and Dias, *Endangerment, Biodiversity, and Culture.*

14. Tsing, *Mushroom at the End;* Haraway, *Staying with the Trouble;* Van Dooren, *Wake of Crows.* For a different approach, see Balmford, *Wild Hope.*

15. Fullilove, *Profit of the Earth,* 10.

16. Fenzi and Bonneuil, "Genetic Resources."

17. On the relationship between plant breeding and political power, see, for example, Bonneuil, "Penetrating the Natives"; Harwood, *Europe's Green Revolution;* Saraiva, *Fascist Pigs;* and contributions to "Autarky/Autarchy."

18. Pistorius and Van Wijk, *Plant Genetic Information;* Flitner, "Genetic Geographies"; Saraiva, "Breeding Europe."

19. Pistorius, *Scientists, Plants and Politics;* Curry, "Working Collections"; Chacko, "Moving, Making, and Saving." See also Plucknett et al., *Gene Banks;* and Farnham, *Saving Nature's Legacy,* ch. 4.

20. Van Dooren, "Banking Seed"; Saraiva, "Breeding Europe"; Fenzi and Bonneuil, "Genetic Resources"; Curry, "Bean Collection."

21. Bonneuil, "Seeing Nature," 11, 12.

22. Brush, *Farmers' Bounty;* Saraiva, "Breeding Europe"; Fenzi, "'Provincialiser' la Révolution Verte"; Curry, "Gene Banks, Seed Libraries."

23. Cullather's revisionist account of the Green Revolution similarly foregrounds the view of peasant farmers espoused by foundation administrators, policy makers, and academics as key to understanding the history of agricultural aid. See *Hungry World.*

24. For example, histories of rice and wheat conservation would share many similar elements, not least because they were also objects of sustained attention from professional breeders, development agencies, and private industry in the twentieth century.

25. Maize in Mexico has been studied more than any other crop and region combination with respect to changes in diversity (Colin Khoury, pers. comm., July 2020). This extensive body of research underscores the centrality of maize in the history of crop diversity research and conservation and the potentially unique possibilities for telling this history through maize.

26. Ranum, Peña-Rosas, and Garcia-Casal, "Global Maize Production"; "World of Corn," National Corn Growers Association, last accessed 2021, http://worldofcorn.com.

27. Dimitri, Effland, and Conklin, *20th Century Transformation;* MacDonald, Korb, and Hoppe, *Farm Size.* See also J. Anderson, *Industrializing the Corn Belt;* and Clampitt, *Midwest Maize.*

28. The history of the seed industry and the place of corn in US foreign policy are discussed in chapters 5–7. On health effects, see Duffey and Popkin, "High-Fructose Corn Syrup"; and Gálvez, *Eating NAFTA.* Popular treatments of this history include Pollan, *Omnivore's Dilemma;* and Woolf, *King Corn.*

29. E. Anderson, "Sources of Effective Germ-Plasm"; Griliches, "Hybrid Corn"; Troyer, "Development of Hybrid Corn." See also Fitzgerald, *Business of Breeding.*

30. For a further analysis of the multiple meanings of Green Revolution, see Picado, "Breve historia."

31. Meanwhile, a focus on the Americas, mine included, ignores the global histories of maize. For more geographically encompassing accounts, see Warman, *Corn and Capitalism;* McCann, *Maize and Grace;* and Byerlee, "Globalization of Hybrid Maize."

32. See the discussion in Benz, "Maize in the Americas," in Staller, Tykot, and Benz, *Histories of Maize,* 9–20.

33. Goodman and Galinat, "History and Evolution"; Tracy, "Vegetable Uses of Maize."

34. Blake, *Maize for the Gods,* 193–97.

35. This identification also features in contemporary Mayan poetry. See Pigott, "Maize and Semiotic Emergence."

36. Mexican agricultural data are available from the government's Statistical Yearbook of Agricultural Production, Food and Fisheries Information Service, last accessed 2021, https://nube.siap.gob.mx/cierreagricola.

37. Appendini and Quijada, "Consumption Strategies"; Massieu Trigo and Lechuga Montenegro, "Maíz en México"; García Urigüen, *Alimentación de los mexicanos.*

38. Gustavo Esteva, "Árboles," in Esteva and Marielle, *Sin maíz,* 17.

39. Polo, quoted in *Voices of Maíz,* 10.

40. As the sources noted here confirm, there is disagreement about these definitions. Some people consider "farmers' variety" and "landrace" synonymous terms. I try to keep them separate, taking into account the possibility that farmers who are not breeders by profession may develop a distinct line in a comparatively short time through their own selective activities. The product would be more akin to a breeders' variety (often called a "modern cultivar") than a landrace yet would likely remain separate within the economic and political spheres that regulate breeders' varieties today. See Berg, "Landraces and Folk Varieties"; and Camacho Villa et al., "Defining and Identifying."

41. I'm avoiding scare quotes here; I address the politics of some of these terms in the chapters where they are introduced.

42. In understanding the origins of "indigenous maize," a comparison is helpful: in her study of Native American DNA, Kim TallBear explains that this ostensibly biological thing—"Native DNA"—was created through "laboratory methods and

devices" and "discourses of race, ethnicity, nation, family and tribe." See "Emergence, Politics and Marketplace," 23–24; and *Native American DNA.*

43. Qureshi, "Dying Americans," 267. See discussion of indigeneity and settler colonialism in chapter 1 of this book.

44. For example, consider the work of Braiding the Sacred, a movement of Indigenous corn cultures, which is documented on their website, last accessed 2021, https://braidingthesacred.org/. See also Gosia Wozniaka, "A New Bill Could Help Protect the Sacred Seeds of Indigenous People," *Civil Eats,* 19 October 2019, https://civileats.com/2019/10/09/a-new-bill-could-help-protect-the-sacred-seeds-of-indigenous-people/.

45. Useful starting places are Cintli Rodríguez, *Our Sacred Maíz;* and *Voices of Maíz.* Other resources, exemplifying diverse approaches, include Asturias, *Hombres de maíz;* Museo Nacional, *Nuestro maíz;* Sandstrom, *Corn Is Our Blood;* and Peña et al., *Mexican-Origin Foods.*

46. K. Whyte, "Food Sovereignty." See also Mihesuah and Hoover, *Indigenous Food Sovereignty.*

47. Masco, "Crisis in Crisis," S73. See also Roitman, *Anti-crisis.*

CHAPTER ONE. COLLECT

1. H. Howard Biggar, "The Corn Work of the Indians of the Middle West," n.d., ARS Records, 170 80–12–07, entry 29, box 2, folder 12, pp. 9–10, 19, 57.

2. Lehmann, "Collecting European Land-Races"; Bonneuil, "Seeing Nature."

3. Bonneuil, "Seeing Nature," 11–12.

4. P. Wolfe, "Settler Colonialism"; Veracini, "Introducing Settler Colonial Studies."

5. On indigeneity and extinction in the US context, see Dippie, *Vanishing American;* and Qureshi, "Dying Americans." See also Byrd, *Transit of Empire;* McGregor, "Doomed Race"; and Brantlinger, *Dark Vanishings.*

6. Biggar, "Corn Work," ARS Records, 81.

7. Two excellent resources for understanding this extended history are Staller, Tykot, and Benz, *Histories of Maize;* and Staller, *Maize Cobs and Cultures.*

8. See, for example, Harrington, "Some Seneca Corn-Foods"; Clarke, "Iroquois Uses of Maize"; and Cushing, *Zuñi Breadstuff.*

9. Lessens, "Master of Millions," 6–7.

10. Historical data are available from USDA, National Agricultural Statistics Service, last accessed 2021, http://quickstats.nass.usda.gov.

11. Lessens, "Master of Millions," ch. 6. See also Fitzgerald, *Business of Breeding.*

12. "How Uncle Sam Helps."

13. See, for example, Hartley, "Seed Corn," 3.

14. Rosenberg, "Rationalization and Reality," 402. See also Danbom, "Agricultural Experiment Station."

15. In 1923 Guy N. Collins at the USDA thought that, in 425 years of corn cultivation, farmers of "the white race" had produced "almost nothing . . . that can not be duplicated among the cultures of the aborigines." See "Agricultural History of Maize," 423.

16. Enfield, *Indian Corn,* 60–62; Sturtevant, *Varieties of Corn,* 37.

17. Scientists' understanding of the origins of Corn Belt Dent has undergone transformation over time, but this basic story remains consistent. See, for example, W. Brown, "Corn Belt Maize"; and Doebley et al., "Origin of Cornbelt Maize."

18. E. Anderson, "Sources of Effective Germ-Plasm"; Wallace and Brown, *Early Fathers,* 88–91.

19. Curran, "Indian Corn." On the development of open-pollinated corn varieties in the United States, see Anderson and Brown, "Common Maize Varieties"; Wallace and Brown, *Early Fathers;* and Troyer, "U.S. Hybrid Corn."

20. This assumption is based on the patchy archive of the Office of Corn Investigations; see lists of its corn varieties "on hand" for 1913 and 1914, BPI Records, 170 25–5–7, box 1, folder: Inventory of Corn 1914.

21. On the professionalization of corn breeding in the United States, see Fitzgerald, *Business of Breeding;* and Fitzgerald, "Farmers Deskilled." On corn and the institutionalization of genetics, see Kimmelman, "Organisms and Interests." For a review of the literature on the professionalization of plant breeding, see Berry, "Historiographies."

22. Troyer and Hendrickson, "Background and Importance," 907, 910.

23. On plant exploration in the United States, see Kloppenburg, *First the Seed,* ch. 3; and Harris, *Fruits of Eden.* The science and politics of plant introduction are also discussed in Fullilove, *Profit of the Earth,* pt. 1; and Cooke, "Who Wants White Carrots?"

24. "How Uncle Sam Helps."

25. See Collins, *New Type,* 7.

26. Enfield, *Indian Corn,* 60.

27. Troyer, "U.S. Hybrid Corn." See also Sturtevant, *Varieties of Corn;* and Montgomery, *Corn Crops.*

28. Some countries with high levels of corn cultivation did not have statistics available. Finch, "Graphic Summary," 533.

29. Montgomery, *Corn Crops,* 6–7. Historical data are available from USDA, National Agricultural Statistics Service.

30. Ten Eyck and Shoesmith, *Indian Corn,* 270–71, 292.

31. Ibid., 270.

32. "Information for Secretary's Report Corn Investigations," 1908, BPI Records, 170 25–5–7, box 1, folder: Sec Reports—Copies of.

33. "Corn Investigations," 1908, BPI Records, 170 25–5–7, box 1, folder: Sec Reports—Copies of.

34. Biggar, "Corn Work," ARS Records, 3, 26.

35. Collins, "Agricultural History of Maize," esp. 423–29.

36. Collins, "Pueblo Indian Maize Breeding," 266, 268.

37. Biggar, "Corn Work," ARS Records, 31.

38. Oscar Will, quoted in Atkinson and Wilson, *Corn in Montana,* 58. Atkinson and Wilson claim the corn was first offered in 1886, but the catalogs indicate 1888.

39. Will and Hyde, *Corn among the Indians,* 27.

40. Atkinson and Wilson, *Corn in Montana,* 58–59. There is confusion between accounts as to whether the corn came from Arikara or Mandan farmers.

41. This story is found in many accounts; see, for example, Curt Eriksmoen, "Oscar Will Helped 'Seed' N.D.," *Bismarck Tribune,* 28 June 2009. On Will's methods, see Atkinson and Wilson, *Corn in Montana.*

42. Will and Spinden, *Mandans.* On George Will, see Wedel, "George Francis Will."

43. Will and Hyde, *Corn among the Indians,* ded., 23, 33.

44. See correspondence between George Will and James Holding Eagle, Will Family Papers, series 1.2, box 3, folder 2.

45. Will and Hyde, *Corn among the Indians,* 87.

46. Gilbert Wilson to George Will, 23 October 1912, Will Family Papers, series 1.2, box 8, folder 2.

47. Wilson, *Agriculture of the Hidatsa,* 4. See also the notes from these interviews, Wilson Papers, box 4.

48. Wilson, *Agriculture of the Hidatsa,* 1; Fenn, *Encounters,* 325. See also Will and Hyde, *Corn among the Indians,* 48.

49. A detailed history of the Mandan, Hidatsa, and Arikara is R. Meyer, *Village Indians.* See also Fenn, *Encounters.*

50. Wilson, *Agriculture of the Hidatsa,* 9, pref.

51. Biggar, "Corn Work," ARS Records, 32.

52. Dippie, *Vanishing American,* 223, 236. See also Hinsley, *Savages and Scientists.*

53. Wilson, *Agriculture of the Hidatsa.*

54. Ibid., pref., 58–63.

55. Ibid., 65.

56. R. Meyer, *Village Indians,* 114. The Fort Berthold reservation continues today, further reduced in size through an executive order of 1880, an agreement of 1886, and the submerging of more than 260 square miles following the construction of the Garrison Dam between 1947 and 1953.

57. McDonnell, *Dispossession;* M. Meyer, *White Earth Tragedy.*

58. Wilson, *Agriculture of the Hidatsa,* 119–20.

59. Will and Hyde, *Corn among the Indians,* 32.

60. James Holding Eagle to George Will, 2 January 1917, Will Family Papers, series 1.2, box 3, folder 2.

61. H. Howard Biggar to George Will, 9 November 1916, Will Family Papers, series 1.2, box 3, folder 10.

62. G. L. Wilson and F. N. Wilson, "Report on Mandan-Hidatsa Fieldwork: Summer of 1912," Wilson Papers, box 4, folder 1, pp. 144–46.

63. See, for example, Dippie, *Vanishing American;* D. Adams, *Education for Extinction;* Qureshi, "Dying Americans"; and Radin, *Life on Ice.*

64. Hinsley, *Savages and Scientists.*

65. Fitzgerald, "Farmers Deskilled." Another popular breeding strategy was to cross two different types and choose the most promising offspring, for example, as the Hershey family did in the case of Lancaster Surecropper, described earlier.

66. On the making of hybrid corn, see Fitzgerald, *Business of Breeding;* and Crow, "90 Years Ago."

67. Crow, "90 Years Ago."

68. Fitzgerald, "Farmers Deskilled."

69. On Wallace and the early years of Pioneer Hi-Bred, see W. Brown, "H.A. Wallace."

70. E. Anderson, "Sources of Effective Germ-Plasm," 355.

71. Fitzgerald, *Business of Breeding,* 220.

72. E. Anderson, "What I Found Out," 7, 8. See also Kleinman, "Edgar Anderson"; and Kleinman, "His Own Synthesis."

73. E. Anderson, "Sources of Effective Germ-Plasm," 355.

74. Wallace and Brown, *Early Fathers,* 84–85.

75. E. Anderson, "Sources of Effective Germ-Plasm," 357–58.

76. Ibid., 360.

77. Gary Nabhan, "Oscar Will's Seeds: Learning from Agricultural History," *Seedhead News* 16 (1986), 3–4, Native Seeds/SEARCH, last accessed 2021, www.nativeseeds.org/blogs/the-seedhead-news/no-16-winter-solstice-1986.

78. For an account of the Wills from the perspective of settler colonial studies, which aligns better with my reading, see Hill, "Seeds as Ancestors."

CHAPTER TWO. CLASSIFY

1. Sauer, "Personality of Mexico," 356–57; West, *Carl O. Sauer's Fieldwork.* On Sauer, see Parsons, "Carl Ortwin Sauer."

2. Lomnitz, "Bordering on Anthropology"; González, "From Indigenismo to Zapatismo"; López Hernández, "Gloria prehispánica." See also Knight, "Racism, Revolution, and Indigenismo"; Dawson, "Models for the Nation"; and Stavenhagen, "Política indigenista."

3. Carl Sauer to Edgar Anderson, 14 November 1942, Sauer Papers, box 6, folder: Anderson, Edgar 1941–43.

4. Paul Mangelsdorf to Merle Jenkins, 13 May 1940, Mangelsdorf Papers, shelf 37.6, box 1, folder: Jenkins, Merle T. See also Sauer, "Plant and Animal Destruction."

5. Chapman and Heald, "Agrobiodiversity Loss," 162. For a similar assessment, see Brush, *Farmers' Bounty,* ch. 7.

6. Bonfil Batalla, *Maíz,* 7. See also Esteva and Marielle, *Sin maíz.*

7. Kato Yamakake et al., *Origen y diversificación.*

8. Rosa, *Memoria sobre el cultivo.*

9. Chávez, *Cultivo del maíz*. See also Ruiz Erdozain, *Estudio sobre el cultivo*.

10. See Vavilov, *Origin and Geography*. On Vavilov and collecting, see Loskutov, *Vavilov and His Institute*.

11. Wilkes, "Maize," 445; Kuleshov, "Maize of Mexico," quotations on 492, 493. See also Vavilov, *Origin and Geography*, 207–38.

12. Elina, Heim, and Roll-Hansen, "Plant Breeding," 166; Flitner, "Genetic Geographies," 178–79.

13. Bukasov, "Cultivated Plants," 470.

14. West, *Carl O. Sauer's Fieldwork*, 17. This interest culminated in Sauer's 1952 synthesis of crop domestication and diffusion as understood through the lens of cultural geography; see *Agricultural Origins and Dispersals*.

15. Parsons, "Carl Ortwin Sauer"; West, "Contribution of Carl Sauer"; West, *Carl O. Sauer's Fieldwork*, 15–22.

16. Sauer to E. Anderson, 10 October 1941, 27 October 1941, Sauer Papers, box 6, folder: Anderson, Edgar 1941–43.

17. Mangelsdorf to Jenkins, 13 May 1940, Mangelsdorf Papers.

18. Entry points into the history of land reform include S. W. Sanderson, *Land Reform in Mexico;* and Escárcega López and Escobar Toledo, *Historia de la cuestión*. See also M. Wolfe, *Watering the Revolution*.

19. Cotter, "Green Revolution." See also Cotter, *Troubled Harvest,* ch. 3. On agricultural research and education in Mexico, see Olea Franco, "One Century."

20. Cotter, "Cultural Wars," 147–50, Rivas's quotation on 147–48.

21. West, *Carl O. Sauer's Fieldwork,* 17–18.

22. Mangelsdorf to Jenkins, 13 May 1940, Mangelsdorf Papers.

23. Ibid.

24. See a summary and further references in Mangelsdorf and Reeves, "Origin of Indian Corn," 62–70.

25. Scientists by this time had identified two species of teosinte, a perennial native to Mexico (then classified as *Euchlaena perennis*) and a more widely spread annual (then classified as *Euchlaena mexicana*). The latter was commonly proposed as the ancestor of corn.

26. Today teosinte is generally accepted as the direct progenitor. A useful review of past debates and recent research is found in Doebley, "Genetics of Maize Evolution."

27. Mangelsdorf, *Corn,* vi–ix. See also Birchler, "Paul C. Mangelsdorf."

28. Mangelsdorf and Reeves, "Origin of Maize"; Mangelsdorf and Reeves, "Origin of Indian Corn."

29. Mangelsdorf and Reeves, "Origin of Maize," 311.

30. Mangelsdorf to C. A. Krug, 28 October 1941, Mangelsdorf Papers, shelf 37.6, box 1, folder: Cutler, Hugh.

31. E. Anderson to Mangelsdorf, 17 December 1940, Mangelsdorf Papers, shelf 37.8, box 1, folder: Anderson, Edgar (2 of 2); Browman, "Necrology."

32. Hugh Cutler to Mangelsdorf, 14 May 1942, Mangelsdorf Papers, shelf 37.8, box 3, folder: Cutler, Hugh.

33. Cutler to family, 28 March 1942, private collection of Bill and Elisabeth Cutler.

34. Cutler to Mangelsdorf, 24 August 1942, Mangelsdorf Papers, shelf 37.8, box 3, folder: Cutler, Hugh.

35. Stocking, *Race, Culture, and Evolution;* Kuper, *Invention of Primitive Society.*

36. Davis, "Ethnobotany."

37. Cutler, misc. notes, 16 June 1942, Mangelsdorf Papers, shelf 37.8, box 3, folder: Cutler, Hugh.

38. Mangelsdorf to Cutler, 23 March 1944, Mangelsdorf Papers, shelf 37.8, box 3, folder: Cutler, Hugh.

39. Brockway, *Science and Colonial Expansion;* Drayton, *Nature's Government;* Schiebinger, *Plants and Empire;* Schiebinger and Swan, *Colonial Botany.*

40. Curry, "Imperilled Crops."

41. Bonneuil, "Seeing Nature." See also Müller-Wille and Rheinberger, *Cultural History of Heredity.*

42. See, for example, Elina, Heim, and Roll-Hansen, "Plant Breeding," 166; and Flitner, "Genetic Geographies," 178–79. See also Saraiva, "Breeding Europe."

43. Flitner, "Genetic Geographies."

44. Cutler to family, 28 March 1942, private collection.

45. Survey Commission, "Agricultural Conditions and Problems," Rockefeller Foundation Archives, RG 1.1, series 323, box 1, folder 2. For a retrospective account by the survey team, see Stakman, Bradfield, and Mangelsdorf, *Campaigns against Hunger.*

46. Paul Mangelsdorf, oral history, 1966, Rockefeller Foundation Archives, RG 13, pp. 49, 51–52.

47. There are many accounts of the Rockefeller Foundation's Mexican agricultural program; a thorough historiographical analysis can be found in Gutiérrez Núñez, "Cambio agrario." For an early actors' account, see Stakman, Bradfield, and Mangelsdorf, *Campaigns against Hunger.* For its place in the history of Mexican agriculture and agronomy, see Hewitt de Alcántara, *Modernizing Mexican Agriculture;* Jennings, *Foundations;* Cotter, *Troubled Harvest;* and Barahona, "Mendelism and Agriculture." On its relationship to Cold War geopolitics and the global Green Revolution, see Perkins, *Geopolitics;* and Cullather, *Hungry World.* For accounts (in addition to Gutiérrez Núñez's) of corn breeding, see Olea Franco, "Introducción del maíz híbrido"; Aboites Manrique, *Mirada diferente;* and Matchett, "Untold Innovation."

48. Olsson emphasizes that Wallace influenced the Rockefeller Foundation plans only indirectly; see *Agrarian Crossings,* 120–21. Earlier accounts assert a more direct connection; see Cotter, *Troubled Harvest,* 139; and Perkins, "Rockefeller Foundation," 8.

49. See the discussion in Cullather, *Hungry World,* ch. 2.

50. Survey Commission, "Agricultural Conditions and Problems," Rockefeller Foundation Archives.

51. Matchett, "Untold Innovation," ch. 4.

52. "Comments by Professor Carl Sauer," 10 February 1941, Rockefeller Foundation Archives, RG 1.1, series 323, box 1, folder 2.

53. Olsson observes that Iowa was in fact *not* the model that Mangelsdorf and those who planned the corn-breeding program had in mind. They instead saw comparisons in the southern United States and therefore called for open-pollinated varieties and other types with reusable seed more suitable to the constraints of underresourced Mexican farmers. See *Agrarian Crossings,* ch. 5.

54. E. Anderson to Sauer, 19 August 1943, Sauer Papers, box 6, folder: Anderson, Edgar, 1941–43.

55. Soto Laveaga, *"Largo Dislocare."*

56. Matchett, "Untold Innovation," 55–61; Matchett, "At Odds over Inbreeding," 357. See also Barahona Echeverría and Gaona Robles, "History of Science"; Aboites Manrique, *Mirada diferente,* 85–100; and Barahona, Pinar, and Ayala, *Genética en México,* ch. 3.

57. Matchett, "At Odds over Inbreeding," 358. See also Barahona, "Mendelism and Agriculture."

58. J. George Harrar, diary, August 15 to November 30, 1943, Rockefeller Foundation Archives, RG 1.1, series 323, box 1, folder 6.

59. "Annual Report, Rockefeller Agricultural Program in Mexico," 1943; "Review of the Work of the Oficina de Estudios Especiales," 1947, both in Rockefeller Foundation Archives, RG 6.13, series 1.1, box 2, folder 21.

60. Edwin Wellhausen, oral history, 1966, Rockefeller Foundation Archives, RG 13, 29, 133; see also Wellhausen, Roberts, and Hernández, *Maize in Central America,* 40–43.

61. Hernández Xolocotzi, "Experiences in the Collection," 4–5. Hernández's career is discussed in chapter 5.

62. Harwood, "Peasant Friendly Plant Breeding"; Matchett, "Untold Innovation," 86–96.

63. [Paul Mangelsdorf], report to Ing. Marte Gómez, 15 April 1946, Mangelsdorf Papers, shelf 37.51, box 1, folder: Rockefeller Foundation—PCM Papers/Reports.

64. Mangelsdorf, "Report on a Trip to Mexico . . . February 14 to 27, 1947," Mangelsdorf Papers, shelf 37.51, box 1, folder: Rockefeller Foundation—PCM Papers/Reports.

65. E. Anderson to Mangelsdorf, 15 June 1946, 9 January 1948, Mangelsdorf Papers, shelf 37.8, box 1, folder: Anderson, Edgar (1 of 2).

66. E. Anderson to Mangelsdorf, n.d., Anderson Papers, RG 3/2/4, series 3, Correspondence, box 11, folder 1. See also Kleinman, "His Own Synthesis"; and Kleinman, "Edgar Anderson."

67. Anderson and Cutler, "Races of Zea Mays."

68. For an overview of this history of the term race *(raza),* see Hartigan, *Care of the Species,* ch. 2.

69. For example, the work of Earnest Hooton provided a model of the measurement and statistical analyses of multiple nonadaptive characteristics to aggregate

humans into groups of common descent. Anderson and Cutler, "Races of Zea Mays," 71, 77. See also Curry, "Taxonomy."

70. Anderson and Cutler, "Races of Zea Mays," 71.

71. E. Anderson, "Maize in Mexico," 148.

72. Wellhausen to Mangelsdorf, 20 August 1948, Mangelsdorf Papers, shelf 37.8, box 9, folder: Wellhausen, E.J.

73. Mangelsdorf, oral history, Rockefeller Foundation Archives, 69–70; J. George Harrar to Warren Weaver, [ca. September 1948], Mangelsdorf Papers, shelf 37.8, box 9, folder: Warren Weaver, 1945–48.

74. Mangelsdorf, oral history, Rockefeller Foundation Archives, 72; Wellhausen, Roberts, and Hernández, *Razas de maíz.*

75. Wellhausen, Roberts, and Hernández, *Razas de maíz;* translations and quotations from their 1952 English edition, *Races of Maize,* on 161–62.

76. Doremus, "Indigenism." For an overview of mestizo identity in Mexico, especially its construction in and through science, see López-Beltrán and García Deister, "Aproximaciones científicas."

77. Wellhausen, Roberts, and Hernández, *Races of Maize,* 203.

78. Ibid., quotation from foreword.

79. Paul Mangelsdorf, "A Proposal for the Collection, Classification and Maintenance of Corn Varieties from Latin America," n.d., Mangelsdorf Papers, shelf 37.12, box 8, folder: Jones DF (1 of 2).

CHAPTER THREE. PRESERVE

1. Ralph Cleland to Paul Mangelsdorf, 20 September 1949, Mangelsdorf Papers, shelf 37.10, box 5, folder: Preservation of Indigenous Strains of Maize—Letters; Cleland to Harry Miller, 17 December 1949, Rockefeller Foundation Archives, RG 1.2, series 300, box 1, folder 2. On the Maize Committee, see Curry, "Breeding Uniformity"; and Méndez Rojas, *"Libros del maíz."*

2. Committee on Preservation of Indigenous Strains of Maize (cited hereafter as Maize Committee), Third Report, December 1952, NAS Archives, Biology and Agriculture Division, folder: Agricultural Board Com on Preservation of Maize: Reports, Progress, 1952–54.

3. Bonneuil, "Seeing Nature," 11.

4. Fullilove, *Profit of the Earth,* 140.

5. Veracini, "Introducing Settler Colonial Studies," 4.

6. Maize Committee, "Proposed Plan for the Collection and Maintenance of Native Races of Maize," 3 February 1951, Rockefeller Foundation Archives, RG 1.2, series 300, box 1, folder 2.

7. Ibid.

8. Harry S. Truman, inaugural address, 20 January 1949, Harry S. Truman Library and Museum, last accessed 2021, www.trumanlibrary.gov/library /public-papers/19/inaugural-address.

9. *National Academy of Sciences,* 77.

10. Early on it was proposed that only ten to twenty plants were required to regenerate a sample, but this number was continually revised upward. Maize Committee, "Proposed Plan," Rockefeller Foundation Archives.

11. Mangelsdorf to Cleland, 3 October 1949, Mangelsdorf Papers, shelf 37.10, box 5, folder: Preservation of Indigenous Strains of Maize—Letters.

12. Vessuri, "Foreign Scientists"; Shepherd, "Imperial Science." For a more general overview of the foundation's work in Latin America through the 1950s, see Cueto, *Missionaries of Science.*

13. Perkins, "Rockefeller Foundation"; Cullather, *Hungry World,* ch. 2.

14. "Interview with R. E. Cleland on the Preservation of Maize Stocks in Latin America," 24 January 1950, Rockefeller Foundation Archives, RG 1.2, series 300, box 1, folder 2.

15. Lorek, "Strange Priests"; Lorek, "Developing Paradise."

16. Lorek, "Strange Priests." See also Rockefeller Foundation, *Annual Report* (1947), 166–67.

17. Rockefeller Foundation, *Annual Report* (1950), 164–69.

18. Mejía Prado, "Víctor Manuel Patiño." On Patiño's collecting for the Rockefeller Program in Columbia, see also Lorek, "Developing Paradise."

19. J. George Harrar to Chester Barnard, 29 August 1951, 15 October 1951, Rockefeller Foundation Archives, RG 1.2, series 300, box 1, folder 2.

20. Barnard to Harrar, 27 September 1951, Mangelsdorf Papers, shelf 37.10, box 5, folder Preservation of Indigenous Strains of Maize—Letters.

21. Bonneuil, "Seeing Nature," 10.

22. Curry, "History of Seed Banking."

23. "Research and Marketing," 294; see also Bowers, "Research and Marketing Act."

24. Burgess, *National Program for Conservation,* 15; Shands, Fitzgerald, and Eberhart, "Plant Germplasm Preservation," 101.

25. Burgess, *National Program for Conservation.*

26. Maize Committee, minutes, 26 October 1951, NAS Archives, Biology and Agriculture Division, folder: Agricultural Board Com on Preservation of Maize: Meetings, 1951–58.

27. Saidel and Plonski, "Shaping Modern Science."

28. "Friedrich Gustav Brieger (depoimento, 1977)," Rio de Janeiro, CPDOC, 2010, last accessed 2021, www.fgv.br/cpdoc/historal/arq/Entrevista475.pdf; Friedrich Brieger to Paul Mangelsdorf, 3 December 1941, Mangelsdorf Papers, shelf 37.8, box 3, folder: Foreign A–E.

29. Cleland to Brieger, 8 January 1951, Mangelsdorf Papers, shelf 37.10, box 5, folder: National Academy of Arts and Sciences, 1954– (1 of 2).

30. Maize Committee, minutes, 26 October 1951, NAS Archives; C. O. Erlanson to Milton Lee, 12 January 1951, Mangelsdorf Papers, shelf 37.10, box 5, folder: Preservation of Indigenous Strains of Maize—Letters.

31. Ralph Cleland to Warren Weaver, 7 May 1951, Mangelsdorf Papers, shelf 37.10, box 7, folder C.

32. "Informe preliminar de Victor Manuel Patiño," 29 September 1952, Rockefeller Foundation Archives, RG 1.2, series 300, box 1, folder 3.

33. Lewis Roberts to Harrar, n.d., "Progress of Project to Collect the Corns of the Andean Region," Rockefeller Foundation Archives, RG 1.2, series 300, box 1, folder 3.

34. Edwin Wellhausen to Alfredo Carballo, 6 May 1952, Rockefeller Foundation Archives, RG 6.13, series 1.1, box 11, folder 137.

35. See correspondence in Rockefeller Foundation Archives, RG 6.13, series 1.1, box 11, folder 137. On Carballo, see Méndez Rojas, "Técnicos o especialistas."

36. Wellhausen to Harrar, 22 December 1952, Rockefeller Foundation Archives, RG 1.2, series 300, box 1, folder 3.

37. Maize Committee, Fourth Report, January 1953, NAS Archives, Biology and Agriculture Division, Folder Agricultural Board Com on Preservation of Maize: Reports, Progress, 1952–54.

38. Friedrich Brieger, "Report on a Collecting Trip to Argentina and Uruguay, July 1952," Rockefeller Foundation Archives, RG 1.2, series 300, box 1, folder 3; Mangelsdorf to J. A. Clark, 9 February 1953, Mangelsdorf Papers, shelf 37.10, box 5, folder: National Academy of Arts and Sciences, 1954– (2 of 2).

39. See correspondence with and about Brieger in Mangelsdorf Papers, shelf 37.12, box 9, folder: Maize Committee; and also shelf 37.10, box 5, folder: National Academy of Arts and Sciences, 1954– (2 of 2).

40. Brieger to Cleland, 10 February 1951, Mangelsdorf Papers, shelf 37.10, box 5, folder: National Academy of Arts and Sciences, 1954– (1 of 2).

41. Maize Committee, Fifth Report, May 1953, NAS Archives, Biology and Agriculture Division, Folder Agricultural Board Com on Preservation of Maize: Reports, Progress, 1952–54.

42. Brieger to Cleland, 10 February 1951, Mangelsdorf Papers.

43. "Program of the Brasilian Corn Center, 1956–1957," [ca. June 1956], Mangelsdorf Papers, shelf 37.12, box 9, folder: Maize Committee.

44. Maize Committee, Seventh Report, October 1953, NAS Archives, Biology and Agriculture Division, Folder Agricultural Board Com on Preservation of Maize: Reports, Progress, 1952–54.

45. Higman, Concise History.

46. Duvick, "William L. Brown."

47. Maize Committee, Third Report, NAS Archives.

48. William Brown, diary of collecting trip to West Indies, 1953, Brown Papers, box 5, folder: ENV.

49. William Brown to Edgar Anderson, 11 February 1952, Anderson Papers, series 3, box 12, folder 1.

50. W. Brown, diary, Brown Papers.

51. W. Brown, Maize in the West Indies, 2–3.

52. TallBear, Native American DNA, 151.

53. W. Brown, *Maize in the West Indies,* 2. On Brown's use of Caribbean maize in breeding, see Brown to H. F. Robinson, 30 December 1958, Brown Papers, box 5, folder 10.

54. William Hatheway to Mangelsdorf, 20 April 1953, 23 April 1953, 28 April 1953, Mangelsdorf Papers, shelf 37.10, box 3, folder: Hatheway, Wm. H.

55. Hatheway to Mangelsdorf, 2 May 1953, Mangelsdorf Papers, shelf 37.10, box 3, folder: Hatheway, Wm. H.

56. Hatheway to Mangelsdorf, 13 May 1953, 18 May 1953, Mangelsdorf Papers, shelf 37.10, box 3, folder: Hatheway, Wm. H.

57. Hatheway to Mangelsdorf, 16 September 1954, 10 October 1954, Mangelsdorf Papers, shelf 37.10, box 3, folder: Hatheway, Wm. H.

58. Hatheway, *Maize in Cuba.*

59. Mangelsdorf to Hatheway, 21 November 1955, Mangelsdorf Papers, shelf 37.12, box 7, folder: Hatheway, W. H.

60. Hatheway to Mangelsdorf, 11 May 1956, Mangelsdorf Papers, shelf 37.12, box 7, folder: Hatheway, Wm. H.

61. Hatheway, *Maize in Cuba,* 8, with references to "mongrel" types throughout.

62. Ibid., 14–16.

63. "Informe preliminar," Rockefeller Foundation Archives; "Llegada de un funcionario de la Fundación Rockefeller," *El Comercio* (Lima, Peru), 4 July 1952; Roberts to Víctor Patiño, 19 August 1952, 21 August 1952; Roberts to Alexander Grobman, 21 August 1952; and additional correspondence in Rockefeller Foundation Archives, RG 1.2, series 300, box 1, folder 3. See also G. W. Gray to Harrar, 24 August 1953, Rockefeller Foundation Archives, RG 1.2, series 300, box 1, folder 4.

64. For a detailed description, see Clark to Harrar, 31 March 1954, Rockefeller Foundation Archives, RG 1.2, series 300, box 1, folder 5.

65. Brieger to Mangelsdorf, 17 August 1955, Mangelsdorf Papers, shelf 37.12, box 9, folder: Maize Committee; Cleland to Brieger, 23 September 1955, Mangelsdorf Papers, shelf 37.12, box 4, folder: Cleland, Ralph E.

66. Mangelsdorf to E. Anderson, n.d., Anderson Papers, series 3, box 14, folder 22.

67. See, for example, Roberts et al., *Maize in Colombia,* 18–20; and Hatheway, *Maize in Cuba,* 13–14.

68. Wellhausen, Fuentes, and Hernández Corzo, *Maize in Central America,* 29.

69. Grobman, Salhuana, and Sevilla, *Maize in Peru,* 38.

70. Brieger et al., *Maize in Brazil,* 4–5, 56–57.

71. Bonneuil, "Seeing Nature," 12.

72. Ramírez et al., *Maize in Bolivia,* 2.

73. Grant et al., *Maize in Venezuela,* 4–5, 11.

74. Clark, "Collection, Preservation, and Utilization." See also Maize Committee, "Collections of Original Strains of Corn I," National Academy of Sciences–National Research Council, 1954, 3–13. Photocopies of this volume and a second of 1955 can be found in the Maize Committee files, NAS Archives.

75. The estimate of two hundred seeds is from M. Goodman, "Evaluation and Critique," 209.

76. Cleland to Weaver, 7 May 1951, Mangelsdorf Papers.

77. Burgess, *National Program for Conservation,* 15–16, 22; Report of the Subcommittee on National Seed Storage, 27 February 1950, ARS Records, 170 80–12–06, box 6, folder: National Seed Storage Laboratory—1957.

78. Maize Committee, minutes, 22 April 1957, NAS Archives, Biology and Agriculture Division, folder: Agricultural Board Com on Preservation of Maize: Executive Com, 1953–60.

CHAPTER FOUR. COPY

1. Otto Frankel, quotation in Bennett, *FAO/IBP Technical Conference,* 9.

2. On gene banks as objects of conservation, see Peres, "Saving the Gene Pool."

3. Fowler, "Svalbard Seed Vault"; see also Fowler, *Seeds on Ice.*

4. Radin and Kowal, *Cryopolitics,* 9, image of the seed vault on 74.

5. This concept was characterized to great influence in Ehrlich, *Population Bomb.*

6. Cullather, "Stretching the Surface"; Connelly, "To Inherit the Earth." Population concerns have a longer history; see Bashford, *Global Population.*

7. See, for example, Frankel, "Development and Maintenance," 91; H. Harlan, *One Man's Life,* 88–89; R. Whyte, *Plant Exploration,* 97; J. Harlan, "Distribution and Utilization," 205; and E. James, "Perpetuation and Protection."

8. Claims for the usefulness of crop wild relatives in plant breeding had been made in the preceding decades and by the 1960s were gaining greater visibility. See Plucknett et al., *Gene Banks;* and Hajjar and Hodgkin, "Use of Wild Relatives."

9. Dodd, "Food and Agriculture Organization."

10. Trentmann, "Coping with Shortage." See also Vernon, *Hunger,* chs. 4 and 5.

11. Staples, "To Win the Peace"; Staples, *Birth of Development;* Marchisio and Di Blase, *Food and Agriculture Organization;* Jarosz, "Political Economy."

12. FAO Subcommittee on Plant and Animal Stocks, minutes, 26–30 May 1947; minutes, 19–23 April 1948, both in NAS Archives, Biology and Agriculture Division, folder: Subcom on Plant and Animal Stocks, 1947–48.

13. R. Whyte, *Plant Exploration,* 90–92.

14. "Recommendation Regarding FAO Activities in the Field of Plant Exploration, Collection and Introduction," *Plant Introduction Newsletter* 1 (November 1957), 1–3. See the discussion of FAO activities in Pistorius, *Scientists, Plants and Politics,* esp. 10–16.

15. *FAO Technical Meeting.*

16. Ibid., 53, 49.

17. Chapman and Heald, "Agrobiodiversity Loss"; see also Brush, *Farmers' Bounty.*

18. J.J. de Jong, "Good Seed Doesn't Cost—It Pays," *Foreign Agriculture* 25, no. 1 (1961), 3; Pistorius and Van Wijk, *Plant Genetic Information,* 92–93.

19. "Seed Program Beats Nature at Its Own Game," February 1965, FAO Feature FR-361/65, Frankel Papers, box 12, folder 171.

20. *FAO Technical Meeting,* 31, 35.

21. On conservation and decolonization, see Adams and Mulligan, *Decolonizing Nature;* Matheka, "Decolonisation"; Schauer, "Imperial Ark"; and De Bont, "World Laboratory."

22. *FAO Technical Meeting,* 53. See also Fenzi and Bonneuil, "Genetic Resources."

23. See, for example, R. Whyte, *Plant Exploration,* 97; J. Harlan, "Distribution and Utilization," 205; and *FAO Technical Meeting,* 9.

24. See, for example, W. Rudolf, "Problems of Collection, Maintenance and Evaluation of Wild Species of Cultivated Plants," *Plant Introduction Newsletter* 5 (June 1959): 4.

25. Escobar, *Encountering Development,* ch. 4.

26. Ibid., ch. 3. See also Rist, *History of Development;* and Cullather, "Development."

27. Cullather, *Hungry World,* 7.

28. Cullather, "Stretching the Surface."

29. Perkins, *Geopolitics.* Here I discuss agricultural aid, but the foundation also (re)engaged with population control in the early 1960s. See Connelly, *Fatal Misconception.*

30. J. George Harrar and Warren Weaver, "Research on Rice," October 1954, Rockefeller Foundation Archives, RG 3.1, series 915, box 3, folder 23. On the Mexican program as a model, see Cullather, *Hungry World,* ch. 2.

31. J. George Harrar, Paul C. Mangelsdorf, and Warren Weaver, "Notes on Indian Agriculture," 11 April 1952, Rockefeller Foundation Archives, RG 1.2, series 460, box 1, folder 4.

32. Perkins, *Geopolitics,* 153–54.

33. Harrar and Weaver, "Research on Rice," Rockefeller Foundation Archives. On the history of IRRI, see Oasa, "International Rice Research Institute"; R. Anderson, "Origins"; Cullather, "Miracles of Modernization"; and Smith, "Imaginaries of Development." An institutional account of IRRI is in Chandler, *Adventure in Applied Science.*

34. Méndez Rojas, "Programa Cooperativo Centroamericano." See also Sterling Wortman, "Report of the Central American Corn Project," 9 August 1954, Rockefeller Foundation Archives, RG 6.13, series 1.1, box 7, folder 93; First Annual Report, September 1954–August 1955, Rockefeller Foundation Archives, RG 6.13, series 1.1, box 6, folder 69; and Rockefeller Foundation, *Annual Report* (1959), 28–29.

35. E.J. Wellhausen, "A Summary of Activities . . . 1959," Rockefeller Foundation Archives, RG 6.13, series 1.1, box 30, folder 332.

36. Rockefeller Foundation, *Annual Report* (1959), 30.

37. Fuente Hernández et al., *Investigación agrícola,* 23–24.

38. Matchett, "Untold Innovation," 183.

39. Perkins, *Geopolitics,* 231.

40. [Edwin Wellhausen], "Inter American Maize Program," n.d., Rockefeller Foundation Archives, RG 6.13, series 1.1, box 30, folder 334. This collection stayed in Chapingo, home of the original Office of Special Studies field station, but the station itself was incorporated into an institutional complex that served as the new INIA headquarters.

41. The history of the site's development as a research center in the 1960s is discussed in Caire-Pérez, "Different Shade of Green"; see also Fuente Hernández et al., *Investigación agrícola,* 29–32.

42. Rockefeller Foundation, *Annual Report* (1959), 30–31.

43. Matchett, "Untold Innovation," 193–98, 205–8.

44. [Wellhausen], "Inter American Maize Program," Rockefeller Foundation Archives.

45. "Colección, clasificación, conservación y evaluación de los linajes de maíz indígena en la América Latina," CIMMYT Archives, Director's Office Files, box 1, folder 29.

46. Rockefeller Foundation, *Annual Report* (1964), 30–31; CIMMYT, *1966–67 Report.* See also Byerlee, *Birth of CIMMYT.*

47. Rockefeller Foundation, *Annual Report* (1963).

48. Perkins, *Geopolitics,* 230–31; Borlaug, "Wheat Breeding," 12.

49. Cullather, "Miracles of Modernization"; Herdt and Capule, *Adoption, Spread, and Production.*

50. William S. Gaud, "The Green Revolution: Accomplishments and Apprehensions," 8 March 1968, AgBioWorld, last accessed 2021, www.agbioworld .org/biotech-info/topics/borlaug/borlaug-green.html.

51. For a persuasive case that the agricultural aid associated with the Green Revolution never aimed to alleviate hunger but instead sought to promote commercialization of agriculture, see Harwood, "Green Revolution."

52. See, for example, Creech and Reitz, "Plant Germ Plasm Now," 5 (corn mentioned on 24); National Research Council, *Genetic Vulnerability,* 60; and TAC Ad Hoc Working Group, "The Collection, Evaluation and Conservation of Plant Genetic Resources," March 1972, Rockefeller Foundation Archives, RG 1.3, series 103D, box 16, folder 101.

53. "Recommendations of the Conference," in Bennett, *FAO/IBP Technical Conference,* 4; *Report of the FAO/IBP,* 1.

54. Fenzi and Bonneuil, "Genetic Resources."

55. Guha, *Environmentalism;* Gottlieb, *Forcing the Spring.* See also Robertson, *Malthusian Moment;* and Egan, "Survival Science."

56. "International Biological Program," 29 November 1963, NAS Archives, IBP, series 1 USNC/IBP: Ad Hoc, folder: Meetings: 3–4 December 1963: General. See also Worthington, *Evolution of IBP;* Kwa, "Representations of Nature"; Schleper, "Life on Earth"; and Schleper, "Conservation Compromises."

57. Worthington, "International Biological Programme," 226.

58. Radin, *Life on Ice,* 6; Radin, "Latent Life." See also Ventura Santos, "Indigenous Peoples."

59. "Survey, Conservation and Utilization of Genetic Resources in Plants Useful to Man," 12 April 1965, Frankel Papers, box 12, folder 171.

60. Ibid. Frankel later remembered the IBP committee that ostensibly generated these materials as "largely a fiction" that nonetheless helped generate action within FAO. See Otto Frankel to Garrison Wilkes, 7 April 1982, Frankel Papers, box 17, folder 245.

61. E.C. Graham, "Committee E. Use and Management of Biological Resources," 15 January 1964, NAS Archives, IBP, series 1 USNC/IBP: Ad Hoc, folder: Task Forces: Use and Management of Biological Resources: 1963–64; "Plant Gene Pools," *IBP News,* no. 5 (March 1966): 48–55.

62. Frankel and Bennett, *Genetic Resources in Plants,* 15–16.

63. Bennett, *FAO/IBP Technical Conference,* 9, 63, 111.

64. Frankel and Bennett, *Genetic Resources in Plants,* 16.

65. See "Editorial" and "Activities of the Crop Ecology and Genetic Resources Branch," *Plant Introduction Newsletter* 22 (July 1969): 1–4, 5–6.

66. See iterations of these plans in *Report of the Third Session;* and *Report of the Fourth Session.*

67. *Report of the Fourth Session.*

68. Chandler, *Adventure in Applied Science,* 103–4.

69. CIMMYT, *1968–69 Report,* 50–51.

70. Sterling Wortman, diary, 17 February 1967, Rockefeller Foundation Archives, RG 12, box 528, folder: Wortman, Sterling L., 1967 (3 of 10).

71. Norman Borlaug to Keith Finlay, 12 January 1970, Rockefeller Foundation Archives, RG 1.3, series 103D, box 18, folder 111.

72. Finlay to Frankel, 26 November 1969, CIMMYT Archives, Director's Office Files, box 27, folder 77; Finlay to R.J. Pichel, 9 December 1969, FAO Archives, RG 10, PL 2/9, box 10AGP50, folder: November 1969–April 1971, vol. 4.

73. Mario Gutiérrez to Finlay, 10 May 1976, Rockefeller Foundation Archives, RG 1.3, series 103D, box 18, folder 115.

74. Sterling Wortman, "World Germplasm Project," 5 March 1969, Rockefeller Foundation Archives, RG 1.3, series 103D, box 16, folder 101.

75. Ibid.

76. Frankel, "Survey."

77. Wortman, "World Germplasm Project," Rockefeller Foundation Archives. See also Curry, "Working Collections."

78. "Meeting on Germ Plasm Preservation, November 18, 1969"; "Germ Plasm Committee Meeting, Wednesday, November 19, 1969," both in Rockefeller Foundation Archives, RG 1.3, series 103D, box 18, folder 111.

79. Maize Germ Plasm Committee, minutes, June 1970, Rockefeller Foundation Archives, RG 1.3, series 103D, box 18, folder 111; William Brown, handwritten notes, June 1970, Brown Papers, box 6, folder 2.

80. Ibid. See also Timothy and Goodman, "Germplasm Preservation."

81. Major Goodman to Lewis Roberts, 22 October 1969, Rockefeller Foundation Archives, RG 1.2, series 100D, box 25, folder 178.

82. Maize Germ Plasm Committee, minutes, Rockefeller Foundation Archives.

83. Ibid.

84. Meeting on the Status of Germ Plasm Collection, Preservation, Evaluation and Utilization, minutes, 27–28 September 1971, Rockefeller Foundation Archives, RG 1.3, series 103D, box 16, folder 101.

85. A critical history of CGIAR remains to be written. For a history of CGIAR's founding written by its second chair, see Baum, *Partners against Hunger*. Other general accounts include Greenland, "International Agricultural Research"; and Alston, Dehmer, and Pardey, "International Initiatives." For an account that contextualizes the emergence of the CGIAR system amid earlier collaborations, see Byerlee and Lynam, "Invention."

86. *Agricultural Development,* foreword, iv–v.

87. Wortman, "Technological Basis," 39.

88. F. F. Hill to McGeorge Bundy, J. George Harrar, David Bell, and William Myers, 1 August 1968, Ford Foundation Archives, FA 624, series IVA, box 26, folder 685; Hill to Lowell Hardin, 23 July 1973, Ford Foundation Archives, FA 624, series IVA, box 26, folder 682.

89. Wortman, "Technological Basis," 20–21, 40.

90. Lowell S. Hardin, "Accelerating Agricultural Modernization in Developing Nations," 1970, CGSpace, http://hdl.handle.net/10947/89, 1, 2.

91. CGIAR Secretariat, "Resolution: Objectives, Composition and Organizational Structure," May 1971, CGSpace, http://hdl.handle.net/10947/5254.

92. CGIAR Secretariat, "CGIAR First Meeting, Washington DC, 19 May 1971: Summary of Proceedings," CGSpace, http://hdl.handle.net/10947/260.

93. CGIAR TAC Secretariat, "Technical Advisory Committee First Meeting, 29 June–2 July 1971: Report," CGSpace, http://hdl.handle.net/10947/1422.

94. See, for example, Franz Kaps, Telex to P. A. Oram, 21 November 1972, World Bank Archives, folder 1759681; Erna Bennett, interview with Gregg Borshmann, November 1994, transcript, ORAL TRC 3179, National Library of Australia; Pistorius, *Scientists, Plants and Politics;* and Pistorius and Van Wijk, *Plant Genetic Information.*

95. TAC Ad Hoc Working Group, "Collection, Evaluation and Conservation," Rockefeller Foundation Archives.

96. See, for example, Harold Graves to John Pino, 21 August 1973, Rockefeller Foundation Archives, RG 1.3, series 103D, box 16, folder 103; and M. Gucovsky, "Report on Third Meeting," 1 May 1972, World Bank Archives, folder 1759681.

97. Sterling Wortman, meeting notes, 20 August 1973, Rockefeller Foundation Archives, RG 1.3, series 103D, box 16, folder 103; see also Wortman, memo, 31 January 1972; and Lewis Roberts, meeting notes, 27 March 1972, both in Rockefeller Foundation Archives, RG 1.3, series 103D, box 16, folder 101.

98. Transcript enclosed in Graves to Pino, Rockefeller Foundation Archives; Lewis Roberts, "Collection, Preservation and Evaluation of the World's Plant

Genetic Resources," 17 September 1973, Rockefeller Foundation Archives, RG 1.3, series 103D, box 16, folder 103.

99. Transcript enclosed in Graves to Pino, Rockefeller Foundation Archives.

100. CGIAR Secretariat, "Summary of Proceedings of CGIAR Meeting, Washington DC," 1–2 November 1973, CGSpace, https://cgspace.cgiar.org/handle /10947/293. See also IBPGR, "Terms of Reference and Operational Rules and Procedures," Rockefeller Foundation Archives, RG 1.3, subseries 103D, box 20, folder 127. Further correspondence related to the initial configuration and funding of IBPGR is found in World Bank Archives, folder 1762066.

101. Roberts to Richard Demuth, 2 June 1976; Pichel to Finlay, 23 December 1975, both in FAO Archives, RG 10, PL 3/11, box 10AGP108, folder: October 1975– December 1979.

102. Robert McK. Bird to Pino, 15 December 1975, Rockefeller Foundation Archives, RG 1.3, series 103D, box 18, folder 114.

103. Major Goodman to Charles Lewis, 14 June 1977, FAO Archives, RG 10, PL 3/11, box 10AGP108, folder: October 1975–December 1979.

104. Ernest Sprague and Finlay, "Current Status of Plant Resources and Utilization," n.d., Rockefeller Foundation Archives, RG 1.3, series 103D, box 18, folder 115.

105. William Brown to Howard Hyland, 8 February 1977, Brown Papers, box 6, folder 12; Report on IBPGR Maize Germplasm Advisory Committee, Second Meeting, 3–4 March 1977, CIMMYT, Mexico, FAO Archives, RG 10, PL 3/11, box 10AGP108, folder: October 1975–December 1979.

106. For a list of the IBPGR's envisioned base collections, see IBPGR, *Annual Report 1977*, 38. See also Peres, "Seed Banking as Cryopower." I am grateful to Sara Peres for many inspiring hours spent interpreting this history.

107. Radin, "Latent Life."

108. Brennan, "Making Data Sustainable." See also Edwards, *Closed World*, 104–6.

CHAPTER FIVE. NEGOTIATE

1. "Mexican Proposal," n.d.; Transcript (first draft), C 81/II/PV/18, both in FAO Archives, LEGL PR 3/11(A), folder: IBPGR-Germplasm Bank, vol. 1.

2. "Mexican Proposal," FAO Archives.

3. Aoki, *Seed Wars*.

4. Chapman and Heald, "Agrobiodiversity Loss," 165–66.

5. Caire-Pérez, "Different Shade of Green," ch. 3. See also Aboites, *Mirada diferente.*

6. Caire-Pérez, "Different Shade of Green," 226–27; Lozano Toledano and Anaya Pérez, "Plan Chapingo." See also McPherson, *Yankee No.*

7. Caire-Pérez, "Different Shade of Green," ch. 5. See also Núñez Gutiérrez, Reyes Canchola, and Ocampo Ledesma, *Huelga nacional.*

8. Pensado, *Rebel Mexico;* Aviña, *Specters of Revolution.*

9. Ortega Paczka, "Reorganización del mejoramiento genético," 370.

10. Barkin, "Persistencia de la pobreza."

11. The agronomist Gilberto Palacios de la Rosa, a maize breeder and director of the National School of Agriculture in the 1960s and early 1970s, played an important role in these shifts, including new directions in maize breeding. See Palacios Rangel and Ocampo Ledesma, *Gilberto Palacios;* and Ortega Paczka, "Don Gilberto Palacios."

12. Ángeles Arrieta, "Maíz y el sorgo," 390–91; Sánchez González and Ordaz Suárez, *Reestudio de las razas,* 11.

13. Ortíz Cereceres, "Antecedentes de la investigación"; Fuente Hernández et al., *Investigación agrícola;* Cotter, *Troubled Harvest,* ch. 7.

14. Wellhausen, "Agriculture of Mexico"; Barkin, "End to Food Self-Sufficiency"; Bartra and Otero, "Agrarian Crisis."

15. Ortega Paczka, "Reorganización del mejoramiento genético."

16. INIA, *Años de investigación agricola;* Fuente Hernández et al., *Investigación agrícola,* ch. 2.

17. Hernández Xolocotzi, introd. to "Xolocotzia," 15–24; Ortega Paczka, "Maíz y sus investigadores"; Cruz León et al., "Obra escrita"; Ortega Paczka, "Vida y aportes."

18. Caire-Pérez, "Different Shade of Green," 165–66, 168–74. Hernández's participation in an influential Mexican botanical commission on *Dioscorea* undoubtedly shaped these views; see Soto Laveaga, *Jungle Laboratories,* ch. 4.

19. Caire-Pérez, "Different Shade of Green," 273, 279–83.

20. Hernández Xolocotzi, Ramos Rodríguez, and Martínez Alfaro, "Etnobotánica," 114; Hernández Xolocotzi, "Concepto de Etnobotánica," 5. See also Toledo, "New Paradigms"; Gliessman, "Agroecology"; and Astier et al., "Back to the Roots."

21. Hernández Xolocotzi and Alanís Flores, "Estudio morfológico." On the purposes of this collecting mission, see Hernández Xolocotzi, *Apuntes sobre la exploración,* 7; and CIMMYT, *1968–69 Report,* 47.

22. Caire-Pérez, "Different Shade of Green," 281–82; Fábregas Puig, "Ecología cultural política," 172. See also Hewitt de Alcántara, *Anthropological Perspectives,* ch. 5.

23. Soto Laveaga, *Jungle Laboratories,* ch. 5, quotations on 131–32.

24. Ángeles Arrieta, "Maíz y el sorgo," 385; Ortega Paczka and Ángeles Arrieta, "Maíz," 75, 76.

25. Ortega Paczka and Ángeles Arrieta, "Maíz," 79. See discussion in Efraím Hernández Xolocotzi, "Exploración etnobotánica."

26. Montes Meneses, "Estrategia para la conservación," 33; Ortega Paczka, "Evaluación de recursos genéticos," 37.

27. INIA, *Unidad de Recursos Genéticos,* 23.

28. Griffin, *Political Economy,* xxi, xi. The quotations are drawn from the prefaces to the first and second editions, respectively (which are both included in the edition cited here).

29. Pearse, "Technology and Peasant Production"; Pearse, *Seeds of Plenty.*

30. Hewitt de Alcántara, *Modernizing Mexican Agriculture,* xiv.

31. Pearse, "Technology and Peasant Production," 129.

32. Hewitt de Alcántara, *Modernizing Mexican Agriculture,* 49–51.

33. S. E. Sanderson, *Transformation of Mexican Agriculture,* ch. 1; Ochoa, *Feeding Mexico,* chs. 7, 8.

34. Hewitt de Alcántara, *Modernizing Mexican Agriculture,* 136; S. E. Sanderson, *Transformation of Mexican Agriculture,* 8.

35. Barkin, "Persistencia de la pobreza."

36. Barkin, "Mexico's Albatross."

37. Ibid., 72.

38. Feder, "New Penetration," 5, 8, 12.

39. Harwood, *Green Revolution.*

40. J. Anderson, *Industrializing the Corn Belt;* USDA, *1969 Census of Agriculture;* historical data are available from USDA, National Agricultural Statistics Service, last accessed 2021, http://quickstats.nass.usda.gov.

41. Hamilton, *Supermarket USA.*

42. McDonald, *Food Power,* 161. See also McGlade, "More a Plowshare."

43. Horsfall, "Fire Brigade Stops," 105–6.

44. Philip Shabecoff, "Corn Crop Down as Blight Spreads," *New York Times,* 3 October 1970, 1; Ullstrup, "Southern Corn Leaf Blight," 42.

45. Mangelsdorf, *Corn,* 239.

46. See different accounts in Duvick, "Cytoplasmic Male-Sterility"; Mangelsdorf, *Corn,* 239–40; and Kloppenburg, *First the Seed,* 105–16.

47. Wade, "Message."

48. National Research Council, *Genetic Vulnerability,* 15, 288–89.

49. Mooney, *Seeds of the Earth,* 11.

50. Mooney, *Seeds of the Earth,* foreword, vi–vii; GRAIN, "A Decade in Review," *Seedling,* 25 February 1993, www.grain.org/en/article/498-a-decade-in-review.

51. Mooney, *Seeds of the Earth.*

52. Lappé and Collins, *Food First.* On Fowler's career, see John Seabrook, "Sowing for Apocalypse," *New Yorker,* 25 August 2007.

53. Erna Bennett, interview with Gregg Borshmann, November 1994, transcript, ORAL TRC 3179, National Library of Australia, 54.

54. Schurman and Munro, "Ideas"; Schurman and Munro, *Fighting for the Future,* 61–65.

55. Bugos and Kevles, "Plants as Intellectual Property"; Fowler, "Plant Patent Act." On the early history of the ICDA Seeds Campaign, see GRAIN, "Seedling Is Not a Full Grown Plant!" originally published in *Seedling,* 15 February 1993, www.grain.org/article/entries/499-seedling-is-not-a-full-grown-plant.

56. Fowler, *Unnatural Selection.*

57. Bugos and Kevles, "Plants as Intellectual Property."

58. Fernandez-Cornejo, *Seed Industry.* Other accounts include Claffey, "Patenting Life Forms."

59. Fernandez-Cornejo, *Seed Industry;* Duvick, "United States."

60. Kloppenburg, *First the Seed,* ch. 6. For an earlier influential account of the relationship between public research and private industry in agriculture, see Hightower, *Hard Tomatoes.*

61. Kloppenburg, *First the Seed,* 128.

62. See Kloppenburg, *First the Seed,* where he argues that, had open-pollinated varieties been pursued as vigorously, the same gains in yield might have been realized. See also a similar analysis by Berlan and Lewontin, "Political Economy."

63. Cited in Fernandez-Cornejo, *Seed Industry,* 10.

64. Kloppenburg, *First the Seed,* 122.

65. My account of the Mexican seed industry in this period draws from L. Goodman, *Improved Seed Industry;* Suárez, "Semillas"; and Barkin and Suárez, *Fín del principio.* See also Luna Mena et al., "Perspectivas de desarrollo."

66. See L. Goodman, *Improved Seed Industry,* 140–42, tables 4.3, 4.4.

67. Ibid., 162.

68. DeWalt, "Mexico's Second Green Revolution"; S. E. Sanderson, *Transformation of Mexican Agriculture;* Barkin and DeWalt, "Sorghum"; Barkin and Suárez, *Fín del principio.*

69. S. E. Sanderson, *Transformation of Mexican Agriculture,* table 4.7; Esteva, "Food Needs and Capacities," 40–41.

70. See discussion of these data in Barkin, "End to Food Self-Sufficiency," 286; and Montanari, "Conception of SAM," 52.

71. Montanari, "Conception of SAM," 48, 51.

72. Austin and Fox, "State-Owned Enterprises," 68.

73. Ibid., 69; L. Goodman, *Improved Seed Industry,* 170.

74. Barkin and Suárez, *Fín del principio;* Austin and Fox, "State-Owned Enterprises," 69.

75. Octavio Véjar, quoted in Barkin, "SAM and Seeds," 124; see also Barkin and Suárez, *Fín del principio,* 158.

76. L. Goodman, *Improved Seed Industry.*

77. Ibid., 13, 123.

78. Aboites Manrique and Martínez Gómez, "Propiedad intelectual," 239; J. Sanderson, *Plants, People and Practices,* 50.

79. Mooney, "International Non-governmental Organizations," 136–37; Fowler, *Unnatural Selection,* 182.

80. Mooney, "International Non-governmental Organizations," 136–37.

81. Fowler, *Unnatural Selection,* 182.

82. "Mexican Proposal," FAO Archives, 1–2. The proposal had its roots in a document prepared for the Spanish delegation by José Esquinas-Alcázar two years earlier. For actors' perspectives on this history, see Martínez Gómez, *Globalización en la agricultura;* José Ramón López Portillo and Francisco Martínez Gómez, "Annex 5: Some Stories on the Inception of the International Treaty," July 2012, FAO, http://www.fao.org/3/ca4854en/ca4854en.pdf; and Esquinas-Alcázar, Hilmi, and López Noriega, "Brief History."

83. "Mexican Proposal," FAO Archives, 1–4, 12.

84. Transcript (first draft), C 81/II/PV/19, FAO Archives, 2.

85. "Plant Genetic Resources," FAO Conference, twenty-first session, resolution 6/81, 1981, www.fao.org/unfao/govbodies/gsbhome/conference/resolutions/1981/en/.

86. Transcript (first draft), C 81/II/PV/19, FAO Archives, 24.

87. FAO Director General, "Plant Genetic Resources," C 83/25, August 1983, 33. I consulted the copy of this report in FAO Archives, LEGL, PR 3/11(A), folder: IBPGR-Plant Genetic Res-Report by DG, vol. 2.

88. See, for example, the correspondence in FAO Archives, LEGL, PR 3/11(A), folder: IBPGR-Plant Genetic Res-Report by DG, vol. 1.

89. Schurman and Munro, *Fighting for the Future.* See also Krimsky, *Genetic Alchemy;* and Wright, *Molecular Politics.*

90. Hughes, *Genentech;* Klausner, "Then There Were Two." The IPO estimates vary by account.

91. Wright, "Recombinant DNA Technology."

92. Kevles, "Ananda Chakrabarty."

93. See, for example, Mooney, "Law of the Seed."

94. These positions are elaborated in the transcripts of the meeting; see transcripts of 1983 FAO Conference, C 83/II/PV/15 and C 83/II/PV/16, FAO Archives, LEGL, PR 3/11(A), folder: IBPGR-Plant Genetic Res-Report by DG, vol. 2. The statements of the US and Nordic delegations are in the same file.

95. Ibid., 294 (C 83/II/PV/15), 306, 316 (C 83/II/PV/16).

96. Transcript (first draft), C 83/II/PV/17, FAO Archives, LEGL PR 3/11(A) folder: IBPGR-Plant Genetic Res-Report by DG, vol. 2, p. 3.

97. "International Undertaking on Plant Genetic Resources," FAO Conference, twenty-second session, resolution 8/83, November 1983, www.fao.org/unfao/govbodies/gsbhome/conference/resolutions/1983/en/.

98. Teichman, "Mexican State"; Minns, *Politics of Developmentalism,* ch. 4.

99. On the end of SAM and its relation to successor programs, see Austin and Esteva, "Final Reflections."

100. Moncada de la Fuente, "Evolución y perspectivas," 46–47.

101. Fuente Hernández et al., "Desenvolvimiento de la investigación," 15, 21.

102. Caetano de Oliveira and Mendoza Mendoza, "Investigación agropecuaria," 76–79.

103. Ortega Paczka and Ángeles Arrieta, "Maíz," 78; Ortega Paczka, "Reorganización del mejoramiento genético"; Cárdenas Ramos, "Bases para la creación."

104. INIA, *Unidad de Recursos Genéticos,* 8, 9; L. Goodman, *Improved Seed Industry,* 151, 210, 215, 228–29.

105. Gliessman, "Agroecology," 27.

106. See, for example, Cuevas Sánchez, "Recursos fitogenéticos"; and Hernández Xolocotzi and Zárate Aquino, "Agricultura tradicional."

107. Ortega Paczka, "Variedades y razas mexicanas."

108. Major Goodman to Howard Sprague, 18 December 1967, NAS Archives, Biology and Agriculture Division, folder: Agricultural Board Com on Preservation of Maize: General; M. Goodman to Lewis Roberts, 22 October 1969, Rockefeller Foundation Archives, RG 1.2, series: 100D, box 25, folder 178.

109. Transcript of discussion at Plant Breeding Forum, encl. in Gordon McCleary to Plant Breeding Forum, 27 August 1982, Goodman Papers, box 25, folder 2.

110. Quentin Jones, "Progress Report on Latin American Maize Collections"; Jones, "Plan for Preservation and Utilization of the Latin American Maize Accessions," 12 May 1982, both in Maize Crop Advisory Committee, Second Meeting Report, 7 December 1982, Goodman Papers, box 49, folder 4.

111. William Brown to M. Goodman, 21 December 1983, Goodman Papers, box 49, folder 6.

112. The exact number varies by account. For one tally, see "Final Report (11/22/88) for Year 5 of the Latin American Maize Regeneration Project," USDA-SCA 58–7B30–573, Goodman Papers, box 49, folder 2.

113. Mooney, "Law of the Seed," 110–21.

114. Otto Frankel to Garrison Wilkes, 15 August 1989, Frankel Papers, box 17, folder 245; W. Brown to M. Goodman, 21 December 1983, Goodman Papers; Wilkes to Frankel, 7 November 1990, Frankel Papers, box 17, folder 245.

CHAPTER SIX. EVALUATE

1. Major Goodman to John Dudley, 15 June 1981, Goodman Papers, box 25, folder 1.

2. M. Goodman to Duane Hill, 15 February 1984, Goodman Papers, box 25, folder 6.

3. Comment of Robert Hougas in M. Goodman, "Evaluation and Critique," 240.

4. Sutter, "Tropics." Sutter's essay provides an excellent entry point into the further literature.

5. Hayden, *When Nature Goes Public;* Voeks, "Disturbance Pharmacopoeias"; Neimark, "Industrializing Nature"; Raby, *American Tropics.*

6. See, for example, Hayden, "Bioprospecting's Representational Dilemma"; and Osseo-Asare, *Bitter Roots.* For a comparable approach in the history of agriculture, see Fullilove, *Profit of the Earth.*

7. Leonelli, "Data"; Leonelli, "How Data Cross Borders."

8. W. Anderson, "Climates of Opinion"; Osborne, *Nature;* Dunlap, *English Diaspora;* Osborne, "Acclimatizing the World."

9. Bonneuil, "Mendelism."

10. An exception (perhaps among several) was the exploration of tropical maize for the United States associated with the Iowa State College–Guatemala Tropical Research Center in the 1940s. See Melhus, *Plant Research.*

11. Wellhausen, "Improving Corn."

12. Edwin Wellhausen, "Improving North American Corn with Exotic Germ Plasm," n.d., Rockefeller Foundation Archives, RG 6.13, series 1.1, box 61, folder 697.

13. Wellhausen, "Exotic Germplasm."

14. William Brown to H. F. Robinson, 30 December 1958, Brown Papers, box 5, folder 10.

15. William Brown, interview by Donald Duvick, 19 December 1989, 4 January, and 22 March 1990, Brown Papers, Biographical File; W. Brown, notes on Henry Wallace, August 1966, Brown Papers, box 3, folder 3; Sehgal, "Contributions."

16. William Brown, "Current Trends in Corn Breeding," ca. 1968, Brown Papers, box 2, folder 30; W. Brown, "Review of Experience in the Use of Exotic Germplasm," 3 March 1969, Brown Papers, box 2, folder 30; Duvick, "William L. Brown," 11.

17. National Research Council, *Genetic Vulnerability,* 1.

18. On the development of public-private relations in corn breeding, see Fitzgerald, *Business of Breeding.*

19. National Research Council, *Genetic Vulnerability,* ch. 8.

20. William Brown, "Use of Exotic Germplasm in Breeding," 9 March 1972, Brown Papers, box 1, folder 8.

21. For an account of breeding strategies within Pioneer Hi-Bred from the 1930s to the 1960s, including a discussion of recycling, see Donald N. Duvick, "Genesis and Evolution of Plant Breeding in the Hi-Bred Corn Company," Goodman Papers, box 5, folder 11.

22. National Research Council, *Genetic Vulnerability,* ch. 8.

23. Ibid., 114.

24. Charles W. Stuber, "Exotic Sources for Broadening Genetic Diversity in Corn Breeding Programs," in *Thirty-Third Annual Corn and Sorghum Research Conference* (1978), collected in Goodman Papers, box 19, folder 3. See additional materials documenting Stuber's research in the same file.

25. Hallauer and Sears, "Integrating Exotic Germplasm." This project antedated the 1970 blight.

26. National Research Council, *Genetic Vulnerability,* ch. 16.

27. Ad Hoc Subcommittee, *Recommended Actions and Policies,* 31.

28. ARS, *National Plant Germplasm System;* W. Brown, "Conservation of Gene Resources."

29. US Comptroller General, *Report to the Congress,* iii–iv.

30. National Plant Genetic Resources Board, *Plant Genetic Resources.*

31. Ibid., 14.

32. Ibid., 16–17. See also Frey, "National Plant Breeding Study."

33. William Brown, "Symposium: Germplasm, Collection and Multiplication," 8 August 1979, Brown Papers, box 1, folder 3.

34. Charles W. Stuber, "Use of Exotic Sources of Germplasm for Maize Improvement," n.d., Goodman Papers, box 19, folder 3.

35. P. K. Bretting and P. H. Sisco, "Germplasm Enhancement of Maize at North Carolina State University," n.d., Brown Papers, box 2, folder 5.

36. National Plant Genetic Resources Board, *Plant Genetic Resources,* 18.

37. "Germplasm Resources Information (GRIP) Update," Maize Crop Advisory Committee, First Meeting Report, 1981, Goodman Papers, box 49, folder 4.

38. "A National Program for Germplasm Enhancement," encl. in Douglas Dewey to Crop Advisory Committee Chairmen, 20 May 1982, Goodman Papers, box 49, folder 5; see also, in the same file, Dewey to Crop Advisory Committee Chairmen, 17 March 1982.

39. W. Brown to Dewey, 25 May 1982, Goodman Papers, box 49, folder 5.

40. M. Goodman to Dewey, 6 April 1982, Goodman Papers, box 49, folder 5.

41. "Germplasm Resources Information," Goodman Papers, 3.

42. Quentin Jones, "Progress Report on Latin American Maize Collections," 26 November 1982, Maize Crop Advisory Committee, Second Meeting Report, 1982, Goodman Papers, box 49, folder 4.

43. Ibid.; Maize Crop Advisory Committee, "Plan for Preservation and Utilization of the Latin American Maize Accessions," n.d., Goodman Papers, carton 113, folder 20.

44. Ricardo Sevilla to M. Goodman, 1 March 1994, Goodman Papers, box 17, folder 5. See also M. Goodman to Gabriel Motes Llamas, 22 September 1987, Goodman Papers, box 26, folder 6; M. Goodman to Manuel Torregnoza, 13 October 1989, Goodman Papers, box 27, folder 6; and "Report on Trip to Colombia and Peru, 4/7/25–4/2/85," n.d., Goodman Papers, carton 112, folder 28.

45. "Report 10/20/87, Latin American Maize Regeneration Project," 20 October 1987, Goodman Papers, box 38, folder 3.

46. "Final Report (11/22/88) for Year 5 of the Latin American Maize Regeneration Project," 22 November 1988, Goodman Papers, box 38, folder 3.

47. M. Goodman, "Evaluation and Critique," 208.

48. Ibid., appended transcript.

49. See proposal drafts from 1982 in Goodman Papers, box 49, folder 5.

50. Castillo-González and Goodman, "Agronomic Evaluations," 855.

51. Pioneer Hi-Bred, "Seed Company and Federal Government Announce Unique Partnership," n.d., Goodman Papers, box 13, folder 3.

52. W. Brown to A. Forrest Troyer, 12 April 1988, Brown Papers, box 3, folder 9; see also W. Brown, "The Evolutionary Aspects of Plant Breeding," 10 February 1981, Brown Papers, box 1, folder 1.

53. William Brown, [untitled], n.d., Brown Papers, box 1, folder 2; Members of the Seed Industry, "Proposal for Maize Germplasm Support," 2 October 1984, Maize Crop Advisory Committee, Meeting Report, 1984, Goodman Papers, box 49, folder 4.

54. Fernandez-Cornejo, *Seed Industry,* 31.

55. Wilfredo Salhuana, "Proposal for Evaluation and Enhancement in Maize," Maize Crop Advisory Committee, Meeting Report, 1984, Goodman Papers, box 49, folder 4.

56. Members of the Seed Industry, "Proposal for Maize Germplasm Support," Goodman Papers.

57. Maize Crop Advisory Committee, Meeting Report, 1984, Goodman Papers; "Maintenance and Evaluation of Maize Germplasm in Latin America," n.d., Goodman Papers, box 7, folder 8. See also Salhuana, Sevilla, and Eberhart, "Final Report."

58. See summary of the seed bank contents and materials regenerated through LAMP in Wilfredo Salhuana, "Regeneration of Latin American Maize Germplasm," n.d., Goodman Papers, box 15, folder 15.

59. M. Goodman to Peter Day, 24 September 1992, Goodman Papers, box 53, folder 4. The evaluation data for the 12,113 accessions evaluated in stage 1 are found in LAMP, *Catálogo del germoplasma*.

60. Salhuana, Jones, and Sevilla, "Latin American Maize Project"; Salhuana, Sevilla, and Eberhart, "Final Report."

61. Salhuana, Sevilla, and Eberhart, "Final Report," 20, 142.

62. Ibid., 20.

63. Eberhart, "USA USDA-NSSL," 47.

64. William Brown, untitled speech, n.d., Brown Papers, box 1, folder 2. See further documentation of this work in Goodman Papers, box 15, folder 15; and W. Brown to R. D. Havener, 29 January 1981, CIMMYT Archives, N9, box 109, folder 6.2.1.

65. R. L. P[aliwal], "Regeneration of Bank Accessions by Pioneer Group," 14 December 1982, CIMMYT Archives, N9, box 109, folder 6.2.1. See also "Pioneer Hi-Bred Project to Aid CIMMYT Maize Preservation," clipping from *Diversity* 1, no. 1 (1982): 13, annotated copy in the same folder.

66. Suketoshi Taba, "The CIMMYT Germplasm Bank," 1986, Goodman Papers, box 22, folder 1. A publication that indicates the changes in the seed bank's perceived role is CIMMYT, *Seed Conservation and Distribution*. On CGIAR funding, see documentation in CIMMYT Archives, N9, box 109, folder 6.2.1.

67. Quentin Jones, "Plan for Preservation and Utilization of the Latin American Maize Accessions," 12 May 1982, Maize Crop Advisory Committee, Second Meeting Report, Goodman Papers; R. L. Paliwal, memo, 26 April 1982, CIMMYT Archives, N9, box 109, folder 6.2.1.

68. "Comments on the Letter from Dr. Charles M. Murphy," n.d., CIMMYT Archives, N9, box 109, folder 6.2.1.

69. Taba, "Current Activities."

70. "Rescuing Endangered Maize Germplasm in Latin America," CIMMYT news release, 1991, CIMMYT Library, NR-0007, NR002.D91.

71. CIMMYT, "A Proposal for a Project to Regenerate Maize Accessions Stored in National Germplasm Banks in Latin America and the Caribbean," June 1991, CIMMYT Library, P-0001, M-A1D01.D91; Taba and Eberhart, "Cooperation Saves Seed," v.

72. CIMMYT, *Recent Advances*, preface, v.

73. D. C. Hess, "Global Implications of Germplasm Conservation and Utilization," CIMMYT Library, R-0076.

74. Taba, "CIMMYT Germplasm Bank," Goodman Papers; Suketoshi Taba to R. P. Cantrell and R. L. Paliwal, 2 December 1985, CIMMYT Archives, N9, box 109, folder 6.2.1.

75. Taba, "Current Activities," 14.

76. Leonelli and Ankeny, "Re-thinking Organisms."

77. "Filling the Corn Germplasm Enhancement Pipeline," encl. in M. Goodman to Bill Kuhn and Charles Stuber, 10 November 1992, Goodman Papers, box 9, folder 11; M. Goodman, "US Maize Germplasm," 135.

78. Bretting and Sisco, "Germplasm Enhancement of Maize," Brown Papers.

79. Arnel Hallauer to Linda Pollak, 28 January 1993, Goodman Papers, box 15, folder 9.

80. Pollak, "History and Success," 51–52.

81. Salhuana, "U.S. Germplasm Enhancement," 2.

82. Pollak, "History and Success," 54.

83. Ibid.

84. GEM protocols are discussed in Pollak, "History and Success"; and Salhuana and Pollak, "Latin American Maize Project."

85. Salhuana, "U.S. Germplasm Enhancement," 3.

86. Lee and Hardin, "GEM Searches for Treasures," 4.

87. Salhuana, "U.S. Germplasm Enhancement," 5.

CHAPTER SEVEN. GROW

1. Quist and Chapela, "Transgenic DNA."

2. Delborne, "Transgenes and Transgressions"; McAfee, "Beyond Technoscience"; Fitting, *Struggle for Maize;* Kinchy, *Seeds, Science, and Struggle,* chs. 3, 4; Bonneuil, Foyer, and Wynne, "Genetic Fallout."

3. P. Brown, "Maya Mother Seeds"; Brandt, "Zapatista Corn."

4. On the origins of the Zapatista uprising, see Harvey, *Chiapas Rebellion;* and Collier, *Basta.*

5. Esteva and Marielle, *Sin maíz.* See also Ita, "Defensa internacional del maíz"; and Pardo Núñez, "Identidad, organización y estrategia."

6. Brush, *Farmers' Bounty,* ch. 9; Fenzi and Bonneuil, "Genetic Resources." For a critical assessment of in situ conservation methods, see Wood and Lenné, "Conservation of Agrobiodiversity On-Farm."

7. Brush, *Farmers' Bounty,* esp. chs. 9, 11.

8. Niethammer, *Seed Banks Serving People.* On Native Seeds/SEARCH, see Schmidt, "United States."

9. "Interview: Gary Nabhan," *Omni* 16, no. 10 (1994): 68–74, 88, 93–94, on 73.

10. Southwest Traditional Crop Conservancy Garden and Seed Bank, *Seed Sources and Resources as of January 1981* (Tucson: Meals for Millions/Freedom from Hunger Foundation, [1981?]), copy at Native Seeds/SEARCH.

11. Ibid., 1.

12. Ibid.

13. [Kent Whealy, ed.], *True Seed Exchange* ([Princeton, MO]: 1976), 17, Seed Savers Exchange Library.

14. Kent Whealy, ed., *The 1981 Seed Savers Exchange* (Princeton, MO: Seeds Savers Exchange, 1981), 47, Seed Savers Exchange Library. See also Curry, "From Bean Collection."

15. Kent Whealy to Gary Nabhan, 25 June 1981, Native Seeds/SEARCH Records, folder: SEE Correspondence.

16. Niethammer, *Seed Banks Serving People,* 1 (emphasis in original).

17. Kent Whealy, "1981 Harvest Tucson Intro," Ceres Trust, last accessed 2021, https://cerestrust.org/1981-harvest-tucson-intro/; Gary Paul Nabhan, "Heirloom Counterculture: The Origin of Heritage Food Revivals," *Heirloom Gardener,* Summer 2013, 22–26.

18. Niethammer, *Seed Banks Serving People,* 1–3.

19. Whealy to Nabhan, Cynthia Anson, Mahina Drees, and others, n.d., Native Seeds/SEARCH Records, folder: Seed Exchange; Whealy, "1981 Harvest Tucson Intro"; Nabhan, "Heirloom Counterculture."

20. Hunn, "Ethnobiology in Four Phases"; Zent, "Genealogy of Scientific Representations."

21. Belasco, *Appetite for Change,* esp. pt. 1; see also Guthman, *Agrarian Dreams,* ch. 1.

22. Whealy, *1981 Seed Savers Exchange,* Seed Savers Exchange Library, 44–49; see also correspondence in Native Seeds/SEARCH Records, folder: Seed Exchange.

23. Southwest Traditional Crop Conservancy Garden and Seed Bank, *Seed Sources and Resources 1982* (Tucson: Meals for Millions, [1982?]), 1. Copy at Native Seeds/SEARCH.

24. Ibid., 2; Gary Nabhan, Cynthia Anson, and Mahina Drees, *Kaicka: Seed Saving the Papago-Pima Way: A Guide for Desert Gardeners and Farmers,* with Danny Lopez (Tucson: Meals for Millions/Freedom from Hunger, 1981), Nabhan Papers, box 2, folder 56.

25. *Seedhead News,* no. 1 (1983): 1. Here and throughout I consulted the digitized collection of *Seedhead News* maintained by Native Seeds/SEARCH, last accessed 2021, www.nativeseeds.org/blogs/the-seedhead-news. On the organization's founding, see Kevin Dahl and Gary Nabhan, "Role of Grassroots Activities in the Maintenance of Biological Diversity," OTA Report 1985, Nabhan Papers, box 7, folder 30, pp. 29–33; and contributions to *Seedhead News,* no. 118, special tribute to Native Seeds/SEARCH cofounder Barney Burns, 1945–2014 (Fall 2014). SEARCH is an acronym for Southwestern Endangered Aridland Resource Clearing House, but this phrase appears not to have been widely used by the organization.

26. I consulted past seed listings for Native Seeds/SEARCH at their administrative offices in Tucson, Arizona, in November 2018.

27. "We've Come Full Circle: Native Seeds/SEARCH Annual Report, 1983–1984," *Seedhead News,* no. 6 (1984): 4.

28. Native Seeds/SEARCH, seed listing, 1985, Native Seeds/SEARCH Records.

29. See documentation for Native Seeds/SEARCH, accession ZM01–002, Native Seeds/SEARCH Records; and US National Plant Germplasm System PI 218186, USDA ARS Germplasm Information Resources Network, last accessed 2021, https://npgsweb.ars-grin.gov/gringlobal/accessiondetail?id=1177853.

30. Native Seeds/SEARCH, accession ZM05–004 (Z08–013), Native Seeds/SEARCH Records.

31. Native Seeds/SEARCH, accession ZM01–005 (Z01–011), Native Seeds/SEARCH Records.

32. "1986–87 Annual Report," *Seedhead News,* no. 18 (1987): 3–5; "1988–89 Annual Report," no. 25 (1989): 3–5; Linda Parker, "Seed Collection Curation at Native Seeds/SEARCH," *Seedhead News,* nos. 32–33 (1991): 21

33. "1987–88 Annual Report," *Seedhead News,* no. 22 (1988): 3–5.

34. "1989–90 Annual Report," *Seedhead News,* no. 29 (1990): 3–5.

35. Gary Nabhan and Karen Reichhardt, "In Situ Conservation of Native Crop Diversity in the U.S./Mexico Borderlands," *Seedhead News,* no. 7 (1984): 1–3.

36. Carol Brandt, "Conserving Traditional Zuni Crops," *Seedhead News,* nos. 32–33 (1991): 15.

37. Kevin Lee López, "Native American Seedbank Workshop," *Seedhead News,* no. 36 (1992): 11.

38. The history of RAFI, including its predecessors and its spinoffs (which include Rural Advancement Foundation International and ETC Group), is difficult to trace. One useful resource is Jaap Hardon, Camila Montecinos, and Tim Roberts, "ETC Group External Review: Report," ETC Group, 18 April 2005, www.etcgroup .org/sites/www.etcgroup.org/files/publication/pdf_file/etc_external_review.pdf.

39. Mooney and Fowler, *Community Seed Bank Kit.* I consulted a copy of this publication archived with the Goodman Papers, box 6, folder 12.

40. "Practical Application: Building the Bank," in Mooney and Fowler, *Community Seed Bank Kit.*

41. Vernooy, Shrestha, and Sthapit, "Origins and Evolution," 12.

42. Frankel, "Genetic Conservation," 57. See also the discussion in Brush, *Farmers' Bounty,* ch. 9.

43. Iltis, "Freezing the Genetic Landscape," 200. See another early suggestion for in situ genetic preserves in Wilkes and Wilkes, "Green Revolution."

44. Brush, *Farmers' Bounty,* 159.

45. See Winkelmann, *New Maize Technology;* and CIMMYT, *Puebla Project.*

46. Clawson and Hoy, "Nealtican, Mexico."

47. Ortega Paczka, "Variación en maíz."

48. Altieri, "Agroecology."

49. Hecht, "Evolution of Agroecological Thought."

50. Hernández Xolocotzi, "Agroecosistema," xvi–xvii. See also Astier et al., "Back to the Roots."

51. Altieri and Merrick, "In Situ Conservation," 90, 91.

52. Brush, *Farmers' Bounty,* preface; Brush, Carney, and Huamán, "Dynamics."

53. Brush, Bellon Corrales, and Schmidt, "Agricultural Development"; Bellon, "Ethnoecology of Maize Variety."

54. Brush, "Farmer-Based Approach," 159; see also Brush, "Reconsidering the Green Revolution."

55. Brush, "Rethinking," 26–27.

56. Altieri, "Crop Genetic Resource Conservation," 78, 79.

57. See, for example, contributions to Glaeser, *Green Revolution Revisited.*

58. Moock and Rhoades, *Diversity.* See also Collinson, *Farming Systems Research;* Cernea and Kassam, *Researching the Culture;* and Price and Palis, "Bringing Farmer Knowledge."

59. See, for example, Collinson, "Low Cost Approach"; Rhoades and Booth, "Farmer-Back-to-Farmer"; and Chambers and Ghildyal, *Agricultural Research.*

60. Byerlee et al., *Planning Technologies,* 1.

61. Johnson, "Individuality and Experimentation"; Conklin, "Hanunóo Agriculture."

62. Howes and Chambers, "Indigenous Technical Knowledge"; Biggs, "Informal R&D"; Biggs and Clay, "Sources of Innovation"; Richards, *Indigenous Agricultural Revolution.*

63. Buckles and Tripp, "Gorras y sombreros." A useful critical review of the literature on participatory agricultural research through the end of the 1980s, written by someone engaged in participatory projects, is Bentley, "Facts, Fantasies and Failures." Analyses of breeding-related participatory research include Weltzien et al., *Technical and Institutional Issues;* and Cleveland and Soleri, "Extending Darwin's Analogy."

64. In addition to accounts detailed in Eyzaguirre and Iwanaga, *Participatory Plant Breeding,* see contributions to Moock and Rhoades, *Diversity.* See also Lightfoot, "Indigenous Research."

65. Blanco Rosas, *Proyecto Sierra.*

66. Ibid., 44–48; Rice, Smale, and Blanco, "Farmers' Use," 13.

67. Blanco Rosas, *Proyecto Sierra,* 49–51, 56.

68. Rice, Smale, and Blanco, "Farmers' Use," 20. On the disappointments of participatory breeding, see Harwood, *Europe's Green Revolution,* 144–48.

69. For pro–free-trade arguments from this period, see Levy and Van Wijnbergen, "Maize"; and Janvry, Sadoulet, and Gordillo de Anda, "Mexico's Maize Producers." For a critical assessment of NAFTA, see Calva, "Posibles efectos." For continuing studies of NAFTA's effects on agriculture, see Nadal, *Environmental and Social Impacts;* Yúnez-Naude, "Lessons from NAFTA"; Rosenzweig, "Debate"; and Fox and Haight, *Subsidizing Inequality.*

70. See, for example, Levy and Van Wijnbergen, "Maize."

71. Harvey, "Rebellion in Chiapas."

72. Boyce, "Ecological Distribution," 280.

73. Fox, "EPA Okays *Bt* Corn"; C. James, "Global Status."

74. See discussion of these concerns in contributions to the special issue (on recombinant DNA), *Trends in Ecology and Evolution* 3, no. 4 (1988).

75. Serratos, Wilcox, and Castillo, *Gene Flow.*

76. Alvarez-Morales, "Implementation of Biosafety Regulations."

77. Schurman and Munro, *Fighting for the Future.*

78. Antonio Turrent and José Antonio Serratos, "Context and Background on Maize and Its Wild Relatives in Mexico," background chapter for the Article 13 Initiative on Maize and Biodiversity, Secretariat of the Commission for Environmental Cooperation of North America, 2004, last accessed 2021, www.researchgate.net/publication/303571977.

79. Fitting, "Importing Corn." On biosecurity in Mexico, see Wanderer, *Life of a Pest.*

80. Robert Bye, Calvin O. Qualset, Alfonso Delgado, and Paul Gepts, "First Annual Scientific Progress Report to the McKnight Foundation," 1995, Goodman Papers, box 51, folder 1.

81. "Conservation of Genetic Diversity in Improvement of Crop Production in Mexico: A Farmer-Based Approach," 15 August 1994, Goodman Papers, box 51, folder 1.

82. Project MILPA, "Conservation of Genetic Diversity and Improvement of Crop Production in Mexico: A Farmer-Based Approach," 1996, Goodman Papers, box 51, folder 2.

83. "History," Collaborative Crop Research Program, last accessed 2021, www.ccrp.org/about/history.

84. The history of Project MILPA can be traced through successive project reports; see Goodman Papers, box 51, folders 1 through 6.

85. Fitting, *Struggle for Maize.* See also Otero, "Neoliberal Globalization"; and Luckstead, Devadoss, and Rodriguez, "Free Trade Agreement."

86. Bellon and Berthaud, "Transgenic Maize."

87. Rincón Sánchez and Hernández Casillas, "Conservación de recursos fitogenéticos," table 3.4.

88. Quist and Chapela, "Transgenic DNA." The concept of introgression was first developed by the botanist Edgar Anderson, whose research features in chapter 2 of this book.

89. Kinchy, *Seeds, Science, and Struggle,* chs. 3 and 4. See also Delborne, "Transgenes and Transgressions"; McAfee, "Beyond Techno-science"; and Bonneuil, Foyer, and Wynne, "Genetic Fallout."

90. Fitting, *Struggle for Maize,* ch. 1; Ita, "Defensa internacional del maíz"; Pardo Núñez, "Identidad, organización y estrategia."

91. Nadal, *Environmental and Social Impacts,* 5; Sámano Rentería, "Movimiento; Otero, "Neoliberal Globalization."

92. "Conclusiones del primer seminario en defensa del maíz," Red en Defensa del Maíz, 23 January 2002, http://redendefensadelmaiz.net/documentos-de-la-red/conclusiones-del-primer-seminario-en-defensa-del-maiz.

93. The network eventually fractured, reflecting divisions among Indigenous and peasant groups (and also NGOs) about how to respond to the disastrous effects of free trade. See Pardo Núñez, "Identidad, organización y estrategia."

94. "Conclusiones del Primer Seminario," Red en Defensa del Maíz.

95. Fitting discusses the portrayal of a static, unchanging Mexican peasantry as the problem of "peasant essentialism" and identifies it at work in both pro- and anti-GM activities. See *Struggle for Maize*.

96. Brandt, "Zapatista Corn."

97. Ibid.

98. See, for example, Doebley, "Molecular Evidence"; and Serratos, Wilcox, and Castillo, *Gene Flow*.

99. McAfee, "Beyond Techno-science," 152; CIMMYT, "Initial Tests Find Mexican Landraces in CIMMYT Gene Bank Free of Promoter Associated with Transgenes," news release, 16 October 2001, SeedQuest, last accessed 2021, www .seedquest.com/News/releases/2002/february/4117b.htm; CIMMYT, "Further Tests at CIMMYT Find No Presence of Promoter Associated with Transgenes in Mexican Landraces in Gene Bank or from Recent Field Collections," news release, 14 December 2001, SeedQuest, last accessed 2021, https://www.seedquest.com /News/releases/2002/january/4117.htm.

100. Delborne, "Transgenes and Transgressions."

101. Although there were many confirmatory studies, the authors had difficulty publishing them; see Bonneuil, Foyer, and Wynne, "Genetic Fallout."

102. CEC, *Maize and Biodiversity,* list of contributors on 10.

103. Ibid., 14. See also the discussion in McAfee, "Beyond Techno-science."

104. CEC, *Maize and Biodiversity,* 26–31.

105. Kinchy, *Seeds, Science, and Struggle,* 71–72; Fitting, *Struggle for Maize,* 70.

106. The manifesto is discussed in Fitting, *Struggle for Maize,* 72–73.

107. Kato Yamakake et al., *Origen y diversificación,* 12; Proyecto Global de Maíces Nativos, "Informe de Gestión y Resultados," first version, March 2011, Comisión Nacional para el Conocimiento y Uso de la Biodiversidad, www .biodiversidad.gob.mx/media/1/genes/files/InformedeGestion_V1.pdf. See also Foyer, "Ver su riqueza."

108. Kinchy, *Seeds, Science, and Struggle,* ch. 2.

109. González-Santos et al., "Model for the Conservation."

110. Ibid., 336, 337; see also Vera Sánchez, González Santos, and Aragón Cuevas, "Community Seed Banks." Further discussion of *custodios* in different regions is found in contributions to Córdova Téllez et al., *Resultados en conservación.*

111. Aragón Cuevas et al., "In Situ Conservation"; Aragón Cuevas, *Bancos comunitarios;* Aragón Cuevas, "Mexico."

112. See, for example, Soleri and Cleveland, "Hopi Crop Diversity"; Soleri, "Developing Methodologies"; and Soleri and Cleveland, "Farmers' Genetic Perceptions."

113. Soleri et al., "Understanding the Potential Impact," 159. See also Soleri, Cleveland, and Aragón Cuevas, "Transgenic Crops." This was followed by a response and counterresponse: Ortiz-García et al., "Transgenic Maize in Mexico"; and Soleri, Cleveland, and Aragón Cuevas, "Response from Soleri."

114. For the most recent version of seed-saving instructions from Native Seeds/SEARCH, see Hought and Kruse-Peeples, *Saving Seed.*

115. Aguirre Gómez and García Leaños, *Selección para el mejoramiento.*

116. Montenegro de Wit, "Are We Losing Diversity?"; Nazarea, Rhoades, and Andrews-Swann, *Seeds of Resistance.*

CODA

1. These critiques are discussed in the introduction to this book. For reviews of crop diversity research and explanations of the trends seen in empirical studies, see Van de Wouw et al., "Genetic Erosion in Crops"; and Thormann and Engels, "Genetic Diversity and Erosion."

2. See a discussion of duplication as a concern for seed banks in Van Hintum and Knüpffer, "Duplication"; and Van Hintum, "Duplication." See also the emphasis on "redundant" duplication—as opposed to safety duplication—in the FAO's "The State of the World's Plant Genetic Resources for Food and Agriculture," first report, 1997, www.fao.org/3/a-w7324e.pdf; and second report, 2010, www.fao.org/3/i1500e/i1500e.pdf.

3. Fullilove, *Profit of the Earth,* 211.

4. For demonstrations of this process of continuous development in maize, see Louette, Charrier, and Berthaud, "In Situ Conservation"; and Perales, Brush, and Qualset, "Dynamic Management."

5. See, for example, Bellon and Risopoulos, "Small-Scale Farmers." See also Cleveland and Soleri, "Extending Darwin's Analogy."

6. Greg Schoen, "The Origins and Journey of 'Carl's Glass Gems' Rainbow Corn," *Mother Earth News,* 13 December 2013, www.motherearthnews.com/homesteading-and-livestock/sustainable-farming/glass-gem-corn-seed-zwfz1212zrob; "The Story of Glass Gem Corn: Beauty, History, and Hope," Native Seeds/SEARCH (blog), 10 June 2013, www.nativeseeds.org/blogs/blog-news/the-story-of-glass-gem-corn-beauty-history-and-hope; "In Memory: Carl Barnes," Seed Broadcast (blog), 19 April 2016, https://seedbroadcast.blogspot.com/2016/04/in-memory-carl-barnes-man-who-saved.html.

7. "Carl Barnes: 'CORNS,'" in Kent Whealy, ed., *The Seed Savers Exchange,* Fall Harvest ed. (Princeton, MO: Seed Savers Exchange, 1982), 39, Seed Savers Exchange Library.

8. Schoen, "Origins and Journey."

9. These issues are discussed in McCouch et al., "Genomics of Gene Banks." In 2021 genomic sequencing is widely regarded as having the potential to transform these long-standing concerns.

10. Historical debates over synonyms, methods of identifying varieties, and the legal ownership of seeds offer insight to these issues. See, for example, Berry, "Plant Breeding Industry"; Holmes, "Changing Techniques"; and Charnley, *"Cui Bono."*

11. This situation is often linked to a decline in public plant breeding vis-à-vis opportunities in private industry. For a review of approaches to restoring breeding efforts that target local needs, including in industrialized settings, see Brouwer, Murphy, and Jones, "Plant Breeding."

12. The chef Dan Barber points to some of these individuals in the United States in "Save Our Food: Free the Seed," *New York Times,* 7 June 2019, www.nytimes .com/interactive/2019/06/07/opinion/sunday/dan-barber-seed-companies.html.

13. Mihesuah and Hoover, *Indigenous Food Sovereignty.* For the perspectives of Indigenous and Native farmers in the United States and Mexico, see, for example, the interviews in *Voices of Maíz.* For a survey of Native American projects centered on food sovereignty, including many that emphasize Native seed, see Hoover, "You Can't Say" (also collected in Mihesuah and Hoover, *Indigenous Food Sovereignty*).

14. "Vision and Mission," Land Institute, last accessed 2021, https:// landinstitute.org/about-us/vision-mission/.

ARCHIVES AND BIBLIOGRAPHY

All archival materials, including many from digital repositories; sources found in magazines, newspapers, and newsletters; and references to website or database content are cited in full at first mention in the notes rather than in this bibliography.

LIBRARIES AND ARCHIVES

Anderson, Edgar Shannon. Papers. RG 3/2/3. Missouri Botanical Garden Archives, Saint Louis.

ARS (Agricultural Research Service). Records. RG 310. National Archives and Records Administration, College Park, MD.

BPI (Bureau of Plant Industry). Records. RG 54. National Archives and Records Administration, College Park, MD.

Brown, William L. Papers. RG 4/2/1/1. Missouri Botanical Garden Archives, Saint Louis.

CGSpace. CGIAR (Consultative Group on International Agricultural Research). Digital Repository. https://cgspace.cgiar.org.

CIMMYT (International Center for the Improvement of Maize and Wheat). Library and Archives. John Woolson Library, Centro Internacional de Mejoramiento de Maíz y Trigo, El Batán, Mexico.

FAO (Food and Agriculture Organization). Library and Archives. David Lubin Memorial Library, Rome.

Ford Foundation. Archives. Rockefeller Archive Center, Sleepy Hollow, NY.

Frankel, Sir Otto H. Papers. MS 106. Basser Library, Australian Academy of Science, Canberra.

Goodman, Major M. Papers. MC 00184. Special Collections Research Center, North Carolina State University, Raleigh.

Mangelsdorf, Paul C. Papers. HUGFP 37.xx. Pusey Library, Harvard University, Cambridge, MA.

Nabhan, Gary. Papers. MSS 337. Special Collections, University of Arizona, Tucson.

NAS (National Academy of Sciences). Archives. National Academy of Sciences, Washington, DC.

National Library of Australia. Canberra.

Native Seeds/SEARCH. Records. Native Seeds/SEARCH Offices, Tucson, AZ.

Rockefeller Foundation. Archives. Rockefeller Archive Center, Sleepy Hollow, NY.

Sauer, Carl Ortwin. Papers. MSS 77/170. Bancroft Library, University of California, Berkeley.

Seed Savers Exchange Library. Decorah, IA.

USDA Division of Cereal Crops and Diseases. Photograph Collection. Special Collections, USDA National Agricultural Library, Beltsville, MD.

Will Family. Papers. MS 10190. State Historical Society of North Dakota, Bismarck.

Wilson, Gilbert Livingstone. Papers. W557. Division of Anthropology Archives, American Museum of Natural History, NY.

World Bank. Archives. World Bank Group, Washington, DC.

BIBLIOGRAPHY

Aboites Manrique, Gilberto. *Una mirada diferente de la Revolución Verde: Ciencia, nación y compromiso social*. Mexico City: Plaza y Valdés, 2002.

Aboites Manrique, Gilberto, and Francisco Martínez Gómez. "La propiedad intelectual de variedades vegetales en México." *Agrociencia* 39, no. 2 (2005): 237–45.

Adams, David Wallace. *Education for Extinction: American Indians and the Boarding School Experience, 1875–1928*. Lawrence: University of Kansas Press, 1995.

Adams, William M. *Against Extinction: The Story of Conservation*. London: Earthscan, 2004.

Adams, William M., and Martin Mulligan, eds. *Decolonizing Nature: Strategies for Conservation in a Post-Colonial Era*. London: Earthscan, 2003.

Ad Hoc Subcommittee of ARPAC (Agricultural Research Policy Advisory Committee). *Recommended Actions and Policies for Minimizing the Genetic Vulnerability of Our Major Crops*. Special report. Washington, DC: USDA/National Association of State Universities and Land Grant Colleges, 1973.

Agricultural Development: Proceedings of a Conference Sponsored by the Rockefeller Foundation, 1969. New York: Rockefeller Foundation, [1970?].

Aguirre Gómez, José Alfonso, and María de Lourdes García Leaños. *Selección para el mejoramiento de maíz criollo: Manual de capacitación*. Folleto para Productores 4. Mexico City: INIFAP, 2012.

Alston, Julian M., Steven Dehmer, and Philip G. Pardey. "International Initiatives in Agricultural R&D: The Changing Fortunes of the CGIAR." In *Agricultural*

R&D in the Developing World: Too Little, Too Late?, edited by Philip G. Pardey, Julian M. Alston, and Roley R. Piggott, 313–60. Washington, DC: International Food Policy Research Institute, 2006.

Altieri, Miguel A. "Agroecology: A New Research and Development Paradigm for World Agriculture." *Agriculture, Ecosystems and Environment* 27, nos. 1–4 (1989): 37–46.

———. "Rethinking Crop Genetic Resource Conservation: A View from the South." *Conservation Biology* 3, no. 1 (1989): 77–79.

Altieri, Miguel A., and Laura C. Merrick. "In Situ Conservation of Crop Genetic Resources through Maintenance of Traditional Farming Systems." *Economic Botany* 41, no. 1 (1987): 86–96.

Alvarez-Morales, A. "Implementation of Biosafety Regulations in a Developing Country: The Case of Mexico." *African Crop Science Journal* 3, no. 3 (1995): 309–14.

Aoki, Keith. *Seed Wars: Controversies and Cases on Plant Genetic Resources and Intellectual Property*. Durham, NC: Carolina Academic Press, 2008.

Anderson, Edgar. "Maize in Mexico: A Preliminary Survey." *Annals of the Missouri Botanical Garden* 33, no. 2 (1946): 147–247.

———. "The Sources of Effective Germ-Plasm in Hybrid Maize." *Annals of the Missouri Botanical Garden* 31, no. 4 (1944): 355–61.

———. "What I Found Out about the Corn Plant." *Missouri Botanical Garden Bulletin* 57, no. 5 (1969): 6–9.

Anderson, Edgar, and William L. Brown. "The History of the Common Maize Varieties of the United States Corn Belt." *Agricultural History* 26, no. 1 (1952): 2–8.

Anderson, Edgar, and Hugh Cutler. "Races of Zea Mays: I. Their Recognition and Classification." *Annals of the Missouri Botanical Garden* 29, no. 2 (1942): 69–86, 88.

Anderson, J. L. *Industrializing the Corn Belt: Agriculture, Technology, and Environment*. DeKalb: Northern Illinois University Press, 2009.

Anderson, Robert S. "The Origins of the International Rice Research Institute." *Minerva* 29, no. 1 (1991): 61–89.

Anderson, Warwick. "Climates of Opinion: Acclimatization in Nineteenth-Century France and England." *Victorian Studies* 35, no. 2 (1992): 135–57.

Ángeles Arrieta, Hermilo H. "El maíz y el sorgo y sus programas de mejoramiento genético en México." In *Memorias del Tercer Congreso Nacional de Fitogenética*, 382–446. Chapingo, Mexico: Sociedad Mexicana de Fitogenética (SOMEFI), 1968.

Appendini, Kirsten, and Ma. Guadalupe Quijada. "Consumption Strategies in Mexican Rural Households: Pursuing Food Security with Quality." *Agriculture and Human Values* 33 (2016): 439–54.

Aragón Cuevas, Flavio. "Mexico: Community Seed Banks in Oaxaca." In Vernooy, Shrestha, and Sthapit, *Community Seed Banks,* 136–39.

Aragón Cuevas, Flavio, F. Humberto Castro García, José Manuel Cabrera Toledo, and Leodegario Osorio Alcalá. *Bancos comunitarios de semillas para conservar in situ la diversidad vegetal*. Oaxaca: INIFAP, 2011.

Aragón Cuevas, Flavio, Suketoshi Taba, F. Humberto Castro-García, Juan Manuel Hernández-Casillas, José Manuel Cabrera-Toledo, Leodegario Osorio Alcalá, and Nicolás Dillánes Ramírez. "In Situ Conservation and Use of Local Maize Races in Oaxaca, Mexico." In *Latin American Maize Germplasm Conservation: Regeneration, In Situ Conservation, Core Subsets, and Prebreeding,* edited by Suketoshi Taba, 26–37. Workshop Proceedings, 7–10 April 2003. Mexico City: CIMMYT, 2005.

ARS (Agricultural Research Service). *The National Plant Germplasm System.* Program Aid 1188. Washington, DC: USDA, 1977.

Astier, Marta, Jorge Quetzal Argueta, Quetzalcóatl Orozco-Ramírez, María V. González, Jaime Morales, Peter R. W. Gerritsen, Miguel A. Escalona, et al. "Back to the Roots: Understanding Current Agroecological Movement, Science, and Practice in Mexico." *Agroecology and Sustainable Food Systems* 41, nos. 3–4 (2017): 329–48.

Asturias, Miguel Ángel. *Hombres de maíz.* Buenos Aires: Losada, 1973.

Atkinson, Alfred, and M. L. Wilson. *Corn in Montana: History, Characteristics, Adaptation.* Bulletin 107. Bozeman: Montana Agricultural College Experiment Station, 1915.

Austin, James E., and Gustavo Esteva, eds. *Food Policy in Mexico: The Search for Self-Sufficiency.* Ithaca, NY: Cornell University Press, 1987.

———. "Final Reflections." In Austin and Esteva, *Food Policy in Mexico,* 353–73.

Austin, James E., and Jonathan Fox. "State-Owned Enterprises: Food Policy Implementers." In Austin and Esteva, *Food Policy in Mexico,* 61–91.

"Autarky/Autarchy: Genetics, Food Production, and the Building of Fascism." Special issue, *Historical Studies in the Natural Sciences* 40, no. 4 (2010).

Aviña, Alexander. *Specters of Revolution: Peasant Guerrillas in the Cold War Countryside.* Oxford: Oxford University Press, 2014.

Balmford, Andrew. *Wild Hope: On the Front Lines of Conservation Success.* Chicago: University of Chicago Press, 2012.

Barahona, Ana. "Mendelism and Agriculture in the First Decades of the XXth Century in Mexico." In *A Cultural History of Heredity IV: Heredity in the Century of the Gene,* 111–28. Preprint 343. Berlin: Max Planck Institute for the History of Science, 2008.

Barahona, Ana, Susana Pinar, and Francisco J. Ayala. *La genética en México: Institucionalización de una disciplina.* Mexico City: UNAM, 2003.

Barahona Echeverría, Ana, and Ana Lilia Gaona Robles. "The History of Science and the Introduction of Plant Genetics in Mexico." *History and Philosophy of the Life Sciences* 23, no. 1 (2001): 151–62.

Barkin, David. "The End to Food Self-Sufficiency in Mexico." *Latin American Perspectives* 14, no. 3 (1987): 271–97.

———. "Mexico's Albatross: The United States Economy." *Latin American Perspectives* 2, no. 2 (1975): 64–80.

———. "La persistencia de la pobreza en México: Un análisis económico estructural." Translated by Mario Luján. *Desarrollo Económico* 10, no. 38 (1970): 263–84.

————. "SAM and Seeds." In Austin and Esteva, *Food Policy in Mexico,* 111–32.

Barkin, David, and Billie R. DeWalt. "Sorghum and the Mexican Food Crisis." *Latin American Research Review* 23, no. 3 (1988): 30–59.

Barkin, David, and Blanca Suárez. *El fin del principio: Las semillas y la seguridad alimentaria.* Mexico City: Centro de Ecodesarrollo, 1983.

Barrow, Mark V., Jr. *Nature's Ghosts: Confronting Extinction from the Age of Jefferson to the Age of Ecology.* Chicago: University of Chicago Press, 2009.

Bartra, Roger, and Gerardo Otero. "Agrarian Crisis and Social Differentiation in Mexico." *Journal of Peasant Studies* 14, no. 3 (1987): 334–62.

Bashford, Alison. *Global Population: History, Geopolitics, and Life on Earth.* New York: Columbia University Press, 2014.

Baum, Warren C. *Partners against Hunger: The Consultative Group on International Agricultural Research.* With Michael L. Lejeune. Washington, DC: World Bank, 1986.

Belasco, Warren J. *Appetite for Change: How the Counter Culture Took on the Food Industry.* 2nd ed. Ithaca, NY: Cornell University Press, 2007.

Bellon, Mauricio R. "The Ethnoecology of Maize Variety Management: A Case Study from Mexico." *Human Ecology* 19, no. 3 (1991): 389–418.

Bellon, Mauricio R., and Julien Berthaud. "Transgenic Maize and the Evolution of Landrace Diversity in Mexico: The Importance of Farmers' Behavior." *Plant Physiology* 134, no. 3 (2004): 883–88.

Bellon, Mauricio R., Alicia Mastretta-Yanes, Alejandro Ponce-Mendoza, Daniel Ortiz-Santamaría, Oswaldo Oliveros-Galindo, Hugo Perales, Francisca Acevedo, and José Sarukhán. "Evolutionary and Food Supply Implications of Ongoing Maize Domestication by Mexican *Campesinos.*" *Proceedings of the Royal Society B* 285, no. 1885 (2018): doc. 20181049.

Bellon, Mauricio R., and Jean Risopoulos. "Small-Scale Farmers Expand the Benefits of Improved Maize Germplasm: A Case Study from Chiapas, Mexico." *World Development* 29, no. 5 (2001): 799–811.

Bennett, Erna, ed. *Record of the FAO/IBP Technical Conference on the Exploration, Utilization and Conservation of Plant Genetic Resources.* Rome: FAO, 1968.

Bentley, Jeffery W. "Facts, Fantasies and Failures of Farmer Participation: Introduction to the Symposium Volume." *CEIBA* 31, no. 2 (1990): 7–27.

Benz, Bruce F. "Maize in the Americas." In Staller, Tykot, and Benz, *Histories of Maize,* 9–20.

Berg, Trygve. "Landraces and Folk Varieties: A Conceptual Reappraisal of Terminology." *Euphytica* 166, no. 3 (2009): 423–30.

Berlan, Jean-Pierre, and Richard C. Lewontin. "The Political Economy of Hybrid Corn." *Monthly Review* 38, no. 3 (1986): 35–47.

Berry, Dominic J. "Historiographies of Plant Breeding and Agriculture." In *Handbook of the Historiography of Biology,* edited by Michael R. Dietrich, Mark E. Borrello, and Oren Harman. Cham, Switzerland: Springer, 2019. https://doi.org/10.1007/978-3-319-74456-8_27-1.

————. "The Plant Breeding Industry after Pure Line Theory: Lessons from the National Institute of Agricultural Botany." *Studies in the History and Philosophy of Science Part C* 46 (2014): 25–37.

Biggs, Stephen. "Informal R&D." *Ceres* 13, no. 4 (1980): 23–26.

Biggs, Stephen D., and Edward J. Clay. "Sources of Innovation in Agricultural Technology." *World Development* 9, no. 4 (1981): 321–36.

Birchler, James A. "Paul C. Mangelsdorf, 1899–1989." *Biographical Memoirs of the National Academy of Sciences.* 2014. www.nasonline.org/publications /biographical-memoirs/memoir-pdfs/mangelsdorf-paul.pdf.

Blake, Michael. *Maize for the Gods: Unearthing the 9,000-Year History of Corn.* Oakland: University of California Press, 2015.

Blanco Rosas, José Luis. *El Proyecto Sierra de Santa Marta: Experimentación participativa para el uso adecuado de recursos genéticos maiceros.* Mexico City: Red de Gestión de Recursos Naturales/Fundación Rockefeller, 1997.

Bonfil Batalla, Guillermo. *El maíz, fundamento de la cultura popular mexicana.* Mexico City: Museo Nacional de Culturas Populares, 1982.

Bonneuil, Christophe. "Mendelism, Plant Breeding and Experimental Cultures: Agriculture and the Development of Genetics in France." *Journal of the History of Biology* 39, no. 2 (2006): 281–308.

————. "Penetrating the Natives: Peanut Breeding, Peasants and the Colonial State in Senegal (1900–1950)." *Science, Technology, and Society* 4, no. 2 (1999): 273–302.

————. "Seeing Nature as a 'Universal Store of Genes': How Biological Diversity Became 'Genetic Resources.'" *Studies in the History and Philosophy of Science Part C* 75 (2019): 1–14.

Bonneuil, Christophe, Jean Foyer, and Brian Wynne. "Genetic Fallout in Biocultural Landscapes: Molecular Imperialism and the Cultural Politics of (Not) Seeing Transgenes in Mexico." *Social Studies of Science* 44, no. 6 (2014): 901–29.

Borlaug, Norman E. "Wheat Breeding and Its Impact on World Food Supply." In *Proceedings of the Third International Wheat Genetics Symposium,* 1–36. Canberra: Australian Academy of Science, 1968.

Bowers, Douglas E. "The Research and Marketing Act of 1946 and Its Effects on Agricultural Marketing Research." *Agricultural History* 56, no. 1 (1982): 249–63.

Boyce, James K. "Ecological Distribution, Agricultural Trade Liberalization, and In Situ Genetic Diversity." *Journal of Income Distribution* 6, no. 2 (1996): 265–86.

Brandt, Marisa. "Zapatista Corn: A Case Study in Biocultural Innovation." *Social Studies of Science* 44, no. 6 (2014): 874–900.

Brantlinger, Patrick. *Dark Vanishings: Discourse on the Extinction of Primitive Races.* Ithaca, NY: Cornell University Press, 2003.

Brennan, Shane. "Making Data Sustainable: Backup Culture and Risk Perception." In *Sustainable Media: Critical Approaches to Media and Environment,* edited by Nicole Starosielski and Janet Walker, 56–76. New York: Routledge, 2016.

Brieger, F. G., J. T. A. Gurgel, E. Paterniani, A. Blumenschein, and M. R. Alleoni. *Races of Maize in Brazil and Other Eastern South American Countries.* Publication 593. Washington, DC: National Research Council, National Academy of Sciences, 1958.

Brockway, Lucile. *Science and Colonial Expansion: The Role of the British Royal Botanic Gardens.* 1979. Reprint, New Haven, CT: Yale University Press, 2002.

Brouwer, Brook O., Kevin M. Murphy, and Stephen S. Jones. "Plant Breeding for Local Food Systems: A Contextual Review of End-Use Selection for Small Grains and Dry Beans in Western Washington." *Renewable Agriculture and Food Systems* 31, no. 2 (2015): 172–84.

Browman, David. "Necrology: Hugh Carson Cutler." *Bulletin of the History of Archaeology* 9, no. 1 (1999): 1–6.

Brown, Peter. "Maya Mother Seeds in Resistance of Highland Chiapas in Defense of Native Corn." In Nazarea, Rhoades, and Andrews-Swann, *Seeds of Resistance,* 151–76.

Brown, William L. "Conservation of Gene Resources in the United States." In *Plant Genetic Resources: A Conservation Imperative,* edited by Christopher W. Yeatman, David Kafton, and Garrison Wilkes, 31–41. Boulder, CO: Westview Press for the American Association for the Advancement of Science.

———. "H. A. Wallace and the Development of Hybrid Corn." *Annals of Iowa* 47 (1983): 167–79.

———. "The Origin of Corn Belt Maize." *Journal of the New York Botanical Garden* 51, no. 610 (1950): 242–46, 255.

———. *Races of Maize in the West Indies.* Publication 792. Washington, DC: National Research Council, National Academy of Sciences, 1960.

Brush, Stephen B. "A Farmer-Based Approach to Conserving Crop Germplasm." *Economic Botany* 45, no. 2 (1991): 153–65.

———. *Farmers' Bounty: Locating Crop Diversity in the Contemporary World.* New Haven, CT: Yale University Press, 2004.

———, ed. *Genes in the Field: On-Farm Conservation of Crop Diversity.* Boca Raton: Lewis; Ottawa: International Development Research Centre; Rome: International Plant Genetic Resources Institute, 2000.

———. "Reconsidering the Green Revolution: Diversity and Stability in Cradle Areas of Crop Domestication." *Human Ecology* 20, no. 2 (1992): 145–67.

———. "Rethinking Crop Genetic Resource Conservation." *Conservation Biology* 3, no. 1 (1989): 19–29.

Brush, Stephen B., Mauricio Bellon Corrales, and Ella Schmidt. "Agricultural Development and Maize Diversity in Mexico." *Human Ecology* 16, no. 3 (1988): 307–28.

Brush, Stephen B., Heath J. Carney, and Zósimo Huamán. "Dynamics of Andean Potato Agriculture." *Economic Botany* 35, no. 1 (1981): 70–88.

Buckles, Daniel, and Robert Tripp. "Gorras y sombreros: Caminos hacia la colaboración entre técnicos y campesinos." In *Gorras y sombreros: Caminos hacia la colaboración entre técnicos y campesinos,* edited by Daniel Buckles, 3–8. Mexico City: CIMMYT, 1993.

Bugos, Glenn E., and Daniel J. Kevles. "Plants as Intellectual Property: American Practice, Law and Policy in World Context." *Osiris* 7 (1992): 74–104.

Bukasov, S. M. "The Cultivated Plants of Mexico, Guatemala and Columbia." Supplement, *Bulletin of Applied Botany, Genetics, and Plant Breeding* (Leningrad) 47 (1930): English summary on 470–553.

Burgess, Sam, ed. *The National Program for Conservation of Crop Germplasm (a Progress Report on Federal/State Cooperation).* Athens, Georgia: University Printing Department, 1971.

Byerlee, Derek. *The Birth of CIMMYT: Pioneering the Idea and Ideals of International Agricultural Research.* Mexico City: CIMMYT, 2016.

———. "The Globalization of Hybrid Maize, 1921–70." *Journal of Global History* 15, no. 1 (2020): 101–22.

Byerlee, Derek, Michael Collinson, D. Winkelmann, S. Biggs, E. R. Moscardi, J. C. Martinez, L. Harrington, and A. Benjamin. *Planning Technologies Appropriate to Farmers: Concepts and Procedures.* 1980. Reprint, Mexico City: CIMMYT, 1988.

Byerlee, Derek, and John K. Lynam. "The Development of the International Center Model for Agricultural Research: A Prehistory of the CGIAR." *World Development* 135 (2020): 105080.

Byrd, Jodi A. *The Transit of Empire: Indigenous Critiques of Colonialism.* Minneapolis: University of Minnesota Press, 2011.

Caetano de Oliveira, Alierso, and Serafín J. Mendoza Mendoza. "La investigación agropecuaria y forestal en México y su vinculación con el estado." In Méndez Ramírez et al., *Investigación agrícola en México,* 67–90.

Caire-Pérez, Matthew. "A Different Shade of Green: Efraím Hernández Xolocotzi, Chapingo, and Mexico's Green Revolution, 1950–1967." PhD diss., University of Oklahoma, 2016.

Calva, José Luis. "Posibles efectos de un Tratado de Libre Comercio México–Estados Unidos sobre el sector agropecuario." *Revista Mexicana de Sociología* 53, no. 3 (1991): 111–24.

Camacho Villa, Tania Carolina, Nigel Maxted, Maria Scholten, and Brian Ford-Lloyd. "Defining and Identifying Crop Landraces." *Plant Genetic Resources* 3, no. 3 (2005): 373–84.

Cárdenas Ramos, Francisco. "Bases para la creación de la Unidad de Recursos Genéticos del INIA." In Cervantes Santana, *Análisis de los recursos,* 467–74.

Castillo-González, F., and M. M. Goodman. "Agronomic Evaluations of Latin American Maize Accessions." *Crop Science* 29, no. 4 (1989): 853–61.

CEC (Commission for Environmental Cooperation). *Maize and Biodiversity: The Effects of Transgenic Maize in Mexico, Key Findings and Recommendations.* Montreal: CEC, 2004.

Cernea, Michael, and Amir H. Kassam, eds. *Researching the Culture in Agri-culture: Social Research for International Development.* Wallingford, CT: CABI, 2006.

Cervantes Santana, Tarcicio, ed. *Análisis de los recursos genéticos disponibles a México*. Chapingo, Mexico: Sociedad Mexicana de Fitogenética, 1978.

Chacko, Xan. "Moving, Making, and Saving Botanic Futures: The History and Practices of Seed Banking." PhD diss., University of California–Davis, 2018.

Chambers, Robert, and B. P. Ghildyal. *Agricultural Research for Resource-Poor Farmers: The Farmer-First-and-Last Model*. New Dehli: Ford Foundation, 1984.

Chandler, Robert F., Jr. *An Adventure in Applied Science: A History of the International Rice Research Institute*. Manila: IRRI, 1992.

Chapman, Susannah, and Paul J. Heald. "Agrobiodiversity Loss and the Construction of Regulatory Frameworks for Crop Germplasm." In *Environmental Resilience and Food Law: Agrobiodiversity and Agroecology*, edited by Gabriela Steier and Alberto Giulio Cianci, 159–82. Boca Raton: CRC Press, 2020.

Charnley, Berris. "*Cui Bono?* Gauging the Successes of Publicly-Funded Plant Breeding in Retrospect." In *Intellectual Property and Genetically Modified Organisms: A Convergence in Laws*, edited by Charles Lawson and Berris Charnley, 7–26. London: Taylor and Francis, 2015.

Chávez, Eduardo. *Cultivo del maíz*. Mexico City: Imprenta y Fototipia de la Secretaría de Fomento, 1913.

CIMMYT. *1966–67 Report*. Mexico City: CIMMYT. http://hdl.handle.net/10883/1344.

———. *1968–69 Report*. Mexico City: CIMMYT. http://hdl.handle.net/10883/1346.

———. *The Puebla Project: Seven Years of Experience, 1967–1973*. El Batán, Mexico: CIMMYT, 1974.

———. *Recent Advances in the Conservation and Utilization of Genetic Resources: Proceedings of the Global Maize Germplasm Workshop*. Mexico City: CIMMYT, 1988.

———. *Seed Conservation and Distribution: The Dual Role of the CIMMYT Maize Germplasm Bank*. Mexico City: CIMMYT, 1986.

Cintli Rodríguez, Roberto. *Our Sacred Maíz Is Our Mother: Indigeneity and Belonging in the Americas*. Tucson: University of Arizona Press, 2014.

Cioc, Mark. *The Game of Conservation: International Treaties to Protect the World's Migratory Animals*. Athens: Ohio University Press, 2009.

Claffey, Barbara A. "Patenting Life Forms: Issues Surrounding the Plant Variety Protection Act." *Southern Journal of Agricultural Economics* 13, no. 2 (1981): 29–37.

Clampitt, Cynthia. *Midwest Maize: How Corn Shaped the U.S. Heartland*. Urbana: University of Illinois Press, 2015.

Clark, J. Allen. "Collection, Preservation, and Utilization of Indigenous Strains of Maize." *Economic Botany* 10, no. 2 (1956): 194–200.

Clarke, John M. "Iroquois Uses of Maize and Other Food." *Education Department Bulletin* 482 (1910): 5–113.

Clawson, David L., and Don R. Hoy. "Nealtican, Mexico: A Peasant Community That Rejected the 'Green Revolution.'" *American Journal of Economics and Sociology* 38, no. 4 (1979): 371–87.

Cleveland, David A., and Daniela Soleri. "Extending Darwin's Analogy: Bridging Differences in Concepts of Selection between Farmers, Biologists, and Plant Breeders." *Economic Botany* 61, no. 2 (2007): 121–36.

Collier, George A. *Basta! Land and the Zapatista Rebellion in Chiapas*. With Elizabeth Lowery Quaratiello. 3rd ed. Oakland: Food First Books, 2005.

Collins, G. N. *A New Type of Indian Corn from China*. Bureau of Plant Industry Bulletin 161. Washington, DC: Government Printing Office, 1909.

———. "Notes on the Agricultural History of Maize." *Agricultural History Society Papers* 2 (1923): 409, 411–29.

———. "Pueblo Indian Maize Breeding." *Journal of Heredity* 5, no. 6 (1914): 255–68.

Collinson, M., ed. *A History of Farming Systems Research*. Wallingford, CT: CABI/FAO, 2000.

———. "A Low Cost Approach to Understanding Small Farmers." *Agricultural Administration* 8 (1981): 433–50.

Conklin, Harold. "Hanunóo Agriculture: A Report on an Integral System of Shifting Cultivation in the Philippines." FAO Development Paper 12. Rome, 1957.

Connelly, Matthew. *Fatal Misconception: The Struggle to Control World Population*. Cambridge, MA: Harvard University Press, 2008.

———. "To Inherit the Earth: Imagining World Population, from the Yellow Peril to the Population Bomb." *Journal of Global History* 1, no. 3 (2006): 299–319.

Cooke, Kathy J. "'Who Wants White Carrots?' Congressional Seed Distribution, 1862 to 1923." *Journal of the Gilded Age and Progressive Era* 17, no. 3 (2018): 475–500.

Córdova Téllez, Leobigildo, Pedro Antonio López, Panuncio Jerónimo Reyes Santiago, Ángel Villegas Monter, Jorge Cadena Iñiguez, Luz María Mera Ovando, Rogelio Lépiz Ildefonso, Rosalinda González Santos, and Oscar Gámez Montiel, eds. *Resultados en conservación, uso y aprovechamiento sustentable de recursos fitogenéticos para la alimentación y la agricultura*. Mexico City: Asociación Nacional para la Innovación y Desarrollo Tecnológico Agrícola, 2015.

Cotter, Joseph. "Cultural Wars and New Technologies: The Discourse of Plant Breeding and the Professionalisation of Mexican Agronomy, 1880–1994." *Science, Technology and Society* 5, no. 2 (2000): 141–68.

———. "The Origins of the Green Revolution in Mexico: Continuity or Change?" In *Latin America in the 1940s: War and Postwar Transitions,* edited by David Rock, 224–41. Berkeley: University of California Press, 1994.

———. *Troubled Harvest: Agronomy and Revolution in Mexico, 1880–2002*. Westport, CT: Praeger, 2003.

Creech, John L., and Louis P. Reitz. "Plant Germ Plasm Now and for Tomorrow." *Advances in Agronomy* 23 (1971): 1–49.

Crow, James F. "90 Years Ago: The Beginning of Hybrid Maize." *Genetics* 148, no. 3 (1998): 923–28.

Cruz León, Artemio, Marcelino Ramírez Castro, Francisco Collazo-Reyes, and Xóchitl Flores Vargas. "La obra escrita de Efraím Hernández Xolocotzi, patrimonio y legado." *Revista de Geografía Agrícola,* nos. 50–51 (2013): 7–29.

Cueto, Marcus, ed. *Missionaries of Science: The Rockefeller Foundation and Latin America*. Bloomington: Indiana University Press, 1994.

Cuevas Sánchez, Jesús Axayacatl. "Recursos fitogenéticos: Bases conceptuales para su estudio y conservación." Departamento de Fitotecnia, Universidad Autónoma Chapingo, 1988.

Cullather, Nick. "Development? It's History." *Diplomatic History* 24, no. 4 (2000): 641–53.

———. *The Hungry World: America's Cold War Battle against Poverty in Asia*. Cambridge, MA: Harvard University Press, 2010.

———. "Miracles of Modernization: The Green Revolution and the Apotheosis of Technology." *Diplomatic History* 28, no. 2 (2004): 227–54.

———. "'Stretching the Surface of the Earth': The Foundations, Neo-Malthusianism and the Modernising Agenda." *Global Society* 28, no. 1 (2014): 104–12.

Curran, William Reid. "Indian Corn: Genesis of Reid's Yellow Dent." *Journal of the Illinois State Historical Society* 11, no. 4 (1919): 576–85.

Curry, Helen Anne. "Breeding Uniformity and Banking Diversity: The Genescapes of Industrial Agriculture, 1935–1970." *Global Environment* 10, no. 1 (2017): 83–113.

———. "From Bean Collection to Seed Bank: Transformations in Heirloom Vegetable Conservation, 1970–1985." *BJHS Themes* 4 (2019): 149–67.

———. "From Working Collections to the World Germplasm Project: Agricultural Modernization and Genetic Conservation at the Rockefeller Foundation." *History and Philosophy of the Life Sciences* 39 (2017): art. 5.

———. "Gene Banks, Seed Libraries, and Vegetable Sanctuaries: The Cultivation and Conservation of Heritage Vegetables in Britain, 1970–1985." *Culture, Agriculture, Food and Environment* 41, no. 2 (2019): 87–96.

———. "The History of Seed Banking and the Hazards of Back Up." Manuscript. January 2021. www.helenannecurry.com/papers.

———. "Imperilled Crops and Endangered Flowers." In *Worlds of Natural History*, edited by H. A. Curry, N. Jardine, J. A. Secord, and E. C. Spary, 460–75. Cambridge: Cambridge University Press, 2018.

———. "Taxonomy, Race Science, and Mexican Maize." *Isis* 112, no. 1 (2021): 1–21.

Cushing, Frank Hamilton. *Zuñi Breadstuff*. 1920. Reprint, New York: Museum of the American Indian, 1974.

Danbom, David B. "The Agricultural Experiment Station and Professionalization: Scientists' Goals for Agriculture." *Agricultural History* 60, no. 2 (1986): 246–55.

Davis, E. Wade. "Ethnobotany: An Old Practice, a New Discipline." In Schultes and Von Reis, *Ethnobotany*, 40–51.

Dawson, Alexander S. "From Models for the Nation to Model Citizens: Indigenismo and the 'Revindication' of the Mexican Indian, 1920–40." *Journal of Latin American Studies* 30, no. 2 (1998): 279–308.

De Bont, Raf. "'Primitives' and Protected Areas: International Conservation and the 'Naturalization' of Indigenous People, ca. 1910–1975." *Journal of the History of Ideas* 76, no. 2 (2015): 215–36.

———. "A World Laboratory: Framing the Albert National Park." *Environmental History* 22, no. 3 (2017): 404–32.

Delborne, Jason A. "Transgenes and Transgressions: Scientific Dissent as Heterogeneous Practice." *Social Studies of Science* 38, no. 4 (2008): 509–41.

DeWalt, Billie R. "Mexico's Second Green Revolution: Food for Feed." *Mexican Studies/Estudios Mexicanos* 1, no. 1 (1985): 29–60.

Dimitri, Carolyn, Anne Effland, and Neilson Conklin. *The 20th Century Transformation of U.S. Agriculture and Farm Policy.* USDA Economic Research Service, Economic Information Bulletin 3. Washington, DC: USDA, June 2005.

Dippie, Brian W. *The Vanishing American: White Attitudes and U.S. Indian Policy.* Middletown, CT: Wesleyan University Press, 1982.

Dodd, Norris E. "The Food and Agriculture Organization of the United Nations: Its History, Organization, and Objectives." *Agricultural History* 23, no. 2 (1949): 81–86.

Doebley, John. "The Genetics of Maize Evolution." *Annual Review of Genetics* 38 (2004): 37–59.

———. "Molecular Evidence for Gene Flow among *Zea* Species." *BioScience* 40, no. 6 (1990): 443–48.

Doebley, John, Jonathan D. Wendel, J. S. C. Smith, Charles W. Stuber, and Major M. Goodman. "The Origin of Cornbelt Maize: The Isozyme Evidence." *Economic Botany* 42, no. 1 (1988): 120–31.

Doremus, Anne. "Indigenism, Mestizaje, and National Identity in Mexico during the 1940s and the 1950s." *Mexican Studies/Estudios Mexicanos* 17, no. 2 (2001): 375–402.

Drayton, Richard. *Nature's Government: Science, Imperial Britain, and the "Improvement" of the World.* New Haven, CT: Yale University Press, 2000.

Duffey, Kiyah J., and Barry M. Popkin. "High-Fructose Corn Syrup: Is This What's for Dinner?" *American Journal of Clinical Nutrition* 88, no. 6 (2008): 1722S–32S.

Dunlap, Thomas R. *Nature and the English Diaspora.* Cambridge: Cambridge University Press, 1999.

Duvick, Donald N. "United States." In *Maize Seed Industries in Developing Countries,* edited by Michael L. Morris, 193–211. Boulder, CO: Rienner, 1998.

———. "The Use of Cytoplasmic Male-Sterility in Hybrid Seed Production." *Economic Botany* 13, no. 3 (1959): 167–95.

———. "William L. Brown, July 16, 1913–March 8, 1991." *Biographical Memoirs of the National Academy of Sciences.* 1996. www.nasonline.org/publications /biographical-memoirs/memoir-pdfs/brown-william-l.pdf.

Eberhart, S. A. "USA USDA-NSSL." In *Latin American Maize Germplasm Conservation: Core Subset Development and Regeneration,* edited by Suketoshi Taba, 47–50. Workshop Proceedings, 1–5 June 1998. Mexico City: CIMMYT, 1999.

Edwards, Paul N. *The Closed World: Computers and the Politics of Discourse in Cold War America.* Cambridge, MA: MIT Press, 1996.

Egan, Michael. "Survival Science: Crisis Disciplines and the Shock of the Environment in the 1970s." *Centaurus* 59 (2017): 26–39.

Ehrlich, Paul. *The Population Bomb.* New York: Buccaneer Books, 1968.

Elina, Olga, Susanne Heim, and Nils Roll-Hansen. "Plant Breeding on the Front: Imperialism, War, and Exploitation." *Osiris* 20 (2005): 161–79.

Enfield, Edward. *Indian Corn: Its Value, Culture, and Uses.* New York: Appleton, 1866.

Escárcega López, Everardo, and Saúl Escobar Toledo. *Historia de la cuestión agraria mexicana: El Cardenismo, un parteaguas histórico en el proceso agrario nacional, 1934–1940.* Mexico City: Siglo XXI, 1990.

Escobar, Arturo. *Encountering Development: The Making and Unmaking of the Third World.* Princeton, NJ: Princeton University Press, 2012.

Esquinas-Alcázar, J., Angela Hilmi, and I. López Noriega. "A Brief History of the Negotiations on the International Treaty on Plant Genetic Resources for Food and Agriculture." In *Crop Genetic Resources as a Global Commons: Challenges in International Law and Governance,* edited by Michael Halewood, Isabel López Noriega, and Selim Louafi, 135–49. London: Routledge, 2012.

Esteva, Gustavo. "Los árboles de las culturas mexicanas." In Esteva and Marielle, *Sin maíz,* 17–28.

———. "Food Needs and Capacities: Four Centuries of Conflict." In Austin and Esteva, *Food Policy in Mexico,* 23–47.

Esteva, Gustavo, and Catherine Marielle, eds. *Sin maíz, no hay país.* Mexico City: Consejo Nacional para la Cultura y las Artes, 2003.

Eyzaguirre, P. and M. Iwanaga, eds. *Participatory Plant Breeding.* Rome: Bioversity, 1996.

Fábregas Puig, Andrés. "La ecología cultural política y el estudio de regiones en México." *Revista de Dialectología y Tradiciones Populares* 64, no. 1 (2009): 167–76.

Farnham, Timothy J. *Saving Nature's Legacy: Origins of the Idea of Biodiversity.* New Haven, CT: Yale University Press, 2007.

Feder, Ernest. "The New Penetration of the Agricultures of the Underdeveloped Countries by the Industrialized Nations and Their Multinational Concerns." Occasional Paper 19. Institute of Latin American Studies, University of Glasgow, 1975.

Fenn, Elizabeth A. *Encounters at the Heart of the World: A History of the Mandan People.* New York: Hill and Wang, 2014.

Fenzi, Marianna. "'Provincialiser' la Révolution Verte: Savoirs, politiques et pratiques de la conservation de la biodiversité cultivée (1943–2015)." PhD diss., École des Hautes Études en Sciences Sociales, Centre Alexandre Koyré, 2017.

Fenzi, Marianna, and Christophe Bonneuil. "From 'Genetic Resources' to 'Ecosystems Services': A Century of Science and Global Policies for Crop Diversity Conservation." *Culture, Agriculture, Food and Environment* 38, no. 2 (2016): 72–83.

Fernandez-Cornejo, Jorge. *The Seed Industry in U.S. Agriculture: An Exploration of Data and Information on Crop Seed Markets, Regulation, Industry Structure, and*

Research and Development. Agriculture Information Bulletin 786. Washington, DC: USDA Economic Research Service, 2004.

Finch, V.C. "A Graphic Summary of World Agriculture." In *USDA Yearbook of Agriculture 1916*, 531–53. Washington, DC: Government Publishing Office, 1917.

Fitting, Elizabeth. "Importing Corn, Exporting Labor: The Neoliberal Corn Regime, GMOs, and the Erosion of Mexican Biodiversity." *Agriculture and Human Values* 23 (2006): 15–26.

———. *The Struggle for Maize: Campesinos, Workers, and Transgenic Corn in the Mexican Countryside.* Durham, NC: Duke University Press, 2011.

Fitzgerald, Deborah. *The Business of Breeding: Hybrid Corn in Illinois, 1890–1940.* Ithaca, NY: Cornell University Press, 1990.

———. "Farmers Deskilled: Hybrid Corn and Farmers' Work." *Technology and Culture* 34, no. 2 (1993): 324–43.

Flitner, Michael. "Genetic Geographies: A Historical Comparison of Agrarian Modernization and Eugenic Thought in Germany, the Soviet Union, and the United States." *Geoforum* 34, no. 2 (2003): 175–85.

Fowler, Cary. "The Plant Patent Act of 1930: A Sociological History of Its Creation." *Journal of the Patent and Trademark Office Society* 82, no. 9 (2000): 621–44.

———. *Seeds on Ice: Svalbard and the Global Seed Vault.* Westport, CT: Prospecta, 2016.

———. "The Svalbard Seed Vault and Crop Security." *BioScience* 58, no. 3 (2008): 190–91.

———. *Unnatural Selection: Technology, Politics and Plant Evolution.* Yverdon, Switzerland: Gordon and Breach, 1994.

Fox, Jeffrey L. "EPA Okays *Bt* Corn; USDA Eases Plant Testing." *Nature Biotechnology* 13 (1995): 1035–36.

Fox, Jonathan, and Libby Haight, eds. *Subsidizing Inequality: Mexican Corn Policy since NAFTA.* Mexico City: Woodrow Wilson Center/University of California–Santa Cruz, 2010.

Foyer, Jean. "Ver su riqueza en los maíces: Un panorama de las iniciativas de conservación de maíces criollos en México." Report for l'ANR BioTEK. 2012. https://halshs.archives-ouvertes.fr/halshs-00994898.

Frankel, O.H. "The Development and Maintenance of Superior Genetic Stocks." *Heredity* 4, no. 1 (1950): 89–102.

———. "Genetic Conservation: Our Evolutionary Responsibility." *Genetics* 78 (1974): 53–65.

———, ed. "Survey of Crop Genetic Resources in their Centres of Diversity." Report AGP: CGR/73/7. Rome: FAO/IBP, 1973.

Frankel, O.H., and E. Bennett, eds. *Genetic Resources in Plants: Their Exploration and Conservation.* IBP Handbook 11. Oxford: Blackwell Scientific, 1970.

Frey, Kenneth J. "National Plant Breeding Study, I." Special Report 98. Iowa Agriculture and Home Economics Experiment Station. 1996.

Friese, Carrie. *Cloning Wild Life: Zoos, Captivity, and the Future of Wild Animals.* New York: New York University Press, 2013.

Frison, Christine, Francisco López, and José T. Esquinas-Alcázar, eds. *Plant Genetic Resources and Food Security: Stakeholder Perspectives on the International Treaty on Plant Genetic Resources for Food and Agriculture*. London: Earthscan/FAO/ Bioversity, 2011.

Fuente Hernández, Juan de la, Margarita González Huerta, María Luisa Jiménez Esquerra, and Rafael Ortega Paczka. "El desenvolvimiento de la investigación agronómica en la década de los ochentas." In Méndez Ramírez et al., *Investigación agrícola en México,* 13–24.

Fuente Hernández, Juan de la, María Luisa Jiménez Esquerra, Margarita González Huerta, Rodolfo Cortés del Moral, and Rafael Ortega Paczka. *La investigación agrícola y el estado méxicano, 1960–1976*. Chapingo, Mexico: Universidad Autónoma Chapingo, 1990.

Fullilove, Courtney. *The Profit of the Earth: The Global Seeds of American Agriculture*. Chicago: University of Chicago Press, 2017.

Galluzzi, Gea, Pablo Eyzaguirre, and Valeria Negri. "Home Gardens: Neglected Hotspots of Agro-biodiversity and Cultural Diversity." *Biodiversity and Conservation* 19, no. 13 (2010): 3635–54.

Gálvez, Alyshia. *Eating NAFTA: Trade, Food Policies, and the Destruction of Mexico*. Oakland: University of California Press, 2018.

García Urigüen, Pedro. *La alimentación de los mexicanos: Cambios sociales y económicos, y su impacto en los hábitos alimenticios*. Mexico City: Canacintra, 2012.

Gissibl, Bernhard, Sabine Höhler, and Patrick Kupper, eds. *Civilizing Nature: National Parks in Global Historical Perspective*. New York: Berghahn Books, 2012.

Glaeser, Bernhard, ed. *The Green Revolution Revisited: Critique and Alternatives*. London: Routledge, 1987.

Gliessman, Steve. "Agroecology: Growing the Roots of Resistance." *Agroecology and Sustainable Food Systems* 37, no. 1 (2013): 19–31.

González, Roberto J. "From Indigenismo to Zapatismo: Theory and Practice in Mexican Anthropology." *Human Organization* 63, no. 2 (2004): 141–50.

González-Santos, Rosalinda, Jorge Cadena-Iñiguez, Francisco J. Morales-Flores, Víctor M. Ruiz-Vera, José Pimentel-López, and Aureliano Peña-Lomelí. "Model for the Conservation and Sustainable Use of Plant Genetic Resources in Mexico." *Wulfenia* 22, no. 2 (2015): 333–53.

Goodman, Louis Wolf. *The Improved Seed Industry: Issues and Options for Mexico*. With Arthur L. Domike and Charles Sands. Washington, DC: Center for International Technical Cooperation, American University, 1982.

Goodman, Major M. "An Evaluation and Critique of Current Germplasm Programs." In *1983 Plant Breeding Research Forum Report,* 195–249. Des Moines, Iowa: Pioneer Hi-Bred International, 1984.

———. "US Maize Germplasm: Origins, Limitations, and Alternatives." In CIMMYT, *Recent Advances,* 130–48.

Goodman, Major M., and Walton C. Galinat. "The History and Evolution of Maize." *Critical Reviews in Plant Sciences* 7, no. 3 (1988): 197–220.

Gottlieb, Robert. *Forcing the Spring: The Transformation of the American Environmental Movement.* Rev. ed. Washington, DC: Island, 2005.

Grant, Ulysses J., William H. Hatheway, David H. Timothy, Clímaco Cassalett D., and Lewis M. Roberts. *Races of Maize in Venezuela.* Publication 1136. Washington, DC: National Research Council, National Academy of Sciences, 1963.

Greenland, D. J. "International Agricultural Research and the CGIAR System: Past, Present, and Future." *Journal of International Development* 9, no. 4 (1997): 459–82.

Griffin, Keith. *The Political Economy of Agrarian Change: An Essay on the Green Revolution.* 2nd ed. London: Macmillan, 1979.

Griliches, Zvi. "Hybrid Corn and the Economics of Innovation." *Science* 132, no. 3422 (1960): 275–80.

Grobman, Alexander, Wilfredo Salhuana, and Ricardo Sevilla. *Races of Maize in Peru.* With Paul C. Mangelsdorf. Publication 915. Washington, DC: National Research Council, National Academy of Sciences, 1961.

Guarino, Luigi, and David B. Lobell. "A Walk on the Wild Side." *Nature Climate Change* 1 (2011): 374–75.

Guha, Ramachandra. *Environmentalism: A Global History.* New York: Longman, 2000.

Guthman, Julie. *Agrarian Dreams: The Paradox of Organic Farming in California.* 2nd ed. Oakland: University of California Press, 2014.

Gutiérrez Núñez, Netzahualcóyotl Luis. "Cambio agrario y Revolución Verde: Dilemas científicos, políticos y agrarios en la agricultura mexicana del maíz, 1920–1970." PhD diss., Colegio de México, 2017.

Hajjar, Reem, and Toby Hodgkin. "The Use of Wild Relatives in Crop Improvement: A Survey of Developments over the Last 20 Years." *Euphytica* 156, nos. 1–2 (2007): 1–13.

Hallauer, Arnel A., and J. H. Sears. "Integrating Exotic Germplasm into Corn Belt Maize Breeding Programs." *Crop Science* 12, no. 2 (1972): 203–6.

Hamilton, Shane. *Supermarket USA: Food and Power in the Cold War Farms Race.* New Haven, CT: Yale University Press, 2018.

Haraway, Donna J. *Staying with the Trouble: Making Kin in the Chthulucene.* Durham, NC: Duke University Press, 2016.

Harlan, Harry V. *One Man's Life with Barley.* New York: Exposition, 1957.

Harlan, Jack R. "Distribution and Utilization of Natural Variability in Cultivated Plants." In *Genetics in Plant Breeding,* no. 9, 191–208. Brookhaven Symposia in Biology. Upton, NY: Brookhaven National Laboratory, 1962.

Harrington, M. R. "Some Seneca Corn-Foods and Their Preparation." *American Anthropologist* 10, no. 4 (1908): 575–90.

Harris, Amanda. *Fruits of Eden: David Fairchild and America's Plant Hunters.* Gainesville: University of Florida, 2015.

Hartigan, John, Jr. *Care of the Species: Races of Corn and the Science of Plant Biodiversity.* Minneapolis: University of Minnesota Press, 2017.

Hartley, C. P. "Seed Corn." *Farmer's Bulletin* 415. Washington, DC: USDA, 1917.

Harvey, Neil. *The Chiapas Rebellion: The Struggle for Land and Democracy.* Durham, NC: Duke University Press, 1998.

———. "Rebellion in Chiapas: Rural Reforms and Popular Struggle." *Third World Quarterly* 16, no. 1 (1995): 39–74.

Harwood, Jonathan. *Europe's Green Revolution and Others Since: The Rise and Fall of Peasant-Friendly Plant Breeding.* Abingdon, UK: Routledge, 2012.

———. "Peasant Friendly Plant Breeding and the Early Years of the Green Revolution in Mexico." *Agricultural History* 83, no. 3 (2009): 384–410.

———. "Was the Green Revolution Intended to Maximise Food Production?" *International Journal of Agricultural Sustainability* 17, no. 4 (2019): 312–25.

Hatheway, William H. *Races of Maize in Cuba.* Publication 453. Washington, DC: National Research Council, National Academy of Sciences, 1957.

Hayden, Cori. "Bioprospecting's Representational Dilemma." *Science as Culture* 14, no. 2 (2005): 185–200.

———. *When Nature Goes Public: The Making and Unmaking of Bioprospecting in Mexico.* Princeton, NJ: Princeton University Press, 2003.

Hecht, Susanna B. "The Evolution of Agroecological Thought." In *Agroecology: The Science of Sustainable Agriculture,* edited by Miguel A. Altieri, 1–20. 2nd ed. Boca Raton: CRC, 2018.

Heise, Ursula K. *Imagining Extinction: The Cultural Meanings of Endangered Species.* Chicago: University of Chicago Press, 2016.

Hellin, Jon, and Mauricio Bellon. "Manejo de semillas y diversidad del maíz." *LEISA* 23, no. 2 (2007): 9–11.

Herdt, R. W. and C. Capule. *Adoption, Spread, and Production Impact of Modern Rice Varieties in Asia.* Los Baños, Philippines: IRRI, 1983.

Hernández Xolocotzi, Efraím. "El agroecosistema, concepto central en el análisis de la enseñanza, la investigación y la educación agrícola en México." In Hernández Xolocotzi, *Agroecosistemas de México,* xv–xix.

———, ed. *Agroecosistemas de México: Contribuciones a la enseñanza, investigación y divulgación agrícola.* 2nd ed. Chapingo, Mexico: Colegio de Postgraduados, 1981.

———. *Apuntes sobre la exploración etnobotánica y su metodología.* Chapingo, Mexico: Colegio de Postgraduados, 1971.

———. "El concepto de etnobotánica." In *La etnobotánica: Tres puntos de vista y una perspectiva,* edited by Alfredo Barrera, 5–8. 1979. Reprint, Chapingo, Mexico: Universidad Autónoma Chapingo, 2001.

———. "Experiences in the Collection of Maize Germplasm." In CIMMYT, *Recent Advances,* 1–8.

———. "Exploración etnobotánica para la obtención de plasma germinal para México." In Cervantes Santana, *Análisis de los recursos,* 3–12.

———. Introduction to "Xolocotzia: Obras de Efraím Hernández Xolocotzi." Special issue, *Revista de Geografía Agrícola* 1 (1985): 15–24.

Hernández Xolocotzi, Efraím, and Glafiro Alanís Flores. "Estudio morfológico de cinco nuevas razas de maíz de la Sierra Madre Occidental de México: Implicaciones filogenéticas y fitogeográficas." *Agrociencia* 5, no. 1 (1970): 3–30.

Hernández Xolocotzi, Efraím, Alberto Ramos Rodríguez, and Miguel Ángel Martínez Alfaro. "Etnobotánica." In *Contribuciones al conocimiento del frijol (Phaseolus) en México,* edited by Mark E. Engelman, 113–38. Chapingo, Mexico: Colegio de Postgraduados, 1979.

Hernández Xolocotzi, Efraím, and Margarita Araceli Zárate Aquino. "Agricultura tradicional y conservación de recursos genéticos *in situ.*" In *Avances en el estudio de los recursos fitogenéticos de México,* edited by Rafael Ortega Paczka, Guadalupe Palomino Hasbach, Fernando Castillo González, Victor A. González Hernández, and Manuel Livera Muñoz, 7–28. Chapingo, Mexico: Sociedad Mexicana de Fitogenética (SOMEFI), 1991.

Hewitt de Alcántara, Cynthia. *Anthropological Perspectives on Rural Mexico.* London: Routledge and Kegan Paul, 1984.

———. *Modernizing Mexican Agriculture: Socioeconomic Implications of Technological Change, 1940–1970.* Geneva: United Nations Research Institute for Social Development, 1976.

Hightower, Jim. *Hard Tomatoes, Hard Times.* Cambridge, MA: Schenkman, 1973.

Higman, B. W. *A Concise History of the Caribbean.* Cambridge: Cambridge University Press, 2011.

Hill, Christina Gish. "Seeds as Ancestors, Seeds as Archives: Seed Sovereignty and the Politics of Repatriation to Native Peoples." *American Indian Culture and Research Journal* 41, no. 3 (2017): 93–112.

Hinsley, Curtis M., Jr. *Scientists and Savages: The Smithsonian Institution and the Development of American Anthropology, 1846–1910.* Washington, DC: Smithsonian Institution Press, 1981.

Hoisington, David, Mireille Khairallah, Timothy Reeves, and Jean-Marcel Ribaut. "Plant Genetic Resources: What Can They Contribute Toward Increased Crop Productivity?" *Proceedings of the National Academy of Sciences* 96, no. 11 (1999): 5937–43.

Holmes, Matthew. "Changing Techniques in Crop Plant Classification: Molecularization at the National Institute of Agricultural Botany during the 1980s." *Annals of Science* 74, no. 2 (2017): 149–64.

Hoover, Elizabeth. "'You Can't Say You're Sovereign if You Can't Feed Yourself': Defining and Enacting Food Sovereignty in American Indian Community Gardening." *American Indian Culture and Research Journal* 41, no. 3 (2017): 31–70.

Horsfall, James G. "The Fire Brigade Stops a Raging Corn Epidemic." In *USDA Yearbook of Agriculture 1975,* 105–14. Washington, DC: Government Publishing Office, 1975.

Hought, Joy, and Melissa Kruse-Peeples. *Saving Seed in the Southwest: Techniques for Seed Stewardship in Aridlands.* Tucson: Native Seeds/SEARCH, 2016.

Howes, Michael, and Robert Chambers. "Indigenous Technical Knowledge: Analysis, Implications and Issues." *IDS Bulletin* 10, no. 2 (1979): 5–11.

"How Uncle Sam Helps Farmers to Grow Better Corn." *Craftsman* 27, no. 1 (1914): 91.

Hughes, Sally Smith. *Genentech: The Beginnings of Biotech.* Chicago: University of Chicago Press, 2011.

Hunn, Eugene. "Ethnobiology in Four Phases." *Journal of Ethnobiology* 27, no. 1 (2007): 1–10.

IBPGR (International Board for Plant Genetic Resources). *Annual Report, 1977.* Rome: IBPGR Secretariat, 1978. www.bioversityinternational.org/e-library /publications/detail/ibpgr-annual-report-1977/.

Iltis, Hugh H. "Freezing the Genetic Landscape." *Maize Genetics Cooperation Newsletter* 48 (1974): 199–200.

INIA (Instituto Nacional de Investigaciones Agrícolas). *INIA: XV años de investigación agrícola.* Mexico City: INIA, 1976.

——. *Unidad de Recursos Genéticos, banco de germoplasma de maíz y sorgo: Informe de actividades desarrolladas durante 1978.* [Chapingo, Mexico: INIA, 1979?].

Ita, Ana de. "La defensa internacional del maíz contra la contaminación transgénica en su centro de origen." *Cotidiano,* no. 173 (2012): 57–65.

James, Clive. "Global Status of Transgenic Crops in 1997." *ISAAA Briefs,* no. 5. Ithaca, NY: International Service for the Acquisition of Agri-biotech Applications, 1997.

James, Edwin. "Perpetuation and Protection of Germ Plasm as Seed." In *Germ Plasm Resources,* edited by Ralph E. Hodgson, 317–26. Publication 66. Washington, DC: American Association for the Advancement of Science, 1961.

Janvry, Alain de, Elisabeth Sadoulet, and Gustavo Gordillo de Anda. "NAFTA and Mexico's Maize Producers." *World Development* 23, no. 8 (1995): 1349–62.

Jarosz, Lucy. "The Political Economy of Global Governance and the World Food Crisis: The Case of the FAO." *Review* (Braudel Center) 32, no. 1 (2009): 37–60.

Jennings, Bruce. *Foundations of International Agricultural Research: Science and Politics in Mexican Agriculture.* Boulder, CO: Westview, 1988.

Johnson, Allen W. "Individuality and Experimentation in Traditional Agriculture." *Human Ecology* 1, no. 2 (1972): 149–59.

Kato Yamakake, Takeo Ángel, Cristina Mapes Sánchez, Luz María Mera Ovando, José Antonio Serratos Hernández, and Robert Arthur Bye Boettler. *Origen y diversificación del maíz: Una revisión analítica.* Mexico City: UNAM/Comisión Nacional para el Conocimiento y Uso de la Biodiversidad, 2009.

Kevles, Daniel J. "Ananda Chakrabarty Wins a Patent: Biotechnology, Law, and Society." *Historical Studies in the Physical and Biological Sciences* 25, no. 1 (1994): 111–35.

Kimmelman, Barbara A. "Organisms and Interests in Scientific Research: R. A. Emerson's Claims for the Unique Contributions of Agricultural Genetics." In *The Right Tools for the Job: At Work in the Twentieth-Century Life Sciences,* edited by Adele H. Clarke and Joan H. Fujimura, 198–232. Princeton, NJ: Princeton University Press, 1992.

Kinchy, Abby. *Seeds, Science, and Struggle: The Global Politics of Transgenic Crops.* Cambridge, MA: MIT Press, 2012.

Klausner, Arthur. "And Then There Were Two." *Nature Biotechnology* 3, no. 7 (1985): 605–12.

Kleinman, Kim. "Edgar Anderson: Interdisciplinary Authority on What Was *Not* Known about Corn." *Endeavour* 23, no. 3 (1999): 114–17.

———. "His Own Synthesis: Corn, Edgar Anderson, and Evolutionary Theory in the 1940s." *Journal of the History of Biology* 32, no. 2 (1999): 293–320.

Kloppenburg, Jack Ralph, Jr. *First the Seed: The Political Economy of Plant Biotechnology, 1492–2000*. 2nd ed. Madison: University of Wisconsin Press, 2004.

Knight, Alan. "Racism, Revolution, and Indigenismo." In *The Idea of Race in Latin America, 1870–1940*, edited by Richard Graham, 71–113. Austin: University of Texas Press, 1990.

Krimsky, Sheldon. *Genetic Alchemy: The Social History of the Recombinant DNA Controversy*. Cambridge, MA: MIT Press, 1982.

Kuleshov, N. N. "The Maize of Mexico, Guatemala, Cuba, Panama and Colombia." In Bukasov, "Cultivated Plants," 492–501.

Kuper, Adam. *The Invention of Primitive Society: Transformations of an Illusion*. New York: Routledge, 1988.

Kwa, Chunglin. "Representations of Nature Mediating between Ecology and Science Policy: The Case of the International Biological Programme." *Social Studies of Science* 17, no. 3 (1987): 413–42.

LAMP (Latin American Maize Program). *Catálogo del germoplasma de maíz*. Beltsville, MD: USDA Agricultural Research Service, 1991.

Lappé, Frances Moore, and Joseph Collins. *Food First: Beyond the Myth of Scarcity*. With Cary Fowler. Boston: Houghton Mifflin, 1977.

Lee, Jill, and Ben Hardin. "GEM Searches for Treasures in Exotic Maize." *Agricultural Research* 45, no. 9 (1997): 4–6.

Lehmann, Christian O. "Collecting European Land-Races and Development of European Gene Banks: Historical Remarks." *Kulturpflanze* 29, no. 1 (1981): 29–40.

Leonelli, Sabina. "Data: From Objects to Assets." *Nature* 574 (2019): 317–21.

———. "How Data Cross Borders: Globalizing Plant Knowledge through Transnational Data Management and Its Epistemic Economy." In *Transnational Transactions: Negotiating the Movement of Knowledge across Borders,* edited by John Krige. Chicago: University of Chicago Press, forthcoming.

Leonelli, Sabina, and Rachel A. Ankeny. "Re-thinking Organisms: The Impact of Databases on Model Organism Biology." *Studies in History and Philosophy of Biological and Biomedical Sciences* 43, no. 1 (2012): 29–36.

Lessens, Kelly J. Sisson. "Master of Millions: King Corn in American Culture." PhD diss., University of Michigan, 2011.

Levy, Santiago, and Sweder van Wijnbergen. "Maize and the Free Trade Agreement between Mexico and the United States." *World Bank Economic Review* 6, no. 3 (1992): 481–502.

Lightfoot, Clive. "Indigenous Research and On-Farm Trials." *Agricultural Administration and Extension* 24 (1987): 79–89.

Lomnitz, Claudio. "Bordering on Anthropology: The Dialectics of a National Tradition in Mexico." *Revue de Synthèse* 4, nos. 3–4 (2000): 345–80.

López-Beltrán, Carlos, and Vivette García Deister. "Aproximaciones científicas al mestizo mexicano." *História, Ciências, Saúde: Manguinhos* 20, no. 2 (2013): 391–410.

López Hernández, Haydeé. "De la gloria prehispánica al socialismo: Las políticas indigenistas del Cardenismo." *Cuicuilco,* no. 57 (2013): 47–74.

Lorek, Timothy Wayne. "Developing Paradise: Agricultural Science in the Conflicted Landscapes of Colombia's Cauca Valley, 1927–1967." PhD diss., Yale University, 2019.

———. "Strange Priests and Walking Experts: Nature, Religion, and Science in Sprouting the Cold War's Green Revolution." In *Itineraries of Science: Science, Technology, and the Environment in Latin America,* edited by Andra B. Chastain and Timothy W. Lorek, 93–113. Pittsburgh: University of Pittsburgh Press, 2020.

Loskutov, Igor G. *Vavilov and His Institute: A History of the World Collection of Plant Genetic Resources in Russia.* Rome: International Plant Genetic Resources Institute, 1999.

Louette, Dominique, André Charrier, and Julien Berthaud. "In Situ Conservation of Maize in Mexico: Genetic Diversity and Maize Seed Management in a Traditional Community." *Economic Botany* 51, no. 1 (1997): 20–38.

Lozano Toledano, Adrián, and Antonio Anaya Pérez. "El Plan Chapingo y su importancia para el campo mexicano." In *La educación superior en el proceso histórico de México,* edited by David Piñera Ramírez, 473–82. Vol. 3. Mexicali, Mexico: Universidad Autónoma de Baja California, 2002.

Luckstead, Jeff, Stephen Devadoss, and Abelardo Rodriguez. "The Effects of North American Free Trade Agreement and United States Farm Policies on Illegal Immigration and Agricultural Trade." *Journal of Agricultural and Applied Economics* 44, no. 1 (2012): 1–19.

Luna Mena, Bethel M., Ma. Alejandra Hinojosa Rodríguez, Óscar J. Ayala Garay, Fernando Castillo González, and J. Apolinar Mejía Contreras. "Perspectivas de desarrollo de la industria semillera de maíz en México." *Revista Fitotecnia Mexicana* 35, no. 1 (2012): 1–7.

MacDonald, James M., Penni Korb, and Robert A. Hoppe. *Farm Size and the Organization of U.S. Crop Farming.* USDA ERS Economic Research Report 152. Washington, DC: USDA Economic Research Service, 2013.

Mangelsdorf, Paul C. *Corn: Its Origin, Evolution and Improvement.* Cambridge, MA: Belknap, 1974.

Mangelsdorf, Paul C., and R. G. Reeves. "The Origin of Indian Corn and Its Relatives." *Texas Agricultural Experiment Station Bulletin* 574 (1939): 1–315.

———. "The Origin of Maize." *Proceedings of the National Academy of Sciences* 24, no. 8 (1938): 303–12.

Marchisio, Sergio, and Antonietta Di Blase. *The Food and Agriculture Organization (FAO).* International Organization and the Evolution of World Society. Vol. 1. Dordrecht, Netherlands: Nijhoff, 1991.

Martínez Gómez, Francisco. *La globalización en la agricultura: Las negociaciones internacionales en torno al germoplasma agrícola.* Mexico City: Plaza y Valdés, 2002.

Masco, Joseph. "The Crisis in Crisis." Supplement, *Current Anthropology* 58, no. 15 (2017): S65–76.

Massawe, Festo, Sean Mayes, and Acga Cheng. "Crop Diversity: An Unexploited Treasure Trove for Food Security." *Trends in Plant Science* 21, no. 5 (2016): 365–68.

Massieu Trigo, Yolanda, and Jesús Lechuga Montenegro. "El maíz en México: Biodiversidad y cambios en el consumo." *Análisis Económico* 17, no. 36 (2002): 281–303.

Matchett, Karin. "At Odds over Inbreeding: An Abandoned Attempt at Mexico/United States Collaboration to 'Improve' Mexican Corn, 1940–1950." *Journal of the History of Biology* 39, no. 2 (2006): 345–72.

———. "Untold Innovation: Scientific Practice and Corn Improvement in Mexico, 1935–1965." PhD diss., University of Minnesota, 2003.

Matheka, Reuben M. "Decolonisation and Wildlife Conservation in Kenya, 1958–68." *Journal of Imperial and Commonwealth History* 36, no. 4 (2008): 615–39.

McAfee, Kathleen. "Beyond Techno-science: Transgenic Maize in the Fight over Mexico's Future." *Geoforum* 39, no. 1 (2008): 148–60.

McCann, James C. *Maize and Grace: Africa's Encounter with a New World Crop.* Cambridge, MA: Harvard University Press, 2005.

McCouch, Susan, Gregory J. Baute, James Bradeen, Paula Bramel, Peter K. Bretting, Edward Buckler, John M. Burke, et al. "Feeding the Future." *Nature* 499 (2013): 23–24.

McCouch, Susan R., Kenneth L. McNally, Wen Wang, and Ruaraidh Sackville Hamilton. "Genomics of Gene Banks: A Case Study in Rice." *American Journal of Botany* 99, no. 2 (2012): 407–23.

McDonald, Bryan L. *Food Power: The Rise and Fall of the Postwar American Food System.* Oxford: Oxford University Press, 2017.

McDonnell, Janet A. *The Dispossession of the American Indian, 1887–1934.* Bloomington: Indiana University Press, 1991.

McGlade, Jacqueline. "More a Plowshare Than a Sword: The Legacy of US Cold War Agricultural Diplomacy." *Agricultural History* 83, no. 1 (2009): 79–102.

McGregor, Russell. "The Doomed Race: A Scientific Axiom of the Late Nineteenth Century." *Australian Journal of Politics and History* 39 (1993): 14–22.

McPherson, Alan. *Yankee No! Anti-Americanism in U.S.–Latin American Relations.* Cambridge, MA: Harvard University Press, 2003.

Mejía Prado, Eduardo. "Víctor Manuel Patiño: La investigación como forma de vida." *Nómadas* 5, no. 1 (1998): 157–73.

Melhus, I. E., ed. *Plant Research in the Tropics.* Ames: Agricultural Experiment Station, Iowa State College, 1949.

Méndez Ramírez, Ignacio, Juan de la Fuente Hernández, Margarita González Huerta, Ma. Luisa Jiménez Esquerra, Rafael Ortega Paczka, Jesús Moncada de la Fuente, Alierso Caetano de Oliveira, et al. *La investigación agrícola en México, en la década de los ochentas.* Chapingo, Mexico: Universidad Autónoma Chapingo, 1991.

Méndez Rojas, Diana Alejandra. "Los libros del maíz: Revolución verde y diversidad biológica en América Latina, 1951–1970." *Letras Históricas* E-ISSN: 2448–8372. www.letrashistoricas.cucsh.udg.mx/index.php/LH/article/view/7281.

———. "El Programa Cooperativo Centroamericano para el Mejoramiento del Maíz: Una historia transnacional de la Revolución Verde desde Costa Rica y Guatemala, 1954–1963." Master's thesis, Instituto Mora, 2018.

———. "¿Técnicos o especialistas? Alfredo Carballo Quirós, la Fundación Rockefeller y la Revolución Verde en Costa Rica, 1949–1962." In *Historias entrelazadas: El intercambio académico en el siglo xx; México, Estados Unidos, América Latina,* edited by Sebastián Rivera Mir. Zinacantepec: Colegio Mexiquense, 2020.

Meyer, Melissa L. *The White Earth Tragedy: Ethnicity and Dispossession at a Minnesota Anishinaabe Reservation, 1889–1920.* Lincoln: University of Nebraska Press, 1994.

Meyer, Roy W. *The Village Indians of the Upper Missouri: The Mandans, Hidatsas, and Arikaras.* Lincoln: University of Nebraska Press, 1977.

Mihesuah, Devon A., and Hoover, Elizabeth. *Indigenous Food Sovereignty in the United States: Restoring Cultural Knowledge, Protecting Environments, and Regaining Health.* Norman: University of Oklahoma Press, 2019.

Minns, John. *The Politics of Developmentalism: The Midas States of Mexico, South Korea, and Taiwan.* London: Palgrave Macmillan, 2006.

Moock, Joyce Lewinger, and Robert E. Rhoades, eds. *Diversity, Farmer Knowledge and Sustainability.* Ithaca, NY: Cornell University Press, 1992.

Mooney, Pat, and Cary Fowler, eds. *The Community Seed Bank Kit.* Durham, NC: Regulator Press for Rural Advancement Fund International, 1986.

Mooney, Patrick R. "International Non-governmental Organizations." In Frison, López, and Esquinas-Alcázar, *Plant Genetic Resources,* 135–48.

———. "The Law of the Seed." *Development Dialogue* 1, no. (1983): 1–172.

———. *Seeds of the Earth: A Private or Public Resource?* London: International Coalition for Development Action, 1979.

Moncada de la Fuente, Jesús. "Evolución y perspectivas de la investigación agrícola en México." In Méndez Ramírez et al., *Investigación agrícola en México,* 25–66.

Montanari, Mario. "The Conception of SAM." In Austin and Esteva, *Food Policy in Mexico,* 48–58.

Montenegro de Wit, Maywa. "Are We Losing Diversity? Navigating Ecological, Political, and Epistemic Dimensions of Agrobiodiversity Conservation." *Agriculture and Human Values* 33, no. 3 (2016): 625–40.

Montes Meneses, Jorge. "Estrategia para la conservación de los recursos genéticos." In Cervantes Santana, *Análisis de los recursos,* 29–35.

Montgomery, E. G. *The Corn Crops: A Discussion of Maize, Kafirs, and Sorghums as Grown in the United States and Canada.* New York: Macmillan, 1916.

Müller-Wille, Staffan, and Hans-Jörg Rheinberger. *A Cultural History of Heredity.* Chicago: University of Chicago Press, 2012.

Museo Nacional de Culturas Populares. *Nuestro maíz: Treinta monografías populares*. Mexico City: Museo Nacional de Culturas Populares, 1982.

Nadal, Alejandro. *The Environmental and Social Impacts of Economic Liberalization on Corn Production in Mexico*. Study for Oxfam Great Britain and World Wildlife Fund International. Gland, Switzerland: Oxfam GB and WWF, 2000.

National Plant Genetic Resources Board. *Plant Genetic Resources: Conservation and Use*. Washington, DC: USDA, 1979.

National Plant Germplasm System: Current Status (1980), Strengths and Weaknesses, Long-Range Plan (1983–1997). Washington, DC: USDA, 1981.

National Research Council. *Genetic Vulnerability of Major Crops*. Washington, DC: National Academy of Sciences, 1972.

Nazarea, Virginia D. *Heirloom Seeds and Their Keepers: Marginality and Memory in the Conservation of Biological Diversity*. Tucson: University of Arizona Press, 2005.

Nazarea, Virginia D., R. E. Rhoades, and J. E. Andrews-Swann, eds. *Seeds of Resistance, Seeds of Hope: Place and Agency in the Conservation of Biodiversity*. Tucson: University of Arizona Press, 2013.

Neimark, Benjamin D. "Industrializing Nature, Knowledge, and Labour: The Political Economy of Bioprospecting in Madagascar." *Geoforum* 43, no. 5 (2012): 980–90.

Niethammer, Carolyn, ed. *Seed Banks Serving People: Highlights of a Workshop, October 9–10, 1981*. With Gary Nabhan, Mahina Drees, and Cynthia Anson. Santa Monica, CA: Meals for Millions/Freedom from Hunger, [1981?].

Núñez Gutiérrez, Hiram Ricardo, Rosaura Reyes Canchola, and Jorge Gustavo Ocampo Ledesma. *La huelga nacional de las escuelas de agricultura en 1967*. Chapingo, Mexico: Centro de Investigaciones Económicas, Sociales y Tecnológicas de la Agroindustria y la Agricultura Mundial, Universidad Autónoma Chapingo, 2008.

Oasa, Edmund Kazuso. "The International Rice Research Institute and the Green Revolution: A Case Study on the Politics of Agricultural Research." PhD diss., University of Hawai'i, 1981.

Ochoa, Enrique C. *Feeding Mexico: The Political Uses of Food since 1910*. Wilmington, DE: Scholarly Resources, 2000.

OECD (Organisation for Economic Co-operation and Development). *Agricultural Outlook, 2018–2027*. Paris: OECD, 2018.

Olea Franco, Adolfo. "La introducción del maíz híbrido en la agricultura mexicana: Una historia de equívocos científicos, intereses comerciales y conflictos sociales." In *Ciencia en los márgenes: Ensayos de historia de las ciencias en México*, edited by Mechthild Rutsch and Carlos Serrano Sánchez, 189–230. Mexico City: UNAM, 1997.

———. "One Century of Higher Agricultural Education and Research in Mexico (1850s–1960s), with a Preliminary Survey on the Same Subjects in the United States." PhD diss., Harvard University, 2001.

Olsson, Tore C. *Agrarian Crossings: Reformers and the Remaking of the US and Mexican Countryside*. Princeton, NJ: Princeton University Press, 2017.

Ortega Paczka, Rafael. "Don Gilberto Palacios de la Rosa: Precursor de la UACh." *Crítica* (2018): 9–14.

———. "Evaluación de recursos genéticos." In Cervantes Santana, *Análisis de los recursos,* 37–48.

———. "El maíz y sus investigadores." *Jornada del Campo,* no. 27 (12 December 2009). www.jornada.com.mx/2009/12/12/maiz.html.

———. "Reorganización del mejoramiento genético del maíz en el INIA." In Hernández Xolocotzi, *Agroecosistemas de México,* 369–90.

———. "Variación en maíz y cambios socio-económicos en Chiapas, Mex., 1946–1971." Master's thesis, Colegio de Postgraduados, Chapingo, 1973.

———. "Variedades y razas mexicanas de maíz y su evaluación en cruzamientos con líneas de clima templado como material de partida para fitomejoramiento." PhD diss., Leningrad, 1985.

———. "Vida y aportes del Maestro Efraím Hernández Xolocotzi." *Revista de Geografía Agrícola,* nos. 50–51 (2013): 31–36.

Ortega Paczka, Rafael, and Hermilo Ángeles Arrieta. "Maíz." In Cervantes Santana, *Análisis de los recursos,* 74–85.

Ortíz Cereceres, Joaquín. "Antecedentes de la investigación agrícola en Mexico y sus repercusiones." *Germen,* no. 3 (1985): 1–17.

Ortiz-García, Sol, Exequiel Ezcurra, Bernd Schoel, Francisca Acevedo, Jorge Soberón, Allison A. Snow, "Transgenic Maize in Mexico." *BioScience* 56, no. 9 (2006): 709.

Osborne, Michael A. "Acclimatizing the World: A History of the Paradigmatic Colonial Science." *Osiris* 15 (2000): 135–51.

———. *Nature, the Exotic, and the Science of French Colonialism.* Bloomington: Indiana University Press, 1994.

Osseo-Asare, Abena Dove. *Bitter Roots: The Search for Healing Plants in Africa.* Chicago: University of Chicago Press, 2014.

Otero, Gerardo. "Neoliberal Globalization, NAFTA, and Migration: Mexico's Loss of Food and Labor Sovereignty." *Journal of Poverty* 15, no. 4 (2011): 384–402.

Palacios Rangel, María Isabel, and Jorge Gustavo Ocampo Ledesma, eds. *Ing. Gilberto Palacios de la Rosa.* Chapingo, Mexico: Universidad Autónoma Chapingo, 2018.

Pardo Núñez, Joaliné. "Identidad, organización y estrategia en dos movimientos que pugnan por la soberanía alimentaria en México." *Desacatos* 55 (2017): 152–71.

Parsons, James J. "Carl Ortwin Sauer, 1889–1975." *Geographical Review* 66, no. 1 (1976): 83–89.

Pearse, Andrew. "Technology and Peasant Production: Reflections on a Global Study." *Development and Change* 8 (1977): 125–59.

———. *Seeds of Plenty, Seeds of Want: Social and Economic Implications of the Green Revolution.* Oxford: Clarendon, 1980.

Peña, Devon G., Luz Calvo, Pancho McFarland, and Gabriel R. Valle. *Mexican-Origin Foods, Foodways, and Social Movements: Decolonial Perspectives.* Fayetteville: University of Arkansas Press, 2017.

Pensado, Jaime M. *Rebel Mexico: Student Unrest and Authoritarian Political Culture during the Long Sixties.* Stanford, CA: Stanford University Press, 2013.

Perales R., Hugo, S. B. Brush, and C. O. Qualset. "Dynamic Management of Maize Landraces in Central Mexico." *Economic Botany* 57, no. 1 (2003): 21–34.

Peres, Sara Marques Mano Ivo. "Saving the Gene Pool: Genebanks and the Political Economy of Crop Germplasm Conservation." PhD diss., University College London, 2017.

———. "Seed Banking as Cryopower: A Cryopolitical Account of the Work of the International Board of Plant Genetic Resources, 1973–1984." *Culture, Agriculture, Food and Environment* 41, no. 2 (2019): 76–86.

Perkins, John H. *Geopolitics and the Green Revolution: Wheat, Genes, and the Cold War.* Oxford: Oxford University Press, 1997.

———. "The Rockefeller Foundation and the Green Revolution, 1941–1956." *Agriculture and Human Values* 7, nos. 3–4 (1990): 6–18.

Picado, Wilson. "Breve historia semántica de la Revolución Verde." In *Agriculturas e innovación tecnológica en la Península Ibérica (1946–1975),* edited by Daniel Lanero and Dulce Freire, 25–50. Madrid: Ministerio de Medio Ambiente y Medio Rural y Marino, 2011.

Pigott, Charles Maurice. "Maize and Semiotic Emergence in a Contemporary Maya Tale: Tec Tun's *U tsikbalo'ob XNuk Nal* [Tales of Old Mother Corn]." *Tapuya: Latin American Science, Technology and Society* 2, no. 1 (2019): 112–26.

Pistorius, Robin. *Scientists, Plants and Politics: A History of the Plant Genetic Resources Movement.* Rome: International Plant Genetic Resources Institute, 1997.

Pistorius, Robin, and Jeroen van Wijk. *The Exploitation of Plant Genetic Information: Political Strategies in Crop Development.* Wallingford, UK: CABI, 1999.

Plucknett, Donald L., Nigel J. H. Smith, J. T. Williams, and N. Murthi Anishetty. *Gene Banks and the World's Food.* Princeton, NJ: Princeton University Press, 1987.

Pollak, Linda M. "The History and Success of the Public-Private Project on Germplasm Enhancement of Maize (GEM)." *Advances in Agronomy* 78 (2003): 45–87.

Pollan, Michael. *Omnivore's Dilemma: A Natural History of Four Meals.* New York: Penguin, 2006.

Price, Lisa Leimar, and Florencia G. Palis. "Bringing Farmer Knowledge and Learning into Agricultural Research: How Agricultural Anthropologists Transformed Strategic Research at the International Rice Research Institute." *Culture, Agriculture, Food and Environment* 38, no. 2 (2016): 123–30.

Quist, David, and Ignacio H. Chapela. "Transgenic DNA Introgressed into Traditional Maize Landraces in Oaxaca, Mexico." *Nature* 414, no. 6863 (2001): 541–43.

Qureshi, Sadiah. "Dying Americans: Race, Extinction, and Conservation in the New World." In *From Plunder to Preservation: Britain and the Heritage of Empire, c. 1800–1940,* edited by Astrid Swenson and Peter Mandler, 267–86. Oxford: Oxford University Press for the British Academy, 2013.

Raby, Megan. *American Tropics: The Caribbean Roots of Biodiversity Science.* Chapel Hill: University of North Carolina Press, 2017.

Radin, Joanna. "Latent Life: Concepts and Practices of Human Tissue Preservation in the International Biological Program." *Social Studies of Science* 43, no. 4 (2013): 484–508.

———. *Life on Ice: A History of New Uses for Cold Blood.* Chicago: University of Chicago Press, 2017.

Radin, Joanna, and Emma Kowal, eds. *Cryopolitics: Frozen Life in a Melting World.* Cambridge, MA: MIT Press, 2017.

Ramírez E., Ricardo, David H. Timothy, Efraín Díaz B., and U. J. Grant. *Races of Maize in Bolivia.* With G. Edward Nicholson Calle, Edgar Anderson, and William L. Brown. Publication 747. Washington, DC: National Research Council, National Academy of Sciences, 1960.

Ranum, Peter, Juan Pablo Peña-Rosas, and Maria Nieves Garcia-Casal. "Global Maize Production, Utilization, and Consumption." *Annals of the New York Academy of Sciences* 1312, no. 1 (2014): 105–12.

"The Research and Marketing Service Act, 1946, U.S.A." *Review of Marketing and Agricultural Economics* 15, no. 8 (1947): 294–95.

Report of the FAO/IBP Technical Conference on the Exploration, Utilization and Conservation of Plant Genetic Resources, 1967. Meeting Report PL/FO: 1967/M/12. Rome: FAO, 1969.

Report of the FAO Technical Meeting on Plant Exploration and Introduction, 1961. Meeting Report PL 1961/8. Rome: FAO, 1961.

Report of the Fourth Session of the FAO Panel of Experts on Plant Exploration and Introduction. Meeting Report AGP 1970/M/3. Rome: FAO, 1970.

Report of the National Academy of Sciences, National Research Council, Fiscal Year 1950–51. Washington, DC: Government Publishing Office, 1953.

Report of the Third Session of the FAO Panel of Experts on Plant Exploration and Introduction. Meeting Report PL: 1969/M/8. Rome: FAO, 1969.

Rhoades, Robert E., and Robert H. Booth. "Farmer-Back-to-Farmer: A Model for Generating Acceptable Agricultural Technology." *Agricultural Administration* 11 (1982): 127–37.

Rice, Elizabeth, Melinda Smale, and José-Luis Blanco. "Farmers' Use of Improved Seed Selection Practices in Mexican Maize: Evidence and Issues from the Sierra de Santa Marta." CIMMYT Economics Working Paper 97–03. Mexico City: CIMMYT, 1997.

Richards, Paul. *Indigenous Agricultural Revolution: Ecology and Food Crops in West Africa.* London: Hutchinson, 1985.

Rincón Sánchez, Froylán, and Juan Manuel Hernández Casillas. "Conservación de recursos fitogenéticos en México." In *Recursos fitogenéticos de México para la alimentación y la agricultura,* edited by P. Ramírez V., R. Ortega P., A. López H., F. Castillo G., M. Livera M., F. Rincón S., and F. Zavala G., 96–111. Chapingo, Mexico: Servicio Nacional de Inspección y Certificación de Semillas/Sociedad Mexicana de Fitogenética (SOMEFI), 2000.

Rist, Gilbert. *The History of Development: From Western Origins to Global Faith.* 4th ed. London: Zed Books, 2014.

Roberts, L. M., U. J. Grant, Ricardo Ramirez E., W. H. Hatheway, and D. L. Smith. *Races of Maize in Colombia.* With Paul C. Mangelsdorf. Publication 510. Washington, DC: National Research Council, National Academy of Sciences, 1957.

Robertson, Thomas. *The Malthusian Moment: Global Population Growth and the Birth of American Environmentalism.* New Brunswick, NJ: Rutgers, 2012.

Rockefeller Foundation. *Annual Report.* New York: Rockefeller Foundation, 1947, 1950, 1959, 1963, 1964. www.rockefellerfoundation.org/annual-reports/.

Roitman, Janet. *Anti-crisis.* Durham, NC: Duke University Press, 2014.

Rosa, Luis de la. *Memoria sobre el cultivo del maíz en México.* Mexico City: Sociedad Literaria, 1846.

Rosenberg, Charles E. "Rationalization and Reality in the Shaping of American Agricultural Research, 1875–1914." *Social Studies of Science* 7 (1977): 401–22.

Rosenzweig, Andrés. "El debate sobre el sector agropecuario mexicano en el Tratado de Libre Comercio de América del Norte." *Estudios y Perspectivas,* no. 30 (2005).

Ruiz Erdozain, Ernesto. *Estudio sobre el cultivo del maíz.* Mexico City: Secretaría de Fomento, 1914.

Saidel, Rochelle G., and Guilherme Ary Plonski. "Shaping Modern Science and Technology in Brazil: The Contribution of Refugees from National Socialism after 1933." *Leo Baeck Institute Year Book* 39, no. 1 (1994): 257–70.

Salhuana, W., and L. Pollak. "Latin American Maize Project (LAMP) and Germplasm Enhancement of Maize (GEM) Project: Generating Useful Breeding Germplasm." *Maydica* 51, no. 2 (2006): 339–55.

Salhuana, Wilfredo. "U.S. Germplasm Enhancement for Maize (GEM)." USDA. 6 December 2005. www.ars.usda.gov/ARSUserFiles/50301000/Reference_Documents/Gem-Executive-report-2005.pdf.

Salhuana, Wilfredo, Quentin Jones, and Ricardo Sevilla. "The Latin American Maize Project: Model for Rescue and Use of Irreplaceable Germplasm." *Diversity* 7, nos. 1–2 (1991): 40–42.

Salhuana, Wilfredo, Ricardo Sevilla, and Steve A. Eberhart. "Final Report: LAMP, Latin American Maize Project." USDA. July 1997. www.ars.usda.gov/ARSUserFiles/50301000/Reference_Documents/LAMP-Final-Report-1997.pdf.

Sámano Rentería, Miguel Ángel. "El movimiento ¡El campo no aguanta más! y el Acuerdo Nacional para el Campo: Situación y perspectiva." *Cotidiano* 19, no. 124 (2004): 64–70.

Sánchez González, José de Jesús, and Lorenzo Ordaz Suárez. *Reestudio de las razas mexicanas de maíz.* Zapopan, Mexico: INIA, 1984.

Sanderson, Jay. *Plants, People and Practices: The Nature and History of the UPOV Convention.* Cambridge: Cambridge University Press, 2017.

Sanderson, Steven E. *The Transformation of Mexican Agriculture: International Structure and the Politics of Rural Change.* Princeton, NJ: Princeton University Press, 1986.

Sanderson, Susan Walsh. *Land Reform in Mexico, 1910–1980.* Orlando: Academic Press, 1984.

Sandstrom, Alan R. *Corn Is Our Blood: Culture and Ethnic Identity in a Contemporary Aztec Village.* Norman: University of Oklahoma Press, 1991.

Saraiva, Tiago. "Breeding Europe: Crop Diversity, Gene Banks, and Commoners." In *Cosmopolitan Commons: Sharing Resources and Risks across Borders,* edited by Nil Disco and Eda Kranakis, 185–212. Cambridge, MA: MIT Press, 2013.

———. *Fascist Pigs: Technoscientific Organisms and the History of Fascism.* Cambridge, MA: MIT Press, 2016.

Sauer, Carl O. *Agricultural Origins and Dispersals.* New York: American Geographical Society, 1952.

———. "The Personality of Mexico." *Geographical Review* 31, no. 3 (1941): 353–64.

———. "Theme of Plant and Animal Destruction in Economic History." *Journal of Farm Economics* 20, no. 4 (1938): 765–75.

Schauer, Jeffrey. "Imperial Ark? The Politics of Wildlife in East and South-Central Africa, 1920–1992." PhD diss., University of California–Berkeley, 2014.

Schiebinger, Londa. *Plants and Empire: Colonial Bioprospecting in the Atlantic World.* Cambridge, MA: Harvard University Press, 2004.

Schiebinger, Londa, and Claudia Swan, eds. *Colonial Botany: Science, Commerce, and Politics in the Early Modern World.* Philadelphia: University of Pennsylvania Press, 2007.

Schleper, Marie Simone. "Conservation Compromises: The MAB and the Legacy of the International Biological Program, 1964–1974." *Journal of the History of Biology* 50, no. 1 (2017): 133–67.

———. "Life on Earth: Controversies on the Science and Politics of Global Nature Conservation." PhD diss., Maastricht University, 2017.

———. *Planning for the Planet: Environmental Expertise and the International Union for Conservation of Nature and Natural Resources, 1960–1980.* New York: Berghahn, 2019.

Schmidt, Chris. "United States: Native Seeds/SEARCH." In Vernooy, Shrestha, and Sthapit, *Community Seed Banks,* 172–75.

Schultes, Richard Evans, and Siri von Reis, eds. *Ethnobotany: Evolution of a Discipline.* London: Chapman and Hall, 1995.

Schurman, Rachel, and William A. Munro. *Fighting for the Future of Food: Activists versus Agribusiness in the Struggle over Biotechnology.* Minneapolis: University of Minnesota Press, 2010.

———. "Ideas, Thinkers, and Social Networks: The Process of Grievance Construction in the Anti-genetic Engineering Movement." *Theory and Society* 35, no. 1 (2006): 1–38.

Sehgal, Suri. "The Contributions of World Maize Expert William L. Brown to Latin America." *Diversity* 7, nos. 1–2 (1991): 43–44.

Sepkoski, David. *Catastrophic Thinking: Extinction and the Value of Diversity from Darwin to the Anthropocene.* Chicago: University of Chicago Press, 2020.

Serratos, J. Antonio, Martha C. Wilcox, and Fernando Castillo, eds. *Gene Flow among Maize Landraces, Improved Maize Varieties, and Teosinte: Implications for Transgenic Maize*. Mexico City: CIMMYT, 1997.

Shands, Henry L., Paul J. Fitzgerald, and Steve A. Eberhart. "Program for Plant Germplasm Preservation in the United States: The U.S. National Plant Germplasm System." In *Biotic Diversity and Germplasm Preservation: Global Imperatives*, edited by Lloyd Knutson and Allan K. Stoner, 97–115. Beltsville Symposia in Agricultural Research 13. Dordrecht, Netherlands: Kluwer, 1989.

Shepherd, Chris J. "Imperial Science: the Rockefeller Foundation and Agricultural Science in Peru, 1940–1960." *Science as Culture* 14, no. 2 (2005): 113–37.

Shiferaw, Bekele, Boddupalli M. Prasanna, Jonathan Hellin, and Marianna Bänziger. "Crops That Feed the World 6: Past Successes and Future Challenges to the Role Played by Maize in Global Food Security." *Food Security* 3, no. 3 (2011): 307–27.

Sistema Alimentario Mexicano. Mexico City: Secretaría de Agricultura y Recursos Hidráulicos, [1980?].

Smith, Elta. "Imaginaries of Development: The Rockefeller Foundation and Rice Research." *Science as Culture* 18, no. 4 (2009): 461–82.

Soleri, Daniela. "Developing Methodologies to Understand Farmer-Managed Maize Folk Varieties and Farmer Seed Selection in the Central Valleys of Oaxaca, Mexico." PhD diss., University of Arizona, 1999.

Soleri, Daniela, and David A. Cleveland. "Farmers' Genetic Perceptions regarding Their Crop Populations: An Example with Maize in the Central Valleys of Oaxaca, Mexico." *Economic Botany* 55, no. 1 (2001): 106–28.

——. "Hopi Crop Diversity and Change." *Journal of Ethnobiology* 13, no. 2 (1993): 203–31.

Soleri, Daniela, David A. Cleveland, Flavio Aragón C., Mario R. Fuentes L., Humberto Ríos L., and Stuart H. Sweeney. "Understanding the Potential Impact of Transgenic Crops in Traditional Agriculture: Maize Farmers' Perspectives in Cuba, Guatemala and Mexico." *Environmental Biosafety Research* 4 (2005): 141–66.

Soleri, Daniela, David A. Cleveland, and Flavio Aragón Cuevas. "Response from Soleri and Colleagues." *BioScience* 56, no. 9 (2006): 709–10.

——. "Transgenic Crops and Crop Varietal Diversity: The Case of Maize in Mexico." *BioScience* 56, no. 6 (2006): 503–13.

Soto Laveaga, Gabriela. *Jungle Laboratories: Mexican Peasants, National Projects, and the Making of the Pill*. Durham, NC: Duke University Press, 2009.

——. "*Largo Dislocare:* Connecting Microhistories to Remap and Recenter Histories of Science." *History and Technology* 34, no. 1 (2018): 21–30.

Stakman, E. C., Richard Bradfield, and Paul C. Mangelsdorf. *Campaigns against Hunger*. Cambridge, MA: Harvard University Press, 1967.

Staller, John, Robert Tykot, and Bruce Benz, eds. *Histories of Maize: Multidisciplinary Approaches to the Prehistory, Linguistics, Biogeography, Domestication, and Evolution of Maize*. London: Routledge, 2016.

Staller, John E. *Maize Cobs and Cultures: History of Zea mays L.* Berlin: Springer, 2010.

Staples, Amy. *The Birth of Development: How the World Bank, Food and Agriculture Organization, and World Health Organization Changed the World, 1945–1965.* Kent, Ohio: Kent State University Press, 2006.

———. "To Win the Peace: The Food and Agriculture Organization, Sir John Boyd Orr, and the World Food Board Proposals." *Peace and Change* 28, no. 4 (2003): 495–523.

Stavenhagen, Rodolfo. "La política indigenista del estado mexicano y los pueblos indígenas en el siglo XX." In *Educación e interculturalidad: Política y políticas,* edited by Bruno Baronnet and Medardo Tapia, 23–48. Cuernavaca, Mexico: Centro Regional de Investigaciones Multidisciplinarias (CRIM)-UNAM, 2013.

Stocking, George W., Jr. *Race, Culture, and Evolution: Essays in the History of Anthropology.* Chicago: University of Chicago Press, 1982.

Sturtevant, E.L. *Varieties of Corn.* USDA Office of Experiment Stations Bulletin 57. Washington DC: Government Publishing Office, 1899.

Suárez, Blanca. "Las semillas, el estado y las transnacionales." *Problemas del Desarrollo* 13, nos. 51–52 (1982–83): 45–102.

Sutter, Paul S. "The Tropics: A Brief History of an Environmental Imaginary." In *The Oxford Handbook of Environmental History,* edited by Andrew C. Isenberg. Oxford Handbooks Online. Oxford: Oxford University Press, 2018. 10.1093 /oxfordhb/9780195324907.013.0007.

Taba, Suketoshi. "Current Activities of CIMMYT's Maize Germplasm Bank." In *The CIMMYT Maize Germplasm Bank: Genetic Resource Preservation, Regeneration, Maintenance, and Use,* edited by Suketoshi Taba, 9–20. Maize Program Special Report. Mexico City: CIMMYT, 1994.

Taba, Suketoshi, and Steve Eberhart. "Cooperation Saves Seed of Thousands of Latin American Maize Landraces." In *Latin American Maize Germplasm Regeneration and Conservation,* edited by Suketoshi Taba, v–x. Workshop Proceedings, 4–6 June 1996. Mexico City: CIMMYT, 1997.

TallBear, Kim. "The Emergence, Politics and Marketplace of Native American DNA." In *Routledge Handbook of Science, Technology, and Society,* edited by Daniel Lee Kleinman and Kelly Moore, 21–37. London: Routledge, 2014.

———. *Native American DNA: Tribal Belonging and the False Promise of Genetic Science.* Minneapolis: University of Minnesota Press, 2013.

Teichman, Judith. "The Mexican State and the Political Implications of Economic Restructuring." *Latin American Perspectives* 73, no. 19 (1992): 88–104.

Ten Eyck, A.M., and V.M. Shoesmith. *Indian Corn.* Bulletin 147. Manhattan: Kansas State Agricultural College, 1907.

Thormann, Imke, and Johannes M.M. Engels. "Genetic Diversity and Erosion: A Global Perspective." In *Genetic Diversity and Erosion in Plants,* edited by M.R. Ahuja and S. Mohan Jain, 1:263–94. Cham, Switzerland: Springer, 2015.

Timothy, D.H., and M.M. Goodman. "Germplasm Preservation: The Basis of Future Feast or Famine; Genetic Resources of Maize: An Example." In *The Plant*

Seed: Development, Preservation, and Germination, edited by Irwin Rubenstein, Ronald L. Phillips, Charles E. Green, and B. G. Gengenbach, 171–200. New York: Academic Press, 1979.

Toledo, Victor Manuel. "New Paradigms for a New Ethnobotany: Reflections on the Case of Mexico." In Schultes and Von Reis, *Ethnobotany,* 75–88.

Tracy, W. F. "Vegetable Uses of Maize (Corn) in Pre-Columbian America." *Hort-Science* 34, no. 5 (1999): 812–13.

Trentmann, Frank. "Coping with Shortage: The Problem of Food Security and Global Visions of Coordination, c. 1890s–1950." In *Food and Conflict in the Age of the Two World Wars,* edited by Frank Trentmann and Flemming Just, 13–48. Basingstoke, UK: Palgrave, 2006.

Troyer, A. Forrest. "Background of U.S. Hybrid Corn." *Crop Science* 39, no. 3 (1999): 601–26.

———. "Development of Hybrid Corn and the Seed Corn Industry." In *Handbook of Maize: Genetics and Genomics,* edited by Jeff Bennetzen and Sarah Hake, 87–114. New York: Springer, 2009.

Troyer, A. Forrest, and Lois G. Hendrickson. "Background and Importance of 'Minnesota 13' Corn." *Crop Science* 47, no. 3 (2007): 905–14.

Tsing, Anna Lowenhaupt. *The Mushroom at the End of the World: On the Possibility of Life in Capitalist Ruins.* Princeton, NJ: University of Princeton Press, 2015.

Ullstrup, A. J. "The Impacts of the Southern Corn Leaf Blight Epidemics of 1970–1971." *Annual Review of Phytopathology* 10 (1972): 37–50.

US Comptroller General. *Report to the Congress of the United States: The Department of Agriculture Can Minimize the Risk of Potential Crop Failures.* CED-81-75. Washington, DC: GAO, 1981.

USDA (US Department of Agriculture). *1969 Census of Agriculture.* Vol. 2. Washington, DC: Government Publishing Office, 1973.

Van de Wouw, Mark, Chris Kik, Theo van Hintum, Rob van Treuren, and Bert Visser. "Genetic Erosion in Crops: Concept, Research Results and Challenges." *Plant Genetic Resources* 8, no. 1 (2010): 1–15.

Van de Wouw, Mark, Theo van Hintum, Chris Kik, Rob van Treuren, and Bert Visser. "Genetic Diversity Trends in Twentieth Century Crop Cultivars: A Meta Analysis." *Theoretical and Applied Genetics* 120, no. 6 (2010): 1241–52.

Van Dooren, Thom. "Banking Seed: Use and Value in the Conservation of Agricultural Diversity." *Science as Culture* 18, no. 4 (2009): 373–95.

———. *Flight Ways: Life and Loss at the Edge of Extinction.* New York: Columbia University Press, 2014.

———. *The Wake of Crows: Living and Dying in Shared Worlds.* New York: Columbia University Press, 2019.

Van Dooren, Thom, Deborah B. Rose, and Matthew Chrulew, eds. *Extinction Studies: Stories of Time, Death, and Generations.* New York: Columbia University Press, 2017.

Van Hintum, Theo J. L. "Duplication within and between Germplasm Collections: III; A Quantitative Model." *Genetic Resources and Crop Evolution* 47 (2000): 507–13.

Van Hintum, Theo J. L., and Helmut Knüpffer. "Duplication within and between Germplasm Collections: I; Identifying Duplication on the Basis of Passport Data." *Genetic Resources and Crop Evolution* 42 (1995): 127–33.

Vavilov, N. I. *Origin and Geography of Cultivated Plants.* Edited by V. F. Dorofeyev. Translated by D. Löve. Cambridge: Cambridge University Press, 1992.

Ventura Santos, Ricardo. "Indigenous Peoples, Postcolonial Contexts and Genomic Research in the Late 20th Century: A View from Amazonia (1960–2000)." *Critique of Anthropology* 22, no. 1 (2002): 81–104.

Veracini, Lorenzo. "Introducing Settler Colonial Studies." *Settler Colonial Studies* 1, no. 1 (2011): 1–12.

Vera Sánchez, Karina Sandibel, Rosalinda González Santos, and Flavio Aragón-Cuevas. "Community Seed Banks in Mexico: An In-Situ Conservation Strategy." In Vernooy, Shrestha and Sthapit, *Community Seed Banks,* 248–53.

Vernon, James. *Hunger: A Modern History.* Cambridge, MA: Harvard University Press, 2007.

Vernooy, Ronnie, Pitambar Shrestha, and Bhuwon Sthapit, eds. *Community Seed Banks: Origins, Evolution and Prospects.* London: Routledge, 2015.

———. "Origins and Evolution." In Vernooy, Shrestha, and Sthapit, *Community Seed Banks,* 11–19.

Vessuri, Hebe M. C. "Foreign Scientists, the Rockefeller Foundation and the Origins of Agricultural Science in Venezuela." *Minerva* 32, no. 3 (1994): 267–96.

Vidal, Fernando, and Nélia Dias, eds. *Endangerment, Biodiversity, and Culture.* London: Routledge, 2016.

Voeks, Robert A. "Disturbance Pharmacopoeias: Medicine and Myth from the Humid Tropics." *Annals of the Association of American Geographers* 94, no. 4 (2004): 868–88.

Voices of Maíz: Collective Work from Abya Yala. Mexico: Voices of Maíz, 2016.

Wade, Nicolas. "A Message from the Corn Blight: The Dangers of Uniformity." *Science* 177, no. 4050 (1972): 678–79.

Wallace, Henry A., and William L. Brown. *Corn and Its Early Fathers.* Ann Arbor: Michigan State University Press, 1956.

Wanderer, Emily Mannix. *The Life of a Pest: An Ethnography of Biological Invasion in Mexico.* Oakland: University of California Press, 2020.

Warman, Arturo. *Corn and Capitalism: How a Botanical Bastard Grew to Global Dominance.* Translated by Nancy L. Westrate. Chapel Hill: University of North Carolina Press, 2003.

Wedel, Waldo R. "George Francis Will, 1884–1955." *American Antiquity* 22, no. 1 (1956): 73–76.

Wellhausen, Edwin J. "The Agriculture of Mexico." *Scientific American* 235, no. 3 (1976): 128–53.

———. "Exotic Germplasm for Improvement of Corn Belt Maize." *Proceedings of the 20th Annual Hybrid Corn Industry Research Conference* (1965): 31–45.

————. "Improving Corn with Exotic Germplasm." *Proceedings of the 11th Annual Hybrid Corn and Sorghum Research Conference* (1956): 85–96.

Wellhausen, E. J., Alejandro Fuentes O., and Antonio Hernández Corzo. *Races of Maize in Central America*. With Paul C. Mangelsdorf. Publication 511. Washington, DC: National Research Council, National Academy of Sciences, 1957.

Wellhausen, E. J., L. M. Roberts, and E. Hernández X. *Races of Maize in Mexico: Their Origin, Characteristics and Distribution*. With Paul C. Mangelsdorf. Cambridge, MA: Bussey Institution of Harvard University, 1952.

————. *Razas de maíz en México: Su origen, características y distribución*. With Paul C. Mangelsdorf. Folleto Técnico 5. Mexico City: Oficina de Estudios Especiales, Secretaría de Agricultura y Ganadería, 1951.

Weltzien, Eva, Margaret E. Smith, Laura S. Meitzner, and Louise Sperling. *Technical and Institutional Issues in Participatory Plant Breeding: From the Perspective of Formal Plant Breeding*. PPB Monograph 1. Cali, Colombia: Participatory Research and Gender Analysis Program, CGIAR, 2003.

West, Robert Cooper. *Carl O. Sauer's Fieldwork in Latin America*. Ann Arbor, MI: Department of Geography, University of Syracuse, 1979.

————. "The Contribution of Carl Sauer to Latin American Geography." *Proceedings of the Conference of Latin Americanist Geographers* 8 (1981): 8–21.

Whyte, Kyle Powys. "Food Sovereignty, Justice, and Indigenous Peoples." In *The Oxford Handbook of Food Ethics,* edited by Anne Barnhill, Tyler Doggett, and Mark Budolfson, 345–66. New York: Oxford, 2018.

Whyte, R. O. *Plant Exploration, Collection, and Introduction*. Rome: FAO, 1958.

Wilkes, Garrison. "Maize: Domestication, Racial Evolution, and Spread." In *Foraging and Farming: The Evolution of Plant Exploitation,* edited by David R. Harris and Gordon C. Hillman, 440–55. London: Hyman, 1989.

Wilkes, H. Garrison, and Susan Wilkes. "The Green Revolution." *Environment: Science and Policy for Sustainable Development* 14 no. 8 (1972): 32–39.

Will, George F., and George E. Hyde. *Corn among the Indians of the Upper Missouri*. Saint Louis: Miner, 1917.

Will, George F., and H. J. Spinden. *The Mandans: A Study of Their Culture, Archaeology and Language*. Papers of the Peabody Museum. Cambridge, MA: Peabody Museum, 1906.

Wilson, G. L. *Agriculture of the Hidatsa Indians: An Indian Interpretation*. Bulletin of the University of Minnesota, Studies in the Social Sciences, no. 9. Minneapolis: University of Minnesota, 1917.

Winkelmann, Donald. *The Adoption of New Maize Technology in Plan Puebla, Mexico*. Mexico City: CIMMYT, 1976.

Wolfe, Mikael D. *Watering the Revolution: An Environmental and Technological History of Agrarian Reform in Mexico*. Durham, NC: Duke University Press, 2017.

Wolfe, Patrick. "Settler Colonialism and the Elimination of the Native." *Journal of Genocide Research* 8, no. 4 (2006): 387–409.

Wood, David, and Jillian M. Lenné. "The Conservation of Agrobiodiversity On-Farm: Questioning the Emerging Paradigm." *Biodiversity and Conservation* 6, no. 1 (1997): 109–29.

Woolf, Aaron, dir. *King Corn.* Westport, NY: Mosaic Films, 2006.

Worthington, E. B. *The Evolution of IBP.* Cambridge: Cambridge University Press, 1975.

———. "The International Biological Programme." *Nature* 208, no. 5007 (1965): 223–26.

Wortman, Sterling. "The Technological Basis for Intensified Agriculture." In *Agricultural Development,* 17–43.

Wright, Susan. *Molecular Politics: Developing American and British Regulatory Policy for Genetic Engineering, 1972–1982.* Chicago: University of Chicago Press, 1994.

———. "Recombinant DNA Technology and Its Social Transformation, 1972–1982." *Osiris* 2 (1986): 303–60.

Yúnez-Naude, Antonio. "Lessons from NAFTA: The Case of Mexico's Agricultural Sector." With Fernando Barceinas Paredes. Report to the World Bank. 2002.

Zent, Stanford. "A Genealogy of Scientific Representations of Indigenous Knowledge." In *Landscape, Process and Power: Re-evaluating Traditional Ecological Knowledge,* edited by Serena Heckler, 19–67. New York: Berghahn, 2009.

INDEX

Navajo, 22–23, 82
Network in Defense of Maize, 217–19,
219*fig*, 223
New England Eight-Rowed, maize variety,
19
Nicaragua, 79
Nigeria, International Institute of Tropical
Agriculture, 106
North American Free Trade Agreement
(NAFTA), 7, 191, 212–17, 217–18
North Carolina State University, US, 163,
172–73, 178, 180, 227
North Central Regional Plant Introduc-
tion Station, Iowa, 227
North Dakota Experiment Station, US, 82

Oaxaca, Mexico, 217, 220, 222
Office of Corn Investigations, USDA, 15,
17–22
Office of Experiment Stations, Mexico,
58–59
Office of Foreign Seed and Plant Introduc-
tion, USDA, 20
Office of Special Studies, Chapingo, Mexico, 57–60, 62–63, 64–66, 68*fig*, 76,
78–79, 91, 105, 166. *see also* National
Agricultural Research Institute,
Chapingo, Mexico
Office of Special Studies, Colombia, 73–74,
76, 78, 89, 91
Oficina de Campos Experimentales, Mexico, 58–59
Oficina de Estudios Especiales, Mexico,
57–60, 62–63, 64–66, 68*fig*, 76, 78–79,
91, 105, 166. *see also* National Agricultural Research Institute, Chapingo,
Mexico
Olmec civilization, 9
Onaveño, maize variety, 199
on-farm conservation, 192–96, 202–7, 215,
216–17, 219–22, 223–24
on-farm experiments, 208–12, 223–24, 233
O'odham huuñ, maize variety, 199
open-pollinated varieties, 34–36, 37, 38, 60,
150–51
organic movement, US, 197
origins of maize, 47–50
Ortega Paczka, Rafael, 157, 204, 205

Oscar H. Will and Company, 38, 227;
catalogue, 14*fig*, 26*fig*

Palacios de la Rosa, Gilberto, 259n11
Panama, 79
Papago, 82
Paraguay, 50, 52
participatory approaches, 208–12, 223–24,
233
passport data, 172, 175, 176, 183
patents, 143, 154, 232
Patiño Rodríguez, Víctor, 74, 78, 89
peas, 167
peasant farmers, 45–46, 129, 130, 131, 133,
203–8, 217–25
The Peasants' Way, 217, 224
Peru, 78, 89, 90, 158, 163, 175, 176, 183,
205–6
pharmaceutical companies, 153
pharmaceuticals, 133, 165
Pioneer Hi-Bred Corn Company, 35, 83,
114, 158, 166–67, 178–81, 181–82, 228
Piracicaba, Brazil, 76, 81, 89, 114–15
plant breeders' rights, 143–44, 154
plant hunting, 55, 56
plant introduction stations, 76, 77
Plant Variety Protection Act (1970), 144
pollination, 34, 35, 139
Polo, Evaristo, 9–10
population growth, 98, 103, 109, 117, 118,
129
potatoes, 32–33, 167, 176, 205–6
poverty, 45, 46, 136, 141
prebreeding, 174, 181
preservation, 65–67, 69–77, 90–95, 109–10,
121–22, 157–58. *see also* collections;
conservation; seed banks
Pride of Dakota, maize variety, 23–24,
24*fig*, 38
primitive maize. *see* farmers' varieties
Productora Nacional de Semillas, Mexico,
146–47, 149–50
Project MILPA (McKnight Integrated
Landrace Preservation Activity), 210*fig*,
215, 215*fig*, 216
Project Sierra de Santa Marta, 209–10, 212,
216
property rights, 126, 140–46, 152–53, 232

Founded in 1893,
UNIVERSITY OF CALIFORNIA PRESS
publishes bold, progressive books and journals
on topics in the arts, humanities, social sciences,
and natural sciences—with a focus on social
justice issues—that inspire thought and action
among readers worldwide.

The UC PRESS FOUNDATION
raises funds to uphold the press's vital role
as an independent, nonprofit publisher, and
receives philanthropic support from a wide
range of individuals and institutions—and from
committed readers like you. To learn more, visit
ucpress.edu/supportus.